X- AND GAMMA-RAY ASTRONOMY

INTERNATIONAL ASTRONOMICAL UNION
UNION ASTRONOMIQUE INTERNATIONALE

SYMPOSIUM No. 55

HELD IN MADRID, SPAIN, 11–13 MAY 1972

X- AND GAMMA-RAY ASTRONOMY

EDITED BY

H. BRADT

Massachusetts Institute of Technology, Cambridge, Mass., U.S.A.

AND

R. GIACCONI

American Science and Engineering, Cambridge, Mass., U.S.A.

D. REIDEL PUBLISHING COMPANY

DORDRECHT-HOLLAND / BOSTON-U.S.A.

1973

Published on behalf of
the International Astronomical Union
by
D. Reidel Publishing Company, P.O. Box 17, Dordrecht, Holland

Sold and distributed in the U.S.A., Canada, and Mexico
by D. Reidel Publishing Company, Inc.
306 Dartmouth Street, Boston,
Mass. 02116, U.S.A.

Library of Congress Catalog Card Number 72–92526

ISBN-13: 978-90-277-0337-8   e-ISBN-13: 978-94-010-2585-0
DOI: 10.1007/978-94-010-2585-0

EDITORS' FOREWORD

The IAU Symposium No. 55 on 'X-Ray and Gamma-Ray Astronomy' has occurred, not entirely by coincidence, at an important moment in the development of these new branches of observational astronomy.

In X-ray astronomy the data from the first X-ray observatory UHURU have contributed to a new view of the X-ray sky and a new conception of the nature and properties of galactic and extragalactic X-ray sources. In gamma-ray astronomy the exciting and often controversial nature of the results underlines the importance of the forthcoming launch of SAS-B, the first orbiting γ-ray observatory.

As Bruno Rossi reminds us (p. 1), the Symposium occurred almost exactly ten years after the first detection of the X-ray star Sco X-1. During this time we have moved from the detection of a handful of the nearest and brightest sources to the detailed study of the nature of stellar sources in the farthest reaches of our own galaxy and in external galaxies of the local group. The detection of pulsating X-ray sources in binary systems permits the measurement of pulsation periods, and orbital parameters with precisions comparable to any yet achieved with traditional observational techniques.

The strong indications that most X-ray sources are extremely compact objects give us confidence that X-ray astronomy will play a significant and possibly decisive role in the study of stars near the end point of stellar evolution.

Still from the point of view of observational advances, the study of extragalactic X-ray sources has revealed that X-ray emission is a common feature of all galaxies. A variety of different processes is observed to take place in different types of galaxies giving rise to a range of several orders of magnitude in intrinsic X-ray luminosity. The discovery that clusters of galaxies are X-ray emitters of finite angular extent and the possible correlation between X-ray luminosity and cluster parameters give us hope that X-ray astronomy will contribute significantly to our understanding of mass distribution in the universe.

From the theoretical point of view this new observational material has caused not only an abundant literature endeavoring to explain the nature of individual X-ray sources; but also, and perhaps more importantly, a rethinking of astrophysical models for celestial objects already known, in light of possible observational consequences. The role of an X-ray emitting phase in the evolution and energetics of stellar and galactic systems is only now beginning to be investigated. It is most interesting to note in this connection that we are only now beginning to realize that a short but intense X-ray emitting phase appears to be a common event in the evolution of stars in close binary systems.

Thus X-ray observations are becoming more closely integrated in the main body of observational astronomy and of astrophysical theories.

The organizing committee of this Symposium endeavored to aid this evolution by inviting several optical and radio astronomers and several astrophysicists to participate in the discussions. From the comments of the participants, and from personal observation, we believe the approach was most stimulating.

The fundamental contributions that each branch of observational astronomy can give in the study of high energy phenomena in galactic and extragalactic objects strongly emphasize the common bond of scientific interest, and of appreciation for the beauty and unlimited variety of nature, which unites us all in the exploration of the universe, no matter how different our techniques.

H. BRADT

R. GIACCONI

ORGANIZING COMMITTEE

R. Giacconi, *Chairman* (U.S.A.)

G. Fazio (U.S.A.)

C. Fichtel (U.S.A.)

H. C. van de Hulst (Netherlands)

H. Friedman (U.S.A.)

M. Oda (Japan)

M. Rees (United Kingdom)

Yash Pal (India)

L. Gratton (Italy)

K. A. Pounds (United Kingdom)

S. L. Mandel'shtam (U.S.S.R.)

ACKNOWLEDGEMENTS

The editors are grateful to numerous people and organizations for their assistance in organizing the Symposium. In particular, we mention the International Astronomical Union, the Committee on Space Research, the organizing committee, especially Drs G. Fazio and Y. Pal for arranging panel discussions, and Dr Z. Niemirowicz of COSPAR. We also thank the following for their assistance in preparing this volume: the invited speakers/authors, their secretaries and colleagues, the reviewers of the manuscripts (Drs K. Brecher, G. Frye, P. Gorenstein, H. Gursky, R. Hjellming, E. Kellogg, W. Kraushaar, J. McClintock, R. Palmieri, D. Schwartz, H. Tananbaum, A. Treves, and W. Tucker), and Mr Ostendorf of Reidel/Dordrecht. Finally, we are grateful to our secretaries, Mrs Polly Sullivan and Miss Diana Valderrama, for their efforts on behalf of the Symposium.

The Symposium was sponsored by Commissions 44 and 48 of the IAU and co-sponsored by COSPAR.

TABLE OF CONTENTS

1. INTRODUCTORY REMARKS

BRUNO B. ROSSI

Massachusetts Institute of Technology, Cambridge, Mass., U.S.A.

It is for me a privilege to chair the first session of this Symposium, which, in a way, may be regarded as a celebration of the 10th anniversary of the birth of X-ray astronomy.

Although my personal contribution has been only marginal, to watch the growth of this new branch of science has been one of the most exhilarating experiences of my whole life as a scientist. The unexpected and astonishing results that have been following one another in quick succession have been for me a striking illustration of how much the imagination of man lags behind the boundless wealth and complexity of nature.

I am sure you all remember the climate of skepticism in which X-ray astronomy began. Which explains why the avowed purpose of the rocket program which opened up this new field (Figure 1.) was an attempt to detect an X-ray fluorescence from the Moon, rather than a search for extra-solar X-ray sources; although the hope of finding such sources was in fact the main motivation of the program. And you remember the surprise, when the rocket did, in fact, detect a galactic X-ray source many orders of magnitude stronger than any one had expected (Figure 2). In retrospect, we see that the little confidence people had in the worth of X-ray astronomy was due primarily to the failure to reckon with the possible existence of hitherto unknown celestial objects whose emission would lie almost entirely in the X-ray band of the spectrum; although it is also true that most astrophysicists had been too conservative in extrapolating toward the high energies the electromagnetic spectra of known peculiar objects, such as supernova remnants.

The comparatively large X-ray fluxes reaching the Earth from celestial sources explains why so many observations of crucial significance could be carried out even before the launching of the first X-ray satellite, at a time when X-ray astronomers had at their disposal only rockets and balloons, with the well-known limitations of these carriers. Just to mention some of the highlights, there was the identification, by the lunar occultation method, of the source in Taurus with the Crab Nebula (Figure 3); the identification of the source in Scorpio with a faint, apparently insignificant star (Figure 4); the measurement of several X-ray spectra; the discovery of a variability of some X-ray sources; the discovery of an X-ray pulsar; the discovery of the first extragalactic sources. Also, a number of high-resolution X-ray pictures of the Sun were taken with the newly-developed grazing-incidence telescopes (Figure 5), which made it possible to plan confidently the use of these powerful instruments for the extra-solar X-ray astronomy as soon as suitable vehicles will become available. Nor must one forget the discovery of a diffuse, nearly isotropic X-ray background and the fairly accurate measurement of its spectrum.

Bradt and Giacconi (eds.), X- and Gamma-Ray Astronomy, 1–6. All Rights Reserved.
Copyright © 1973 by the IAU.

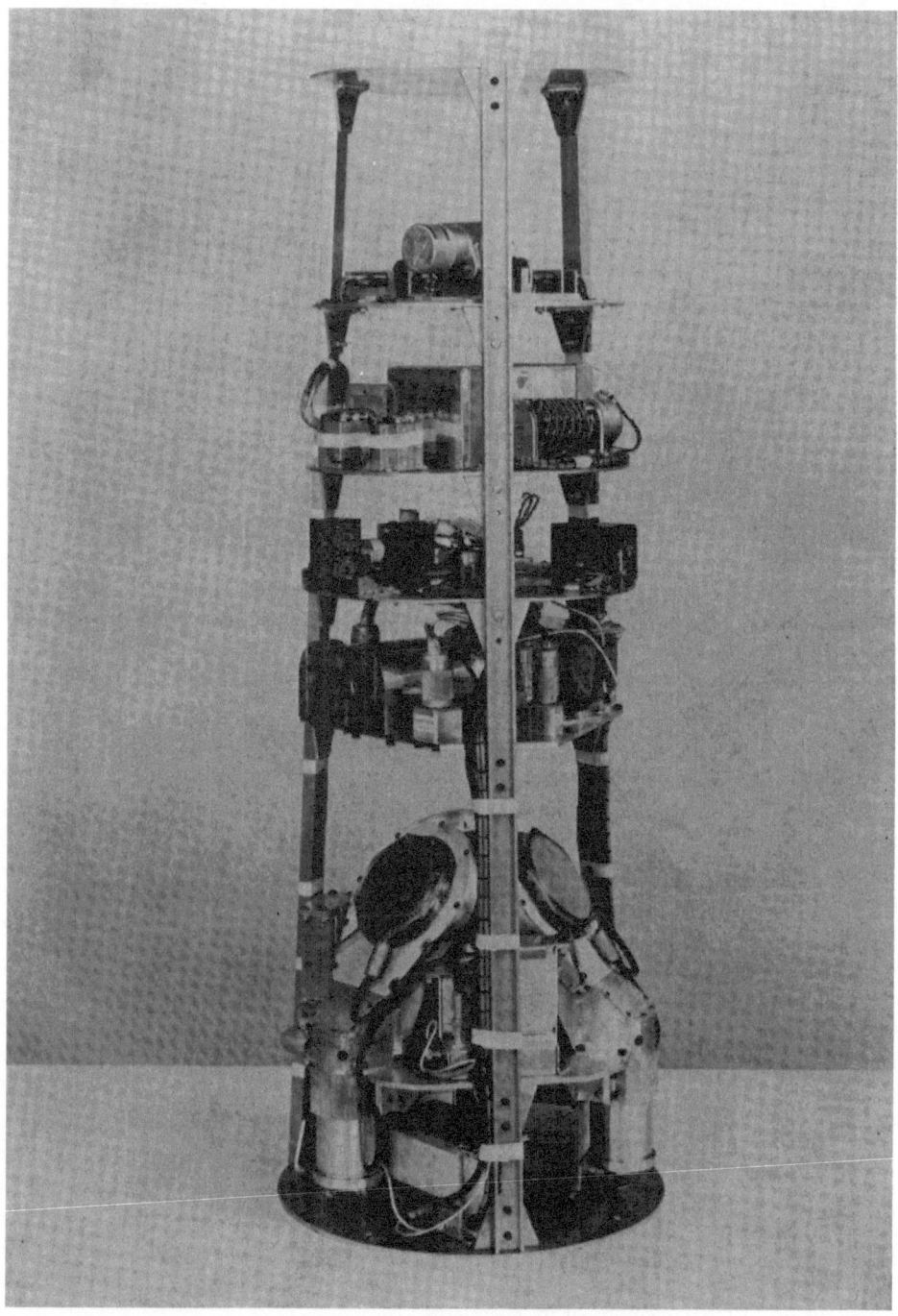

Fig. 1.   Payload of the rocket which detected the first extra-solar X-ray source in 1962.

Rockets and balloons will continue to play an important role in the future. But the performance of UHURU (Figure 6) has made it abundantly clear that the availability of satellites marks the opening of a new era for X-ray astronomy. And here again we find that scientists had been too conservative in their expectations. The main purpose

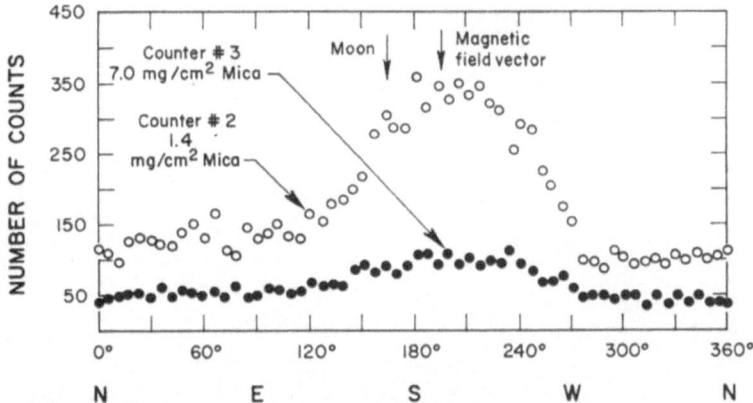

Fig. 2.   Signals from a strong galactic source recorded by the rocket in Figure 1
(*Phys. Rev. Letters* **9**, 439).

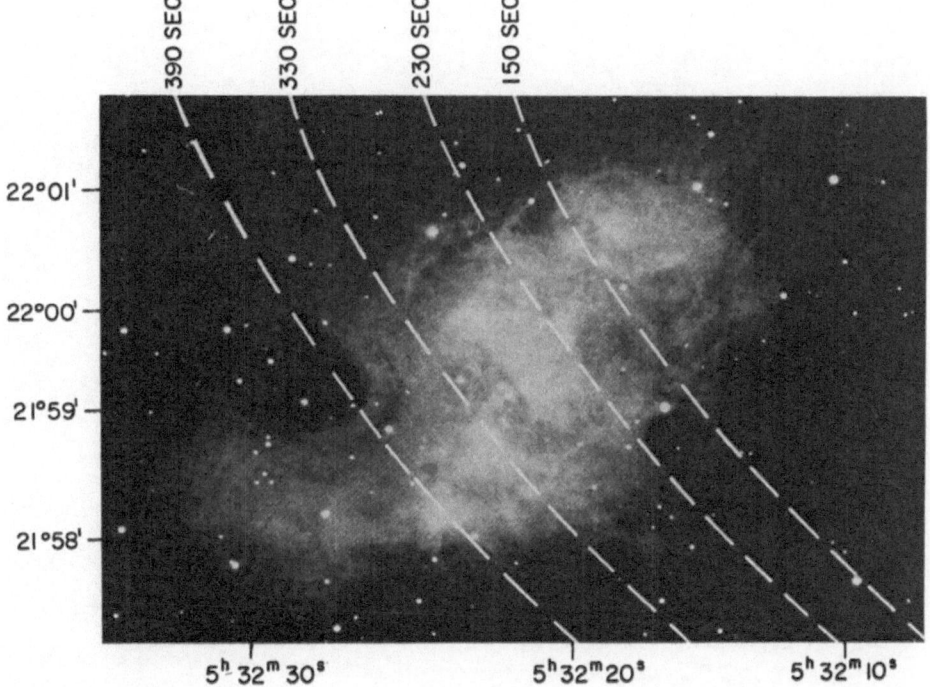

Fig. 3.   Identification of the X-ray source in the Crab Nebula by the lunar occultation method in
1964 (*Science* **146**, 912).

of UHURU, as originally conceived, was to carry out a general X-ray survey of the sky, adding weaker sources to the existing catalogues and determining source positions with good accuracy. UHURU is successfully fulfilling this assignment. But, in so doing, it has discovered new and highly significant features of individual sources; and

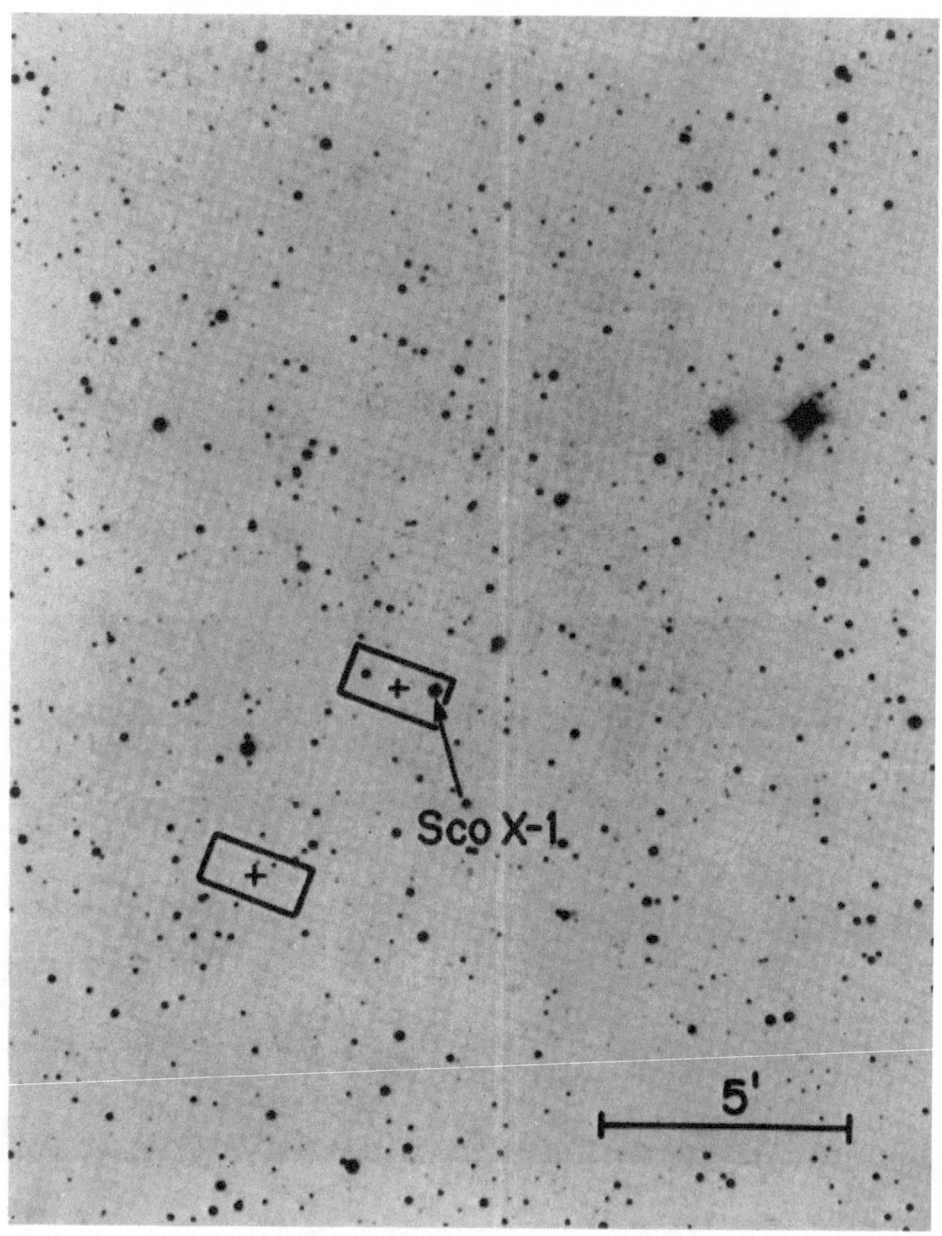

Fig. 4.   Identification of Sco X-1 in 1966 (*Astrophys. J.* **146**, 316).

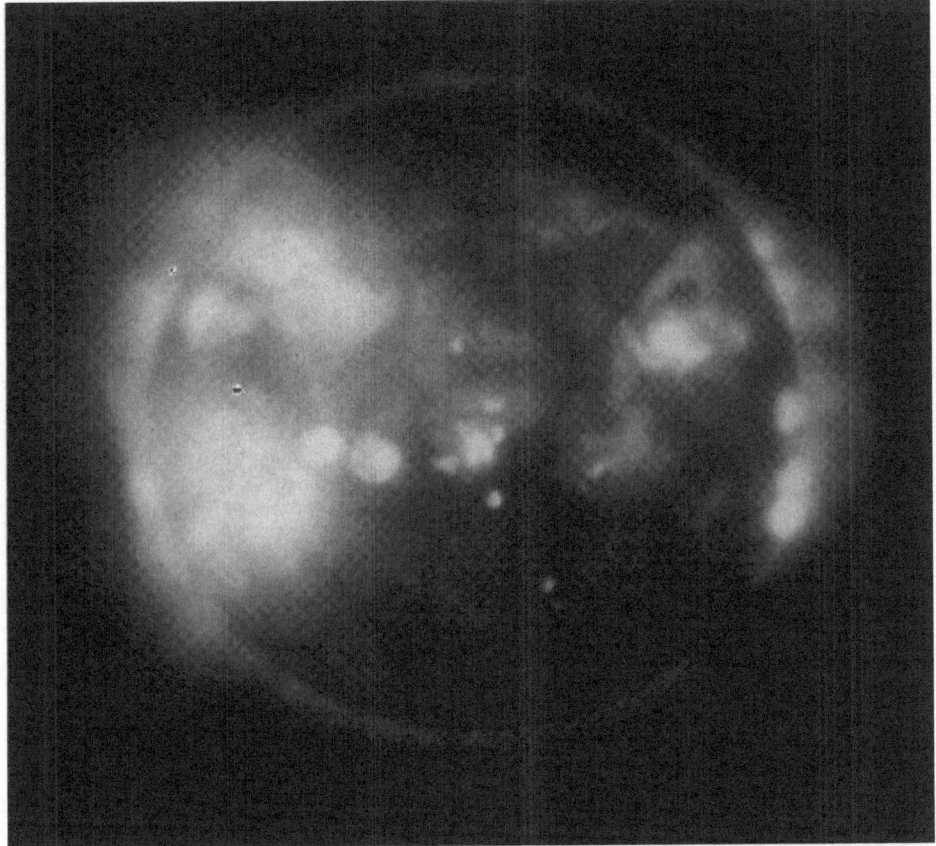

Fig. 5.   Photograph of the Sun in soft X-rays (3–30 Å, 44–55 Å) at the end of a lunar eclipse; note
Moon shadow to left (1970, *Nature* **227**, 818).

this not by virtue of any sophisticated instrumentation, but because of its ability of
scanning the sky at a very slow rate and of going back to any desired source over and
over again, on command from the Earth.

Thus, as you well know, a number of galactic sources were found to undergo large,
short-time fluctuations in their X-ray emission; fluctuations which in some cases are
exactly periodic, in some cases are not. Also long-term periodic changes in the average
X-ray intensity were detected for several sources. Detailed studies of these effects
have already had far-reaching consequences on our thinking about the nature of the
galactic sources. As for the extra-galactic sources several were shown to be extended,
suggesting that the association of galaxies into clusters may result in exceptionally
strong X-ray emission, indicative perhaps of interactions of the galaxies with the
medium between them.

Gamma-ray astronomy has developed at a much slower pace than X-ray astronomy,

Fig. 6.   UHURU: artist's conception.

mainly because of the technical difficulties of detecting very weak γ-ray fluxes. But a promising beginning has been made in this field too.

We still do not have more than tentative theoretical interpretations for the observational results of X-ray and γ-ray astronomy. However it is becoming increasingly clear that the findings of high-energy astronomy will have an important bearing on many of the most crucial problems of contemporary astrophysics, such as the nature and properties of collapsed objects; the possible existence among them of 'black holes'; the physical processes occurring in various kinds of galaxies, particularly in their nuclei; the condition of matter in interstellar and intergalactic space (composition, density, temperature); the related question as to whether our universe is closed or open.

But I do not wish to dwell any longer on past history nor on future expectations, since there are so many new results that we are all anxious to hear about. So let us go on with this morning's program.

PART I

GALACTIC SOURCES

2. UHURU RESULTS ON GALACTIC X-RAY SOURCES

HARVEY D. TANANBAUM

American Science and Engineering, 955 Mass. Ave., Cambridge, Mass., U.S.A.

Abstract. Galactic X-ray sources studied by UHURU can be classified into several categories, including supernova remnants, transient or nova-like sources, Sco X-1 like sources, and pulsating sources. The latter are discussed in some detail with particular emphasis upon the characteristics of the periodic binary X-ray sources, Cen X-3 and Her X-1. It is suggested that all galactic X-ray sources (except S/N remnants) may be binary systems.

1. Introduction

The UHURU satellite has now been scanning the sky for seventeen months. The satellite was built by American Science and Engineering and the Applied Physics Laboratory of Johns Hopkins University for NASA's Goddard Space Flight Center. The original objectives of the satellite were an all sky survey down to a sensitivity of 10^{-4} Sco X-1 designed to locate stronger sources to $1'$, the study of source spectra from 2–20 keV using proportional counters and 8 channels of pulse height analysis, and a search for time variability. Recently the UHURU group at American Science & Engineering generated a catalog of some 125 X-ray sources (Giacconi *et al.*, 1972). The first figure shows the distribution of these sources on the sky plotted in galactic coordinates. Approximately 35 of the sources were known before UHURU and about 90 new sources have now been found. The sources range from an intensity of $\sim 20/000$ cts s^{-1} for Sco X-1 down to about 2 cts s^{-1} for the source identified with M31 (Andromeda). Note that coverage off the galactic plane is incomplete at present – we have analyzed only some 70 days out of 200 spent scanning the sky.

This paper is concerned with the nature of the galactic X-ray sources while the paper by Dr. Kellogg will discuss extragalactic sources. For the galactic sources the most exciting impact of the UHURU satellite has been the discovery of widespread time variability. In several cases, we have found intensity changes of factors of 2 or more in times of seconds or less, requiring the X-ray sources to be compact. With the satellite we have for the first time been able to conduct detailed precise measurements on a number of the galactic sources.

After a brief description of the X-ray emission of our galaxy as a whole, I would like to discuss the evidence for a working hypothesis that the galactic X-ray sources (except for the supernova remnants) are in fact binary systems. The first to consider the possibility of binary systems being X-ray sources were Hayakawa and Matsuoka in 1964 and the others since are too numerous to list here. Along these lines, I will discuss the general properties of sources similar to Sco X-1 (called Sco X-1-like sources) and then in some detail the properties of 3 pulsating sources showing evidence of a binary nature. Then I would like to discuss some new results on 2 objects – GX263+3 (2U0900−40) and the X-ray source in the Small Magellanic Cloud (2U0115−73)

Bradt and Giacconi (eds.), X- and Gamma-Ray Astronomy, 9–28. All Rights Reserved.
Copyright © 1973 by the IAU.

X-RAY SOURCES OBSERVED BY UHURU

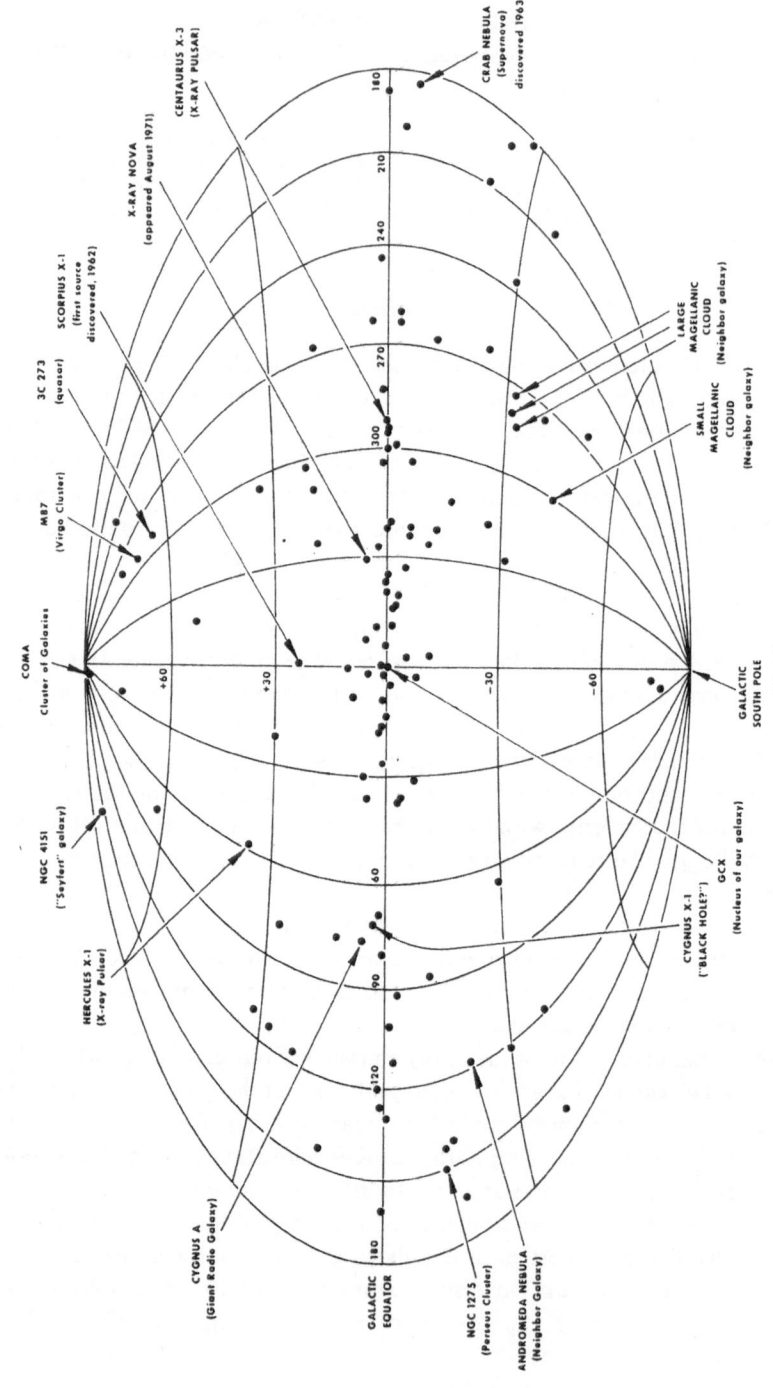

Fig. 1. The X-ray sky as seen by UHURU. Circles indicate X-ray source locations; many of the X-ray sources are labelled. The map is an equal area projection in galactic coordinates.

which both show evidence of a binary nature, yet have some of the properties of the Sco X-1-like objects. These links between the Sco X-1-like objects and the pulsating, binary sources are what lead to the suggestion that all of the galactic X-ray sources (except S/N remnants) may in fact be binaries.

2. The Galaxy as a Whole

When we turn to the X-ray emission from our galaxy we consider first the number versus intensity curve which is shown in Figure 2. We have plotted here for sources within 20° of the galactic plane the number of sources brighter than a given intensity versus intensity with the data corrected for sky coverage. The actual number of sources observed to date within 20° of the plane is 81. The most striking feature of the curve is the difference between the observed slope of the distribution – approximately 0.4 – and a slope of 1 expected for a uniform disk distribution of equal luminosity sources. The much flatter observed slope indicates that we are seeing essentially all of the bright sources in our galaxy and that there is a spread in intrinsic luminosities. Recent work by the Livermore group (Seward *et al.*, 1972) and to a lesser extent by the UHURU group has attempted to determine intrinsic luminosities by measuring low energy cutoffs of sources and assigning distances. This leads to the result that several sources in the direction of the galactic center have a luminosity of order 10^{38} ergs s^{-1}, comparable to the intensities found for 3 sources in the Large Magellanic Cloud (to which

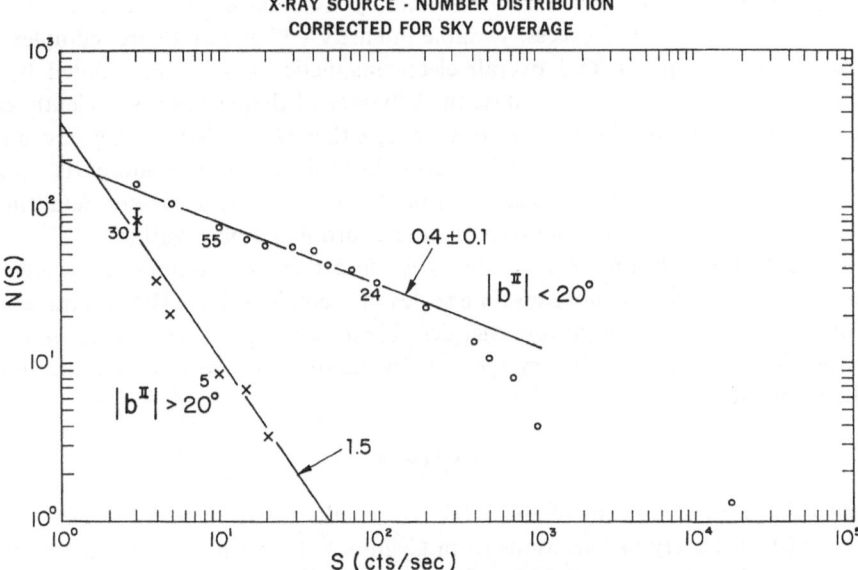

Fig. 2.   X-ray source number distribution as a function of intensity. The number of sources brighter than a given intensity are plotted versus intensity. The sources are divided into 2 groups, those within 20° of the galactic plane and those more than 20° away from the galactic plane.

we know the distance). Many of the other sources can be assigned distances placing them in the various spiral arms of the galaxy. Taken altogether we get a luminosity of order 2×10^{39} ergs s^{-1} for our galaxy, which is comparable to the luminosity which we observe for M31.

3. Classification of X-Ray Sources

Passing on to the nature of individual galactic X ray sources, I will not discuss sources which can be identified with supernova remnants. These will be discussed in a paper by Dr Pounds. Also I will not dwell upon the transient or nova-like sources which have been reported in the literature – Cen X-2, Cen X-4, and more recently from UHURU, 2U1543−47. I would like to point out that the UHURU data on 2U1543−47 show outbursts following the initial appearance of the source and sudden changes in the spectrum. This appears to rule out simple models involving a blast wave or the expansion of a hot gas. Also the absence to date of an optical counterpart, in spite of the small location uncertainty, means that either the object is underluminous for an optical nova or that its distance is of order 10 kpc, producing optical obscuration, and implying an energy of 10^{39} ergs s^{-1} for the X-ray emission.

For the bulk of the galactic sources, based on the observation of large scale time variations, we have previously divided the sources into Sco X-1-like and pulsating. The Sco X-1-like sources show intensity changes of order 50% on time scales of minutes to hours. Their spectra are exponential suggesting thermal bremsstrahlung, and their temperatures vary on the same time scale as the intensity, ranging from 50 to 150 million degrees. Two of the objects, Sco X-1 and Cyg X-2, have similar optical counterparts – blue stars with UV excess, variable intensity – flicker and flare, complex line emission and absorption, and overall electromagnetic emission dominated by the X-rays. Despite several years of study, the behavior of these stars is sufficiently complex that at present we are unable to answer, either way, whether they are binary systems. Sco X-1, GX17+2 and GX9+1 also have similar radio counterparts – highly variable, weak sources. These objects will be discussed in much greater detail in the papers on optical and radio counterparts and coordinated observations.

Turning to the pulsating X-ray sources we find they are dominated by intensity changes of factors of 2 or more on time scales of seconds or less. The spectra tend to be flat and are often cut off at low energies. There is considerable evidence linking a number of these sources to binary systems. In the following sections I present some of the results for these sources.

4. Cygnus X-1

Perhaps the most significant of the UHURU results for the galactic X-ray sources has been the discovery of pulsations from Cygnus X-1, which led to further study of this object and to the present belief that we are dealing with a black hole. I would like to present the data we have on this object and consider the status of the black hole identification. Figure 3 contains data already reported in the literature (Schreier *et al.*,

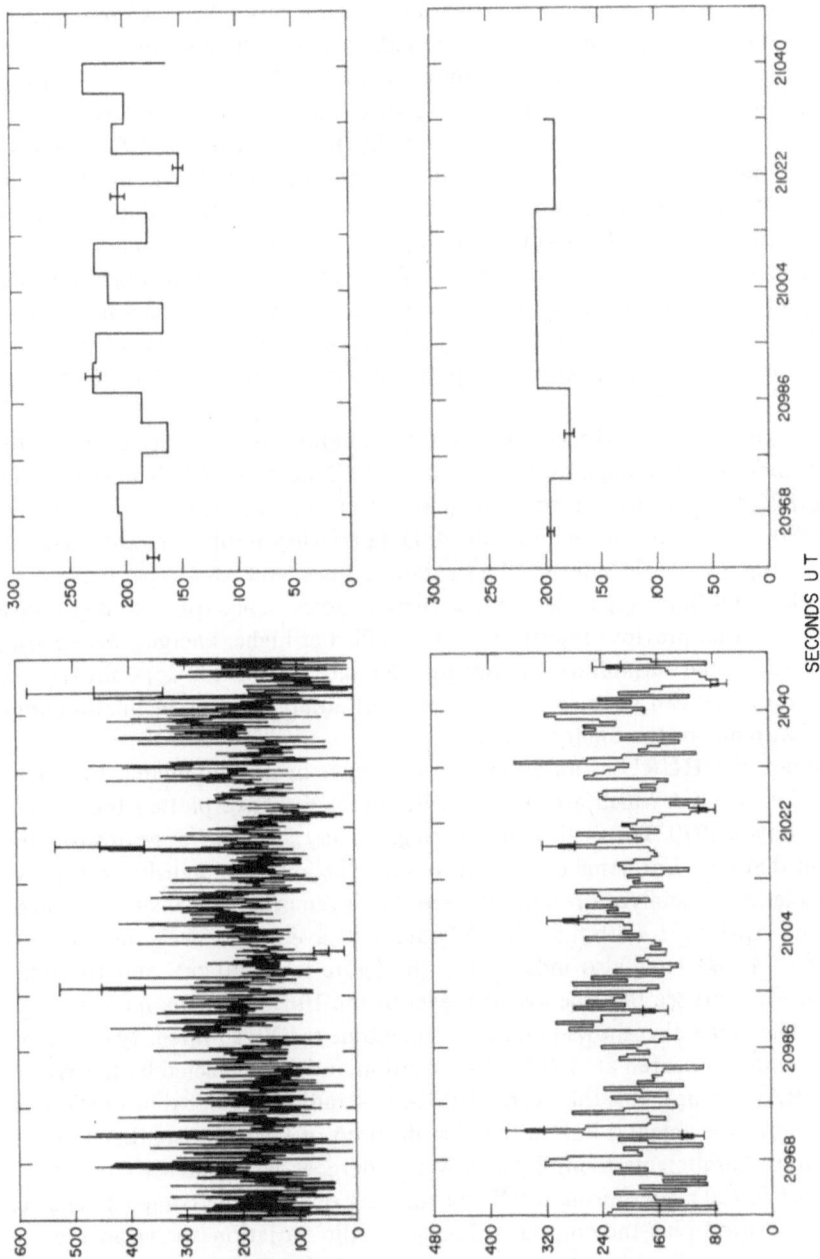

Fig. 3. Observation of Cygnus X-1 on 1971 June 10. Data have been corrected for triangular collimator response. Data are summed over 0.096 s, 0.48 s, 4.8 s, and 14.4 s intervals. Typical 1σ error bars are shown (Schreier *et al.*, 1971).

1971) showing substantial variations in X-ray intensity on time scales from 100 ms to 10's of seconds. Some 80 s of data are shown here summed on 4 time scales from 100 ms up to 14 s. I should point out that similar X-ray variability also reported by scientists at MIT, (Rappaport *et al.*, 1971a), Goddard Space Flight Center (Holt *et al.*, 1971) and Naval Research Laboratory (Shulman *et al.*, 1971) requires us to consider compact objects smaller than 10^9 cm for the X-ray source. With the relatively good X-ray location determined by an MIT rocket flight (Rappaport *et al.*, 1971b) and by UHURU, a radio source was discovered by Braes and Miley (1971) and by Hjellming and Wade (1971). It is this precise radio location that led to the optical identification by Webster and Murdin (1972) and by Bolton (1972) of Cygnus X-1 as a 5.6 day spectroscopic binary system. The central object of this system is a 9th magnitude BO supergiant and conservative mass estimates such as 12 $M_\odot$ lead to a mass in excess of 3 $M_\odot$ for the unseen companion. If the companion is the compact X-ray source then it could be a neutron star more massive than previously considered or it could be a black hole.

We attempted to confirm the identification by looking for a 5.6 day effect on the X-ray light curve. In December 1971 and January 1972 we observed Cygnus X-1 continuously for 35 days. We folded the X-ray data with many different periods including 5.6 days and the results are shown in Figure 4. Data are shown folded modulo 3.0, 5.6, and 6.2 days, the average is indicated by the dotted lines, and 2.0 σ error bars are indicated by the solid lines. We conclude that there is no evidence for a 5.6 day eclipse here, and believe that previous reports of such an effect at higher energies were caused by the large scale time variablility and not by a 5.6 day effect. This does not rule out the identification and can be understood in terms of an appropriate inclination angle for the orbital plane of the binary system.

With the use of UHURU as an observatory we have now accumulated 16 months of data on Cygnus X-1 which are shown in Figure 5. We have plotted the 2–6 keV intensity vs. day of 1970. The vertical lines for a given day show the range of variability observed on that day. For some days we have only the average intensity shown by a dash available in our analyzed results. We see that a remarkable transition occurred in March and April 1971, with the source changing its average 2–6 keV intensity level by a factor of 4. We have also indicated in the figure the 6–10 keV and 10–20 keV X-ray intensities and see that the average level of the 10–20 keV flux increased by a factor of 2. The figure also shows that at the same time the X-ray intensity changed, a weak radio source appeared at the Cyg X-1 location and was detected by the Westerbork and NRAO groups. It is this correlated X-ray – radio behavior that I believe is a most important experimental link in the identification of Cyg X-1 as a black hole.

Reviewing the arguments then, Cygnus X-1 undergoes large intensity changes in times as short as 100 ms requiring the X-ray emitting region to be compact. The very good X-ray position plus the correlated X-ray – radio variation shown in Figure 5 demonstrate the X-ray – radio identification. The optical-radio identification is based on position agreement better than 1″. Then the optical data taken conservatively require at least 3 solar masses in the unseen companion which is the compact X-ray

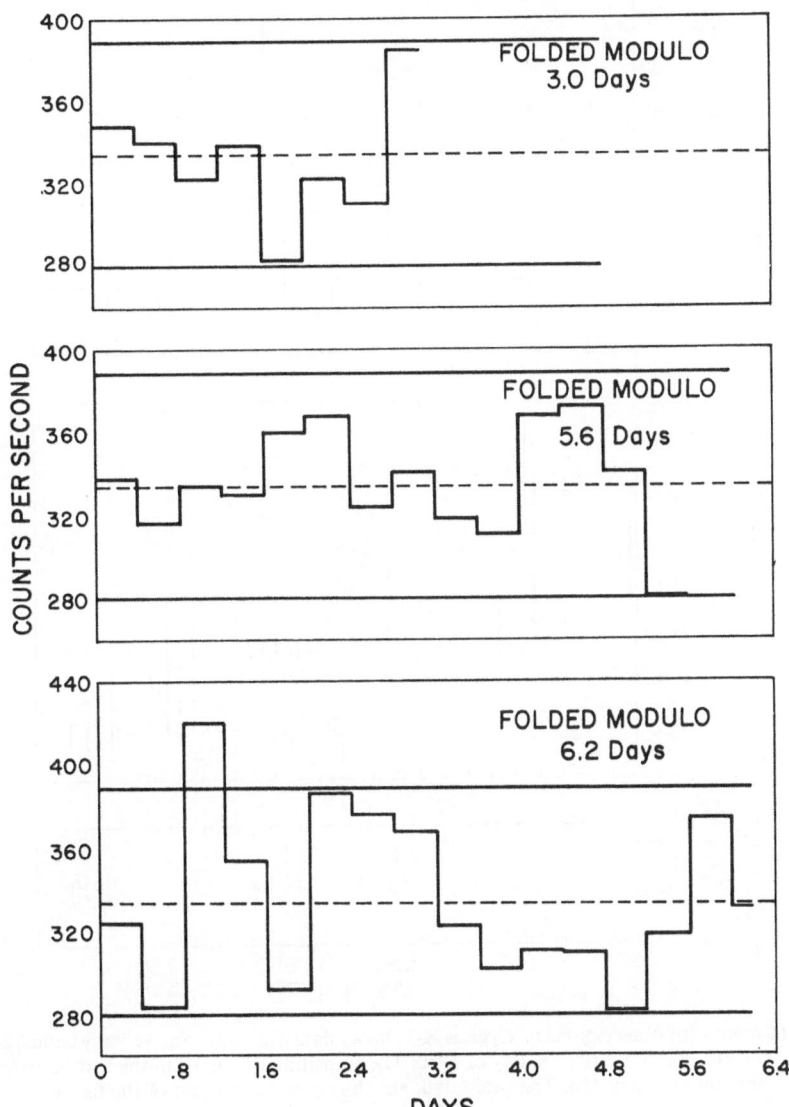

Fig. 4. Cygnus X-1 folded intensity, 17 Dec. 1971 to 21 Jan. 1972. 35 days of 2–6 keV data are shown folded modulo 3.0, 5.6, and 6.2 days. The dotted lines give the average intensity and the solid lines are 2σ error bars.

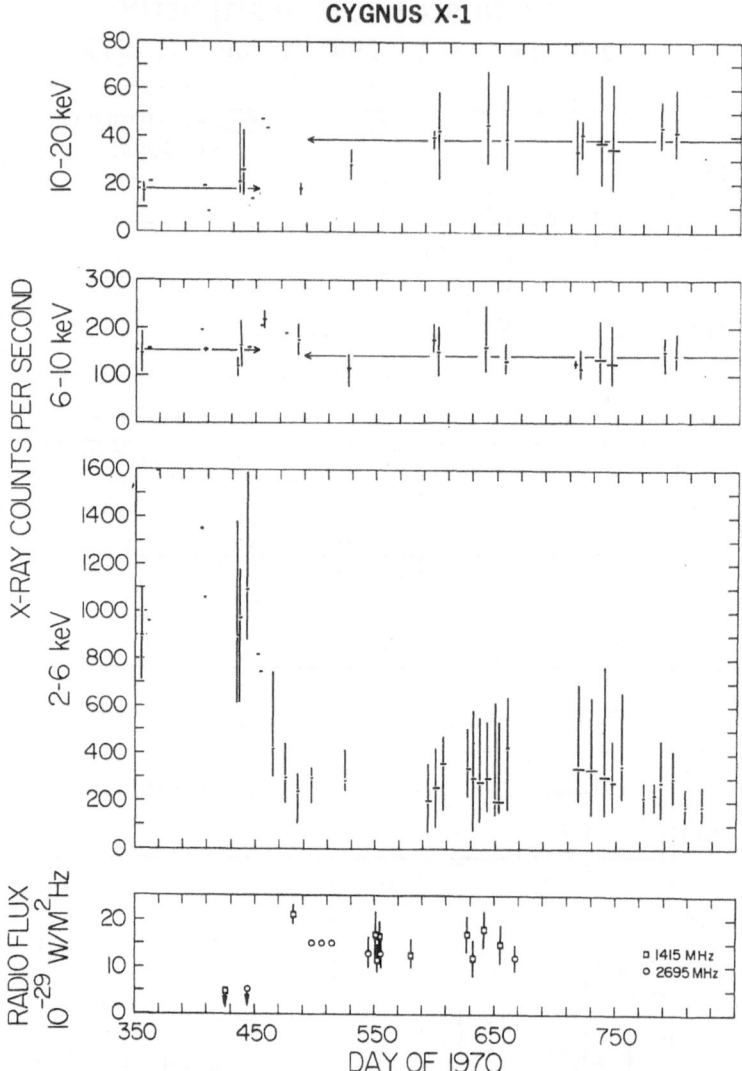

Fig. 5.   16 months of observations of Cygnus X-1. X-ray data are shown for 3 energy bands, 2–6 keV, 6–10 keV, and 10–20 keV plotted vs. day of 1970. The transition discussed in the text occurred in the period near day 450. The radio data are shown at the bottom of the figure.

source. Whether this object could be a massive neutron star or must be a black hole, is a subject more appropriately discussed by the theoreticians.

5. Centaurus X-3 and Hercules X-1

We now come to 2 X-ray sources, Cen X-3 and Hercules X-1, which are identified as

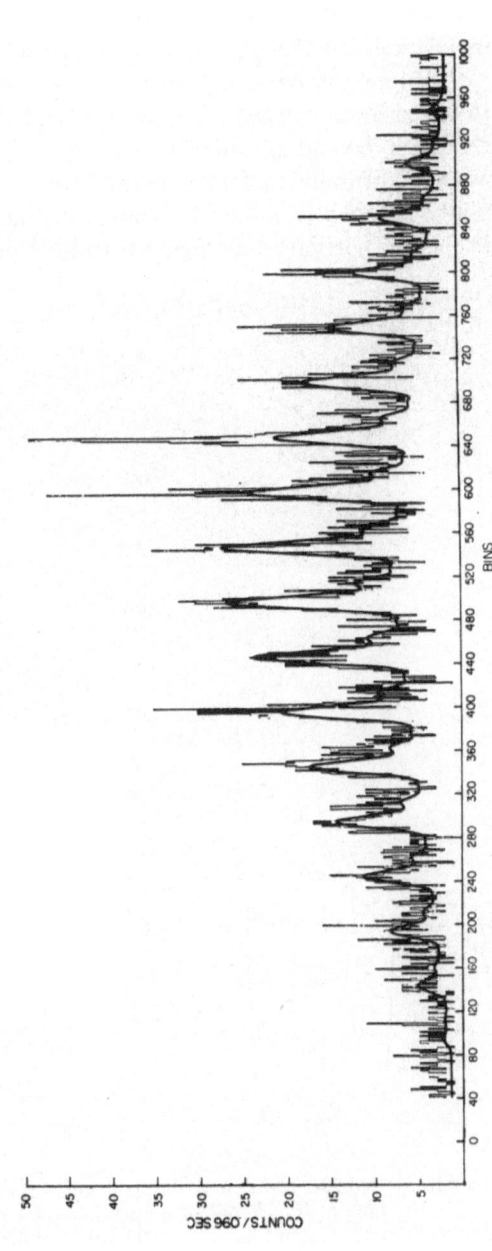

Fig. 6. Counts accumulated in 0.096 s bins from Cen X-3 during a 100-s pass on 7 May 1971. The functional fit obtained by minimizing x^2 is shown as the heavier curve (Schreier *et al.*, 1972, *Astrophys. J. Letters* **172**, L79.)

binaries solely from their X-ray properties. The first source is Cen X-3 which is shown in Figure 6. Here we see a regular pulsing source with a period of 4.8 s. The histogram shows the actual counts observed and the heavier curve is a sine wave plus harmonics fit to the data. The X-ray emission is at least 90% pulsed with the 4.8 s period. Figure 7 shows some of the data that demonstrate that Cen X-3 is an occulting binary system. The bottom portion shows 3 days of intensity data accumulated in May 1971 with a clear cut downward transition followed by an upward transition about a half day later. Many such transitions have been observed and all are fit to a 2.08712±0.00004 day period. By measuring the arrival time of individual 4.8 s pulses we have also determined that the pulsation frequency is Doppler shifted in phase with the 2.087 day occultation cycle as is shown in the top portion of the figure where the pulse arrival

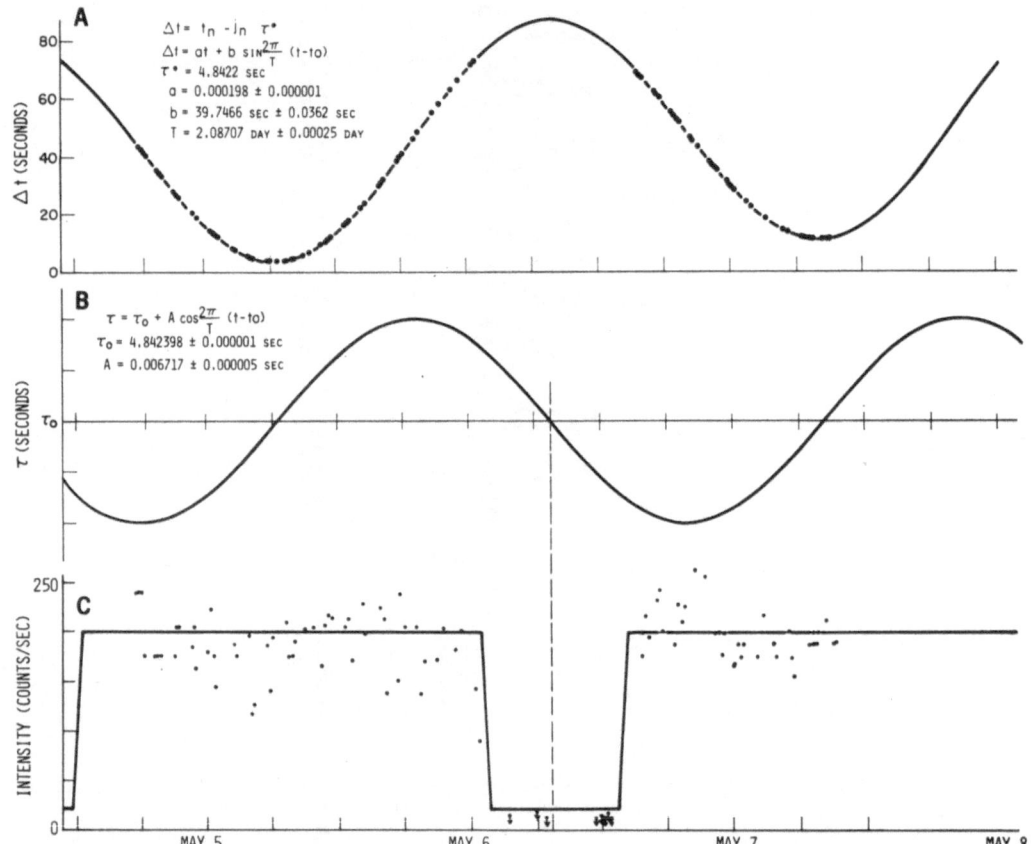

Fig. 7.   Bottom: The intensity observed from Cen X-3 (dots) and the light curve predictions for 5–7 May. Top: The difference Δt between the time of occurrence of a pulse and the time predicted for a constant period, plotted as a function of time. A best fit function and the values of the parameters are given. Center: The dependence of the pulsation period τ on time as derived from the best fit phase function. Note the coincidence of the null points of the period function with the centers of the high and low intensity states (Schreier *et al.*, 1972; see caption Figure 6).

time delays are fit by a sine wave. Under the model of an occulting binary system, we have made very precise determinations of the projected orbital velocity, 415.1 ± 0.4 km s^{-1}, the projected orbital radius $(1.191 \pm 0.001) \times 10^{12}$ cm, and the mass function of the system $(3.074 \pm 0.008) \times 10^{34}$ gm.

Figure 8 is a schematic representation of this system with a compact X-ray object orbiting a central star. We find from the Doppler velocity that the mass of the central star must be at least 15 $M_\odot$ and calculations such as those of R. E. Wilson (1972) for close binaries lead to masses on the order of a tenth of a solar mass for the X-ray emitting object. These calculations assume that the sharpness of the occultation means that the radius of the Roche lobe is greater than or equal to the size of the occulting

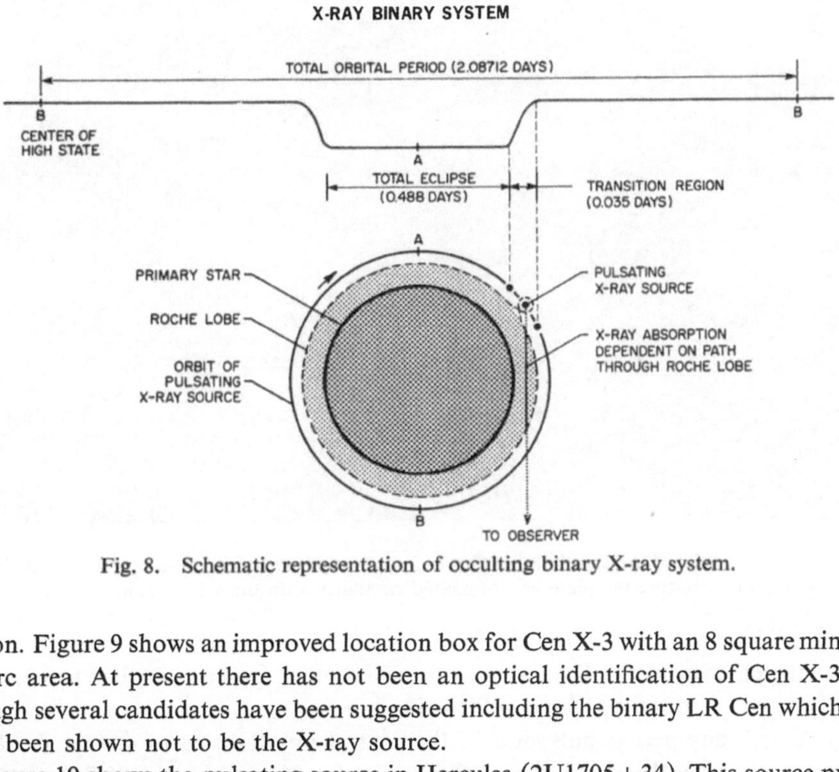

Fig. 8. Schematic representation of occulting binary X-ray system.

region. Figure 9 shows an improved location box for Cen X-3 with an 8 square minutes of arc area. At present there has not been an optical identification of Cen X-3, although several candidates have been suggested including the binary LR Cen which has now been shown not to be the X-ray source.

Figure 10 shows the pulsating source in Hercules (2U1705+34). This source pulses with a 1.24 s period and as for Cen X-3 is essentially totally pulsed. In Figure 11, we see a schematic representation of the light curve for this source. The heavier lines represent all of the data taken on the source from its discovery in November 1971 through March 1972. The source shows a 1.70017 ± 0.00004 day occultation cycle indicated by the dotted curve with many cycles now observed. The data show a further periodicity. For 9 or 10 days the source is intense and pulsing and can be seen following the 1.70 day occultation cycle. Then for 26 days the source is too weak to be observed. We have now observed 6 cycles of this 35.7 periodicity, the most recent being

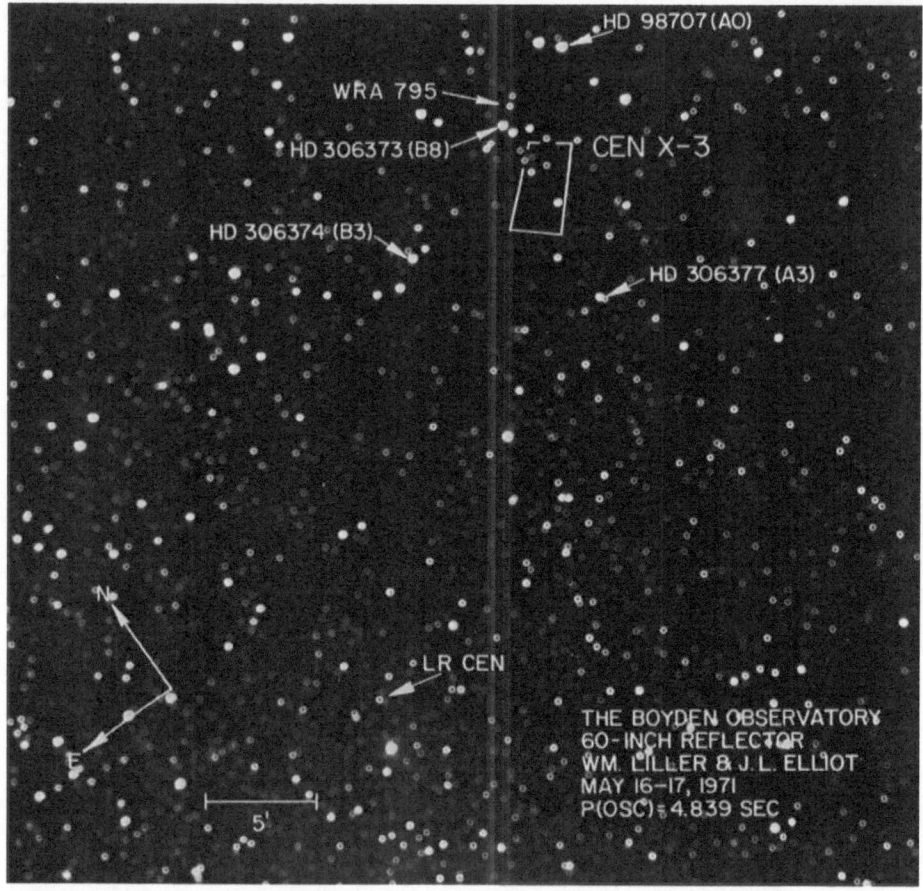

Fig. 9.   90% confidence location box for Cen X-3 superimposed on plate taken by Liller and Elliot.
During the exposure the plate was oscillated circularly with the 4.8 s period of Cen X-3.

during the past few weeks. Several models have been proposed to explain this 35.7 day
cycle; the two currently under consideration are:

(1) A 35.7 day period pulsation of the central object so as to embed the X-ray
source in its atmosphere and absorb the X rays for 26 out of every 35.7 days.

(2) Instabilities in the atmosphere of the central star near its Roche lobe. This model
assumes accretion onto the X-ray star as the energy source; when the Roche lobe is no
longer filled, the energy source is effectively turned off until the atmosphere of the
central star again fills the Roche lobe.

For this X-ray source we again observe a Doppler shift of the 1.24 s pulsations in
phase with the 1.700 day cycle. We find a projected orbit velocity of 169.2 ± 0.4 km s^{-1},
a projected orbital radius of $(3.95 \pm 0.01) \times 10^{11}$ cm, and a mass function of
$(1.69 \pm 0.01) \times 10^{33}$ gm. Since the duration of the occultation is shorter than for Cen

X-3, calculations such as those of Wilson (1972) mentioned earlier, lead to maximum masses from 0.2 to 3.0 $M_\odot$ for the X-ray source and 1.2 to 3.2 $M_\odot$ for the central object assuming 90° inclinations. A search is presently underway for the optical counterpart for this source.

I would also like to point out here that these X-ray measurements of velocities, radii, masses, and pulsation frequencies compare favorably with the precision of most ground based measurements.

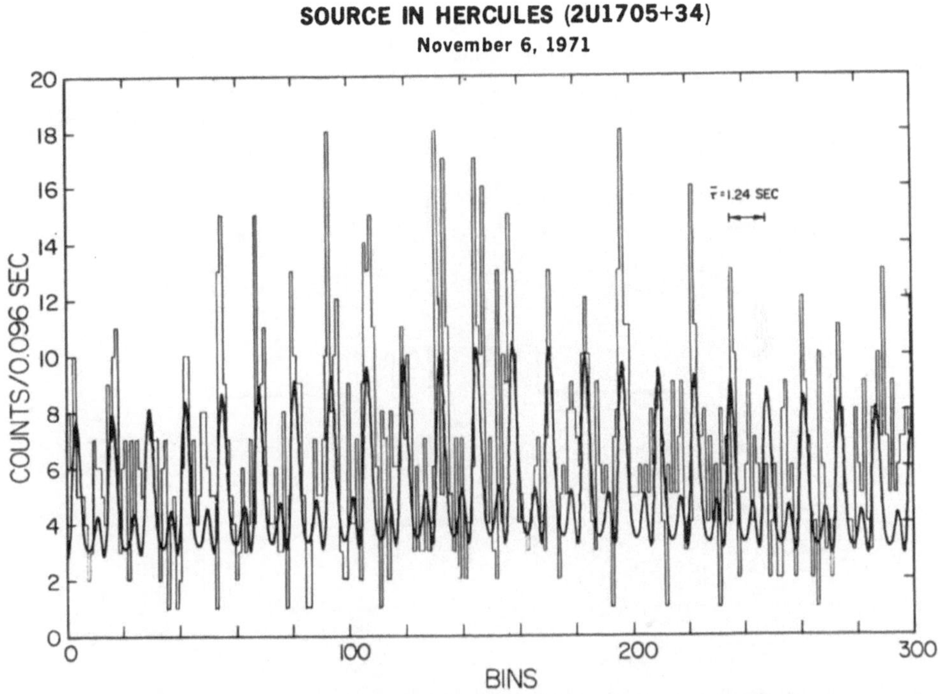

SOURCE IN HERCULES (2U1705+34)
November 6, 1971

Fig. 10.   Counts accumulated in 0.096-s bins from Hercules X-1 during the central 30 s of a 100-s pass on 6 Nov. 1971. The heavier curve is a minimum x^2 fit to the pulsations of a sine function, its first and second harmonics plus a constant, modulated by the triangular response of the collimator (Tananbaum *et al.*, 1972, *Astrophys. J. Letters* **174**, L143.)

6. GX263+3 (2U0900−40)

Still another candidate for identification with a binary system is the X-ray source GX263+3 (2U0900−40). Figure 12 shows the location information for this source; the outer box is the 2U catalog location, the inner box is a 4 sq. minutes of arc location obtained recently using additional UHURU data. The 6.7 magnitude B0 supergiant originally suggested by Kellogg and Murray (1971) and by Bradt and Kunkel (1971) for this identification is shown in the center. This star has been studied by Brucato and Kristian (1972) who found it to have a variable velocity suggesting a binary and to be

HERCULES X-1

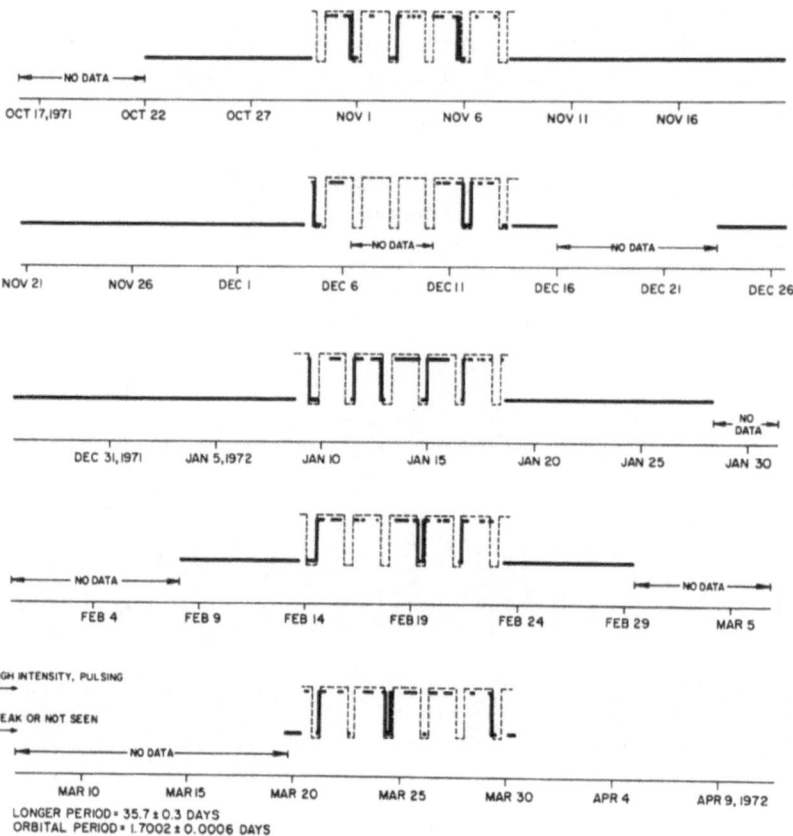

Fig. 11. The heavier line represents schematically the observations of Hercules X-1 as either high intensity pulsed sighting or as weak or not observed sightings. Short term breaks are due to blocking of the source by the Earth or due to dropouts in Quick Look data transmission. Intervals where the satellite was oriented so as not to scan the Hercules source are marked 'no data'. The lighter dashed curve shows a single 1.700-day period fit to all of the observations. Each row in the figure has a duration of 35.7 days and the regular occurrence of the 9-day high intensity pulsed intervals demonstrates the 35.7 day cycle. (Tananbaum *et al.*, 1972 – see caption Figure 10).

at a distance of about 1.3 kpc. Very recently the star was found to be a 7.0 day spectroscopic binary by Hiltner and Osmer (1972) and is reported to have emission lines similar to Cygnus X-1. We are presently observing the source to search for any 7.0 day effects on the X-ray emission.

As an X-ray source, the 2U catalog reported observations of 2U0900−40 ranging from 25 to 75 counts s^{-1}. Analysis of some additional data gives us the results shown in Figure 13 where we have a half day's data with X-ray intensity from 2–6 keV plotted

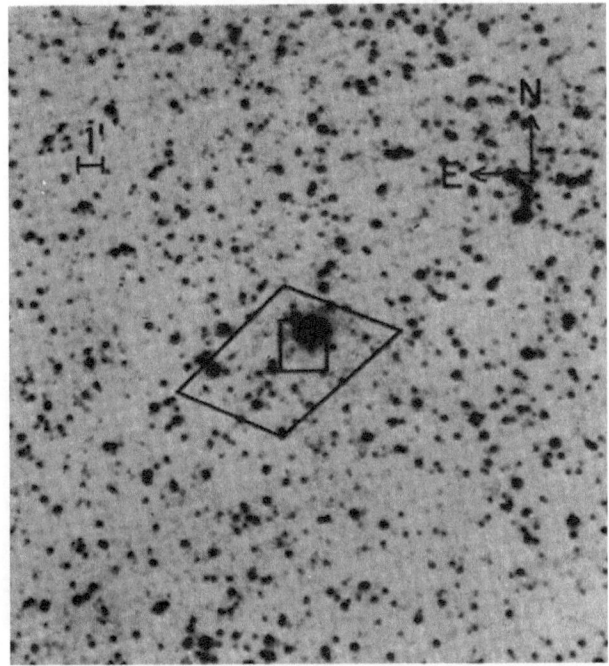

Fig. 12. 90% confidence location box for 2U0900–40 (GX263 + 3). The outer box is the published 2U location and the inner box a more recent determination from additional UHURU data. The 6.7 magnitude B0 supergiant suggested for the identification is the bright star shown in the center.

vs. time for this X-ray source. We see that the source intensity changed by a factor of 30 in 2 hr time and that other intensity variations also occurred. We have also studied the energy spectrum for this source and find it variable but very cut off at low energies at the source and very flat. In this respect the source resembles Cen X-3 and Herc X-1. From the optical identification using a distance of 1.3 kpc, we find a 2–20 keV peak luminosity of 8×10^{36} ergs s^{-1}.

We have also looked for faster time scale intensity variations and Figure 14 contains 20 s of data obtained at the flare peak. We see about a 30% intensity pulse over a time of 2 s. Such behavior is observed only when the source is brightest. Attempting to classify this X-ray source I would have to say that the dominant time scale of intensity variability, minutes to hours, is more typical of Sco X-1-like sources, while the range of intensity, 30, and spectral shape are similar to the pulsating sources.

7. Small Magellanic Cloud (2U0115 – 73)

The last source I would like to discuss is the one in the Small Magellanic Cloud (2U0115 – 73). This source has previously been reported by us as variable in intensity (Leong *et al.*, 1971) and in Figure 15 we show 8 days of data obtained in January 1971

2U0900-40 (GX263+3) 2-6 keV

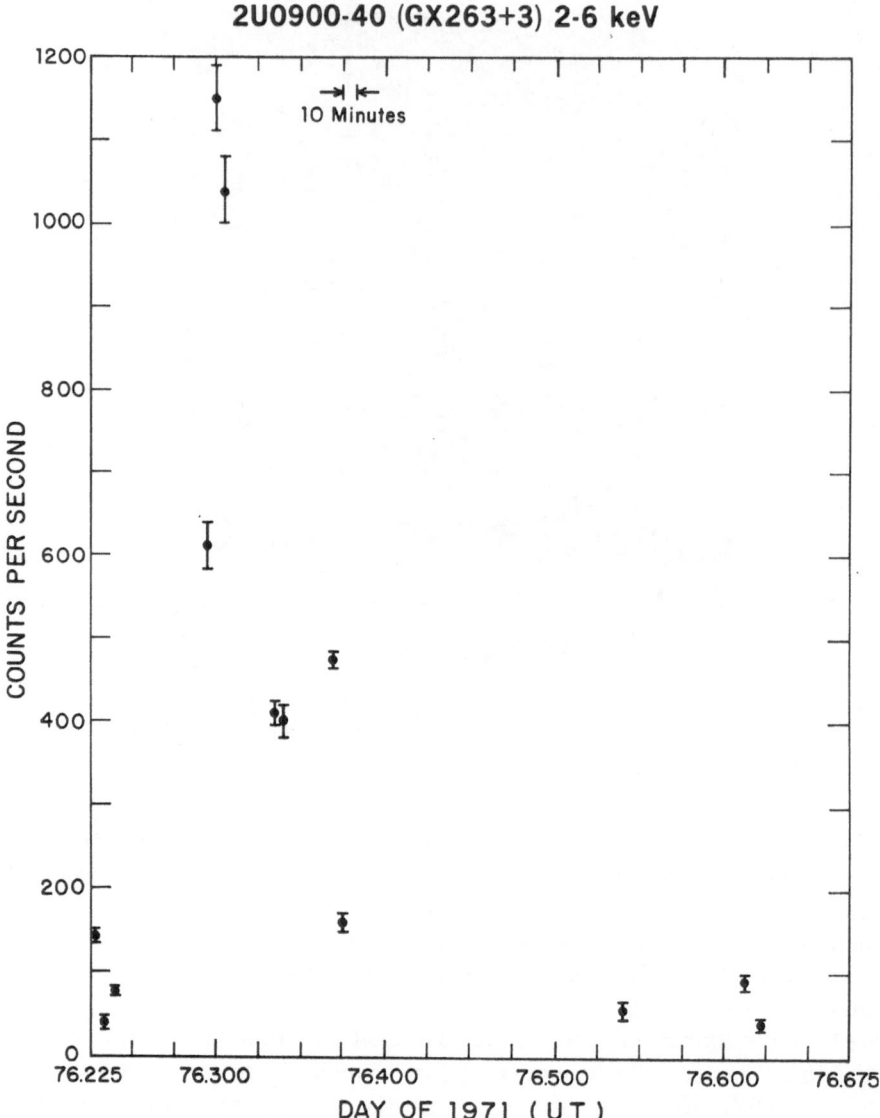

Fig. 13. X-ray intensity for 2U0900−40 plotted as a function of time for ½ day. Data have been corrected for aspect and 1σ error bars are shown.

on this source. Here again the data show highs, lows and transitions, suggesting an occulting system – the period is 3.9 days and the duration of the occultation is 0.6 days. This source is very weak (1/1000 of Sco X-1 in observed intensity) and we can therefore not yet say whether it pulsates on a time scale of seconds. Its spectrum is cut off at low energies at 1.8 keV and a power law fit has an energy index of about 0.15.

In this instance we know the distance to the source, getting a 2–10 keV luminosity of 10^{38} ergs s^{-1} (and using Livermore data (Price *et al.*, 1971) on this source at 40 keV we get $\sim 10^{39}$ ergs s^{-1}). In this respect, the 10^{38} ergs s^{-1} luminosity is comparable to that we found for several of the Sco X-1 like objects towards the center of our own galaxy. Knowledge of the distance to the SMC and thereby the intrinsic source luminosity demonstrates that a binary system can in fact produce luminosities of

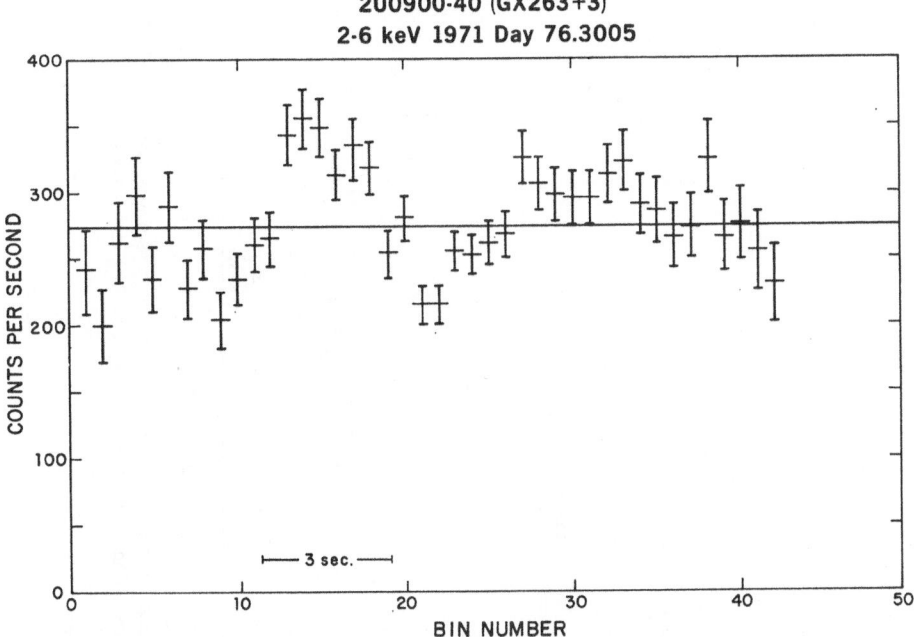

Fig. 14.   20 s of data for 2U0900–40 taken at the peak of the intensity shown in Figure 13. Data have been corrected for the triangular response of the collimator. The solid line shows the average counting level and significant variations are seen.

10^{38} ergs s^{-1} and that the Sco X-1-like sources are not excluded from being binaries by their high luminosities. The SMC source is also significant since the occultations show that we are observing for the first time in X-rays a stellar system in another galaxy. The source is also interesting because when it is bright it accounts for at least 98% of the 2–6 keV X-ray emission of the SMC.

8. Conclusion

From these results it should be obvious that there is still a wealth of information to be obtained from the detailed UHURU data on individual galactic sources. It seems that the division of the sources into Sco X-1-like and pulsating may not hold up, since the

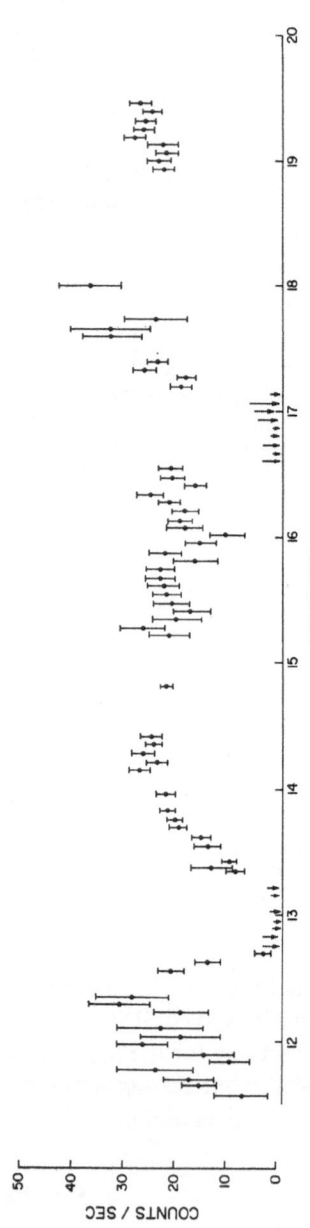

Fig. 15. Light curve for the X-ray source in the Small Magellanic Cloud. The 2–6 keV intensity is plotted for 8 days in January 1971. The data show highs, lows, and transitions suggesting a 3.9-day occulting binary system.

sources GX2 63+3 and SMC seem to have some properties of each class. It is already clear that at least some of the galactic X-ray sources are binary systems and perhaps all of them (except SN remnants) are. Binary models such as those suggested by Pringle and Rees (1972) to be discussed in following papers consider how differences in the nature of the secondary – its mass, radius, magnetic field, spin rate, orbital radius, and in the mass accretion rate could give rise to the variety of X-ray phenomena observed. An important question raised in a recent review by Burbidge (1972) is what makes these few binary systems X-ray sources while most others are not.

Several significant results come from the identification of galactic X-ray sources with binary systems:
- We have an energy mechanism to produce the X-rays – accretion;
- We can determine distances from optical data on the central stars and thereby determine absolute luminosities;
- We can study the final states of stellar evolution – white dwarfs, neutron stars and black holes;
- We can study mass transfer in binaries;
- We can use the X-ray source to probe the central star atmosphere;
- We can determine the masses of both objects in a pulsating binary system if we make an optical identification. The information available is the same as for a double-line spectroscopic binary.

Acknowledgements

I wish to acknowledge the assistance I have had from the X-ray astronomy group at American Science and Engineering. In particular Dr E. Schreier and Mr R. Levinson of AS&E and Mr W. Forman and Miss C. Jones of the Harvard College Observatory contributed very substantially to the analysis of all of the data. I have also benefited from discussions with Dr R. Giacconi and Dr W. Tucker regarding the contents of the paper. This research was sponsored under NASA Contract NAS5-11422.

References

Bolton, C. T.: 1972, *Nature* **235**, 271.
Bradt, H. and Kunkel, W.: 1971, private communication.
Braes, L. and Miley, G. K.: 1971, *Nature* **232**, 246.
Brucato, R. J. and Kristian, J.: 1972, *Astrophys. J. Letters* **173**, L105.
Burbidge, G.: 1972, *Comm. Astrophys. Space Phys.*, to be published.
Giacconi, R., Murray, S., Gursky, H., Kellogg, E., Schreier, E., and Tananbaum, H.: 1972, *Astrophys. J.*, to be published.
Hayakawa, S. and Matsuoka, M.: 1964, *Prog. Theor. Phys. Suppl.* **30**, 204.
Hiltner, W. A. and Osmer, P.: 1972, *IAU Circular*, No. 2398 (and private communication).
Hjellming, R. M. and Wade, C. M.: 1971, *Astrophys. J. Letters* **168**, L21.
Holt, S., Boldt, E., Schwartz, D., Serlemitsos, P., and Bleach, R.: 1971, *Astrophys. J. Letters* **166**, L65.
Kellogg, E. and Murray, S.: 1971, private communication
Leong, C., Kellogg, E., Gursky, H., Tananbaum, H., and Giacconi, R.: 1971, *Astrophys. J. Letters* **170**, L67.

Price, R. E., Groves, D. J., Rodrigues, S. M., Seward, F., Swift, C. D., and Toor, A.: 1971, *Astrophys. J. Letters* **168**, L7.

Pringle, J. E. and Rees, M. J.: 1972, *Astron. Astrophys.*, to be published.

Rappaport, S., Doxsey, R., and Zaumen, W.: 1971a, *Astrophys. J. Letters* **168**, L43.

Rappaport, S., Zaumen, W., and Doxsey, R.: 1971b, *Astrophys. J. Letters* **168**, L17.

Schreier, E., Gursky, H., Kellogg, E., Tananbaum, H., and Giacconi, R.: 1971, *Astrophys. J. Letters* **170**, L21.

Seward, F. D., Burginyon, G. A., Grader, R. J., Hill, R. W., and Palmieri, T. M.: 1972, to be published.

Shulman, S., Fritz, G., Meekins, J., and Friedman, H.: 1971, *Astrophys. J. Letters* **168**, L49.

Webster, L. and Murdin, P.: 1972, *Nature* **235**, 37.

Wilson, R. E.: 1972, *Astrophys. J. Letters* **174**, L27.

3. OBSERVATIONS OF COSMIC X-RAY SOURCES BY THE MIT INSTRUMENT ON THE OSO-7*

G. W. CLARK

Dept. of Physics and Center for Space Research, Massachusetts Institute of Technology, Cambridge, Mass., U.S.A.

Abstract. The Seventh Orbiting Solar Observatory, OSO-7, carries on board a 1–60 keV X-ray detector for the purpose of measuring the positions, spectra and time variations of X-ray sources. The instrument is described herein, and several topics under investigation and results in publication are mentioned.

A 1–60 keV X-ray detector** has been operating successfully in the wheel section of the Seventh Orbiting Solar Observatory since it was launched on September 29, 1971. The instrument was prepared at MIT for the purposes of measuring the positions, spectra and time variation of X-ray sources.

The detectors are two banks of gas proportional counters behind tubular collimators that define two circular fields of view, one with a 1° FWHM centered 15° above the wheel plane, and the other with a 3° FWHM centered 15° below the wheel plane. Figure 1 shows the arrangement of the detectors and collimators and their energy response functions. As the wheel rotates with a period of about 2 s the fields of view sweep out two circular scan bands in the sky. The counts are accumulated in 256 azimuth bins, the contents of which are telemetered sequentially with a period of 191 s. As the spin axis is precessed to maintain it at an angle near 90° from the Sun direction, the scan bands gradually sweep over the sky. Figure 2 is a summary of the sky exposure which was obtained during the first six months in orbit.

More complete descriptions of the instrument and its performance characteristics have been submitted for publication (Clark *et al.*, 1972a). Results have been obtained so far on the following topics:

(1) Measurement of the position and spectrum of Hercules X-1*** (see Figures 3 and 4) (Clark *et al.*, 1972b).

(2) 20-day observation of Scorpio X-1 with simultaneous radio observations (see Figure 5).

(3) The spectrum and time variations of a transient X-ray source in Lupus.

(4) Evidence of the interstellar absorption of 1–3 keV background radiation at the galactic equator near $l^{II} = 145°$.

(5) Upper limits on the X-ray luminosity of three extragalactic supernovae near the times of their optical discovery.

---

* Supported in part by Contract NGL 22-009-115 from the National Aeronautics and Space Administration.
** Participating scientists at MIT are H. V. Bradt, C. Cañizares, D. Hearn W. H. G. Lewin, T. H. Markert, H. W. Schnopper, and G. Sprott.
*** Note added in proof: Her X-1 has been identified with the variable HZ Herculis which lies within 0.2° of the center of the OSO-7 error circle (Liller, 1972).

Bradt and Giacconi (eds.), X- and Gamma-Ray Astronomy, 29–35. All Rights Reserved.
Copyright © 1973 by the IAU.

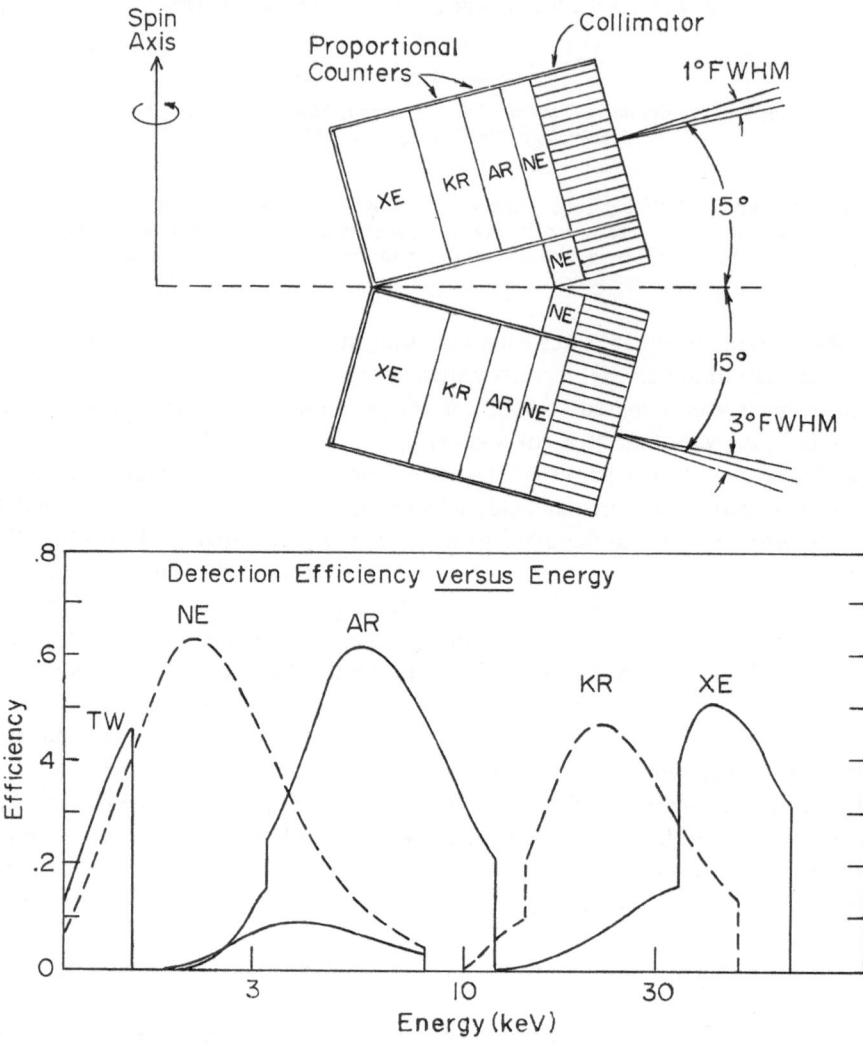

Fig. 1.   Schematic diagram of the 1–60 keV X-ray detectors on the OSO-7 and plots of the detection efficiencies of the proportional counters.

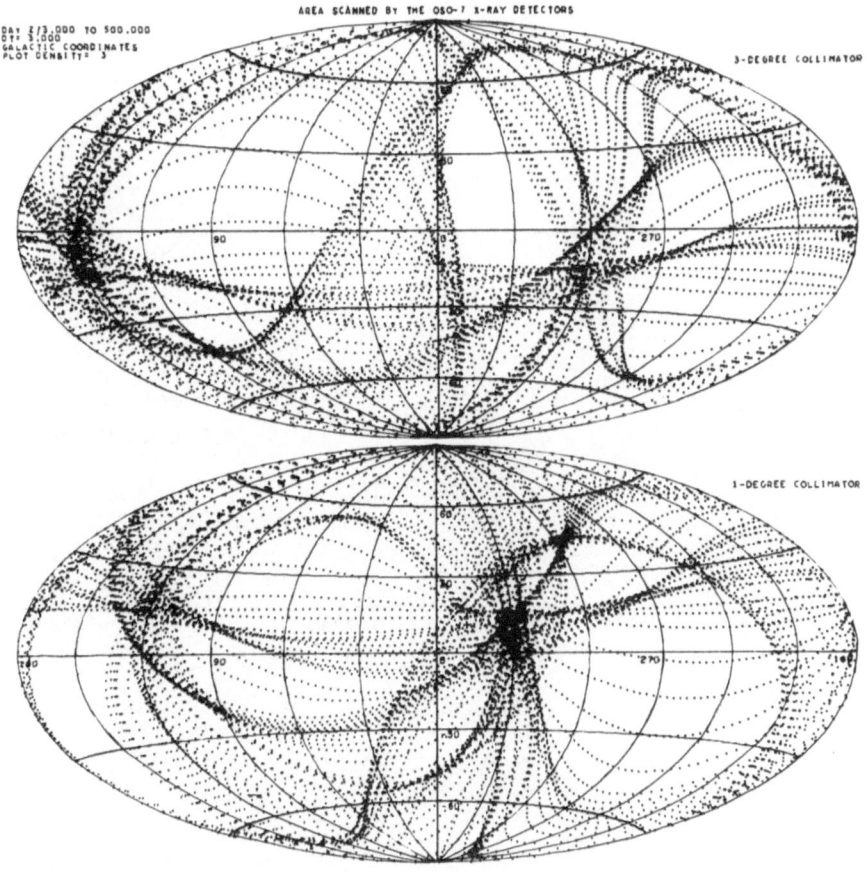

Fig. 2. Maps of sky exposure in galactic coordinates during the first six months of observations.

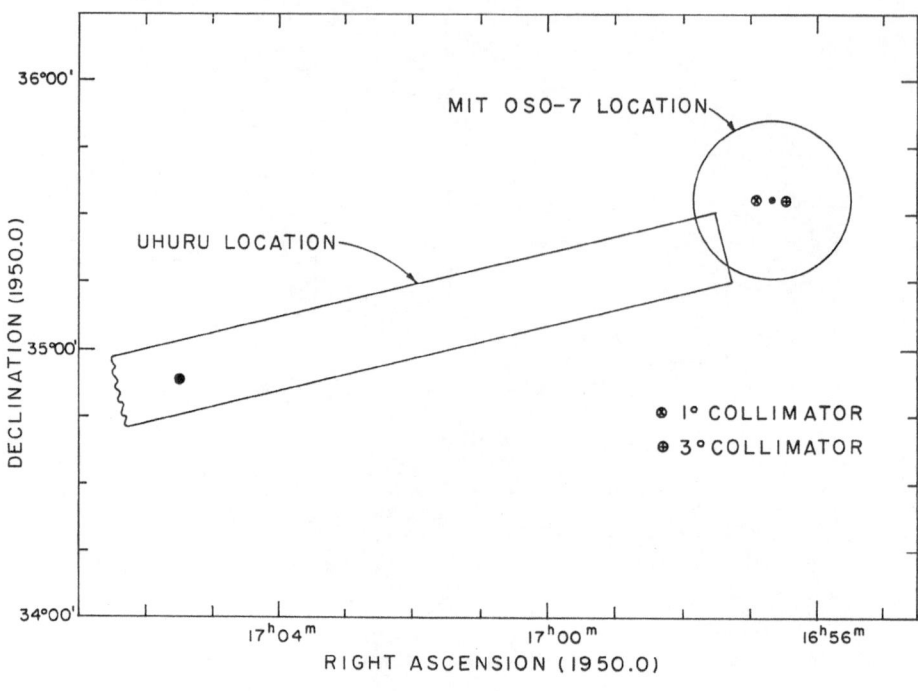

Fig. 3.   Estimated position of Hercules X−1.

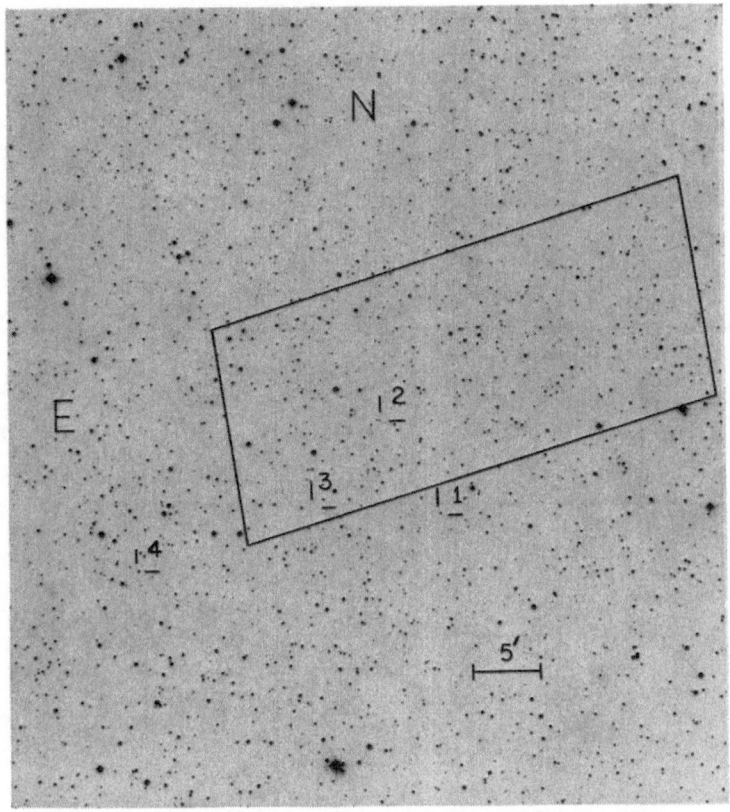

Fig. 4. Enlargement of a Palomar Sky Survey print (National Geographic Society) of the Her X-1 region. The parallelogram represents the approximate intersection of the OSO-7 error box and the extended UHURU error box. Four radio sources (2695 MHz) are indicated (from Doxsey *et al.*, 1972). [*Editor's note in proof*: The recently proposed and well substantiated optical candidate, HZ Herculis, (Davidsen *et al.* and Forman *et al.*, in press) is the moderately bright star 4.0 mm north and 19.8 mm east of the southwest corner of the parallelogram. It is the end of the handle of a 'dipper' composed of fainter stars.]

G. W. CLARK

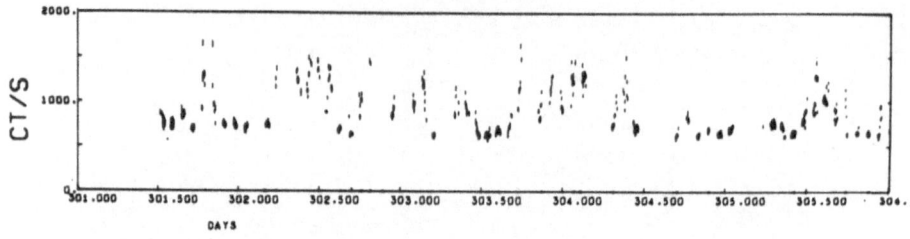

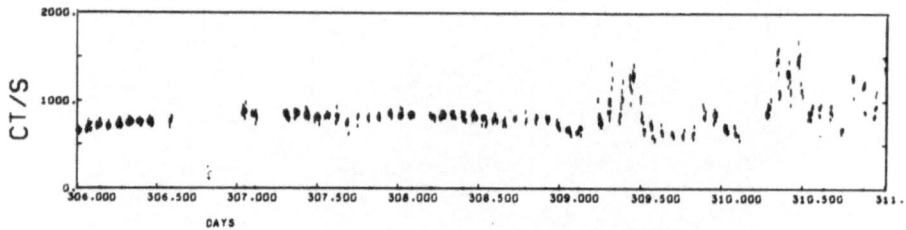

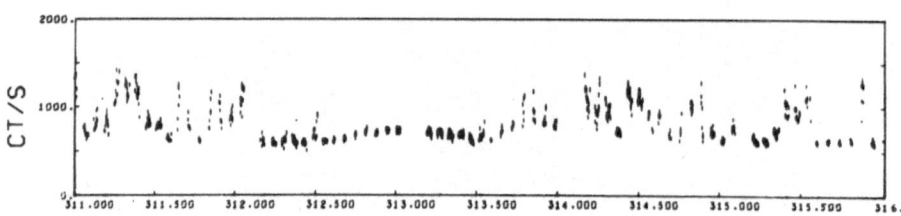

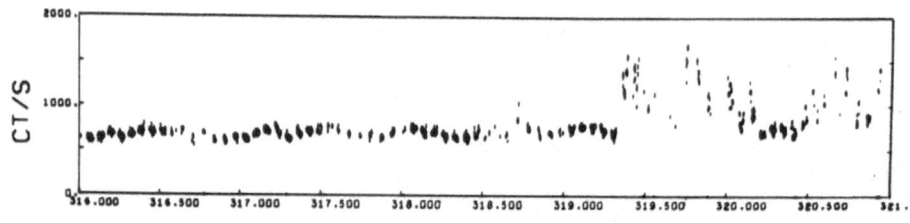

Fig. 5.   Twenty-day observation of Sco X-1 as recorded in the 3° AR detector.

Acknowledgements

The instrument was built at the Center for Space Research of MIT under the supervision of R. Rasche and R. Taylor. The concept of the multichamber detector is due to J. Stein. The data processing system was prepared by P. Northridge, T. Thorsos and S. Watt.

Reference

Clark, G. W., Bradt, H. V., Lewin, W. H. G., Markert, T. H., Schnopper, H. W., and Sprott, G. F.: 1972a, *Astrophys. J.*, in press.
Clark, G. W., Bradt, H. V., Lewin, W. H. G., Markert, T. H., Schnopper, H. W., and Sprott, G. F.: 1972b, *Astrophys. J. Letters*, in press.
Doxsey, R., Murthy, G. T., Rappaport, S., Zaumen, W., and Spencer, J.: 1972, *Astrophys. J. Letters* **176**, L15.
Liller, W.: 1972, IAU Circular 2415 and 2427.

4. BINARY SYSTEMS AS X-RAY SOURCES: A REVIEW*

ROBERT P. KRAFT

*Lick Observatory, Board of Studies in Astronomy and Astrophysics,
University of California, Santa Cruz, Calif., U.S.A.*

Abstract. The observational properties of Sco X-1, Cyg X-2, and Cen X-3 are reviewed in connection with the hypothesis that X-ray power is derived from gravitational energy released when matter is accreted onto the surface of one component in a mass transfer binary star. Evolutionary mechanisms for producing suitable types of binaries are considered. The following boundary conditions on possible evolutionary models are also treated briefly: (1) a quite significant fraction of hard X-ray sources are associated with the nuclear bulge of the galaxy; (2) mass-transfer binaries such as U Gem stars are not hard X-ray sources; (3) counts of binary stars lead to a considerably larger number of X-ray source candidates than are actually observed.

1. Introduction

As a result of the first optical studies of compact X-ray sources, speculation arose that X-ray power might be derived from gravitational energy released when matter was accreted onto the surface of one component in a mass-transfer binary star. The hypothesis was generated by the observation that the optical spectrum of Sco X-1 (Sandage *et al.*, 1966) is similar to that of an old nova or U Geminorum star; it is well known (cf. Kraft, 1962, 1964) that such objects are close binaries in which one component overflows its lobe of the inner zero-velocity surface and transfers mass to its collapsed, probably white dwarf, companion. If the energy of infall of atoms onto a star of radius R and mass $\mathfrak{m}$ is converted into heating the gas, the temperature will be of order

$$T \sim 10^7 \, (\text{K}) \, \frac{\mathfrak{m}/\mathfrak{m}_\odot}{R/R_\odot},\tag{1}$$

so that for objects with $\mathfrak{m}/R$ values in the range 1 to 10 or more (in solar units) *i.e.*, for degenerate objects like white dwarfs, thermal bremsstrahlung with $T > 10^7 \, \text{K}$ could conceivably be emitted (Chodil *et al* 1965; Grader *et al.*, 1966; Peterson and Jacobson, 1966). This idea was given early treatment by Cameron and Mock (1967), Shklovskii (1967) and Prendergast and Burbidge (1968); the last named carried out the most extensive calculations.

In the Prendergast-Burbidge model, matter flows through the inner Lagrangian point into a disk surrounding a white dwarf; owing to radial viscous dissipation, the disk spreads out. The outer part speeds up and moves out while the inner part slows down and moves inward. Detailed hydrodynamical and radiative transfer

---

* Contributions from the Lick Observatory, No. 368. This paper was prepared for, but not given at the symposium. In the author's absence, Dr E. van den Heuvel kindly consented to present his own review of similar material.

Bradt and Giacconi (eds.), X- and Gamma-Ray Astronomy, 36–50. All Rights Reserved.

equations are set up with prescribed initial values of tangential velocity, density and temperature, along with mass-flux and angular momentum as parameters. The calculations show that X-rays can be emitted from the inner part of the disk and must be transferred through the outer cooler part which emits the optical radiation. X-ray luminosities of the order of 10^{35-36} erg s^{-1} were achieved for accretion onto white dwarfs with a mass transfer rate of 2×10^{19} gm s^{-1} (3×10^{-6} m$_\odot$ yr^{-1}). This transfer rate is one to three orders of magnitude larger than rates measured or inferred for U Gem systems (cf. Crawford and Kraft, 1956; Krzeminski and Smak, 1971), an interesting point to be dealt with later.

The reasonable success of this model spurred an interest in the detection of binary motion both in Sco X-1 and in the less-certainly identified Cyg X-2. Extensive photometric and spectroscopic observations of the former (Westphal et al., 1968; Hiltner et al., 1970) showed, however, no certain evidence either of eclipses or of binary motion; erratic changes of brightness of a magnitude or more and of radial velocity by several hundred km s^{-1} were found but contained no evidence of periodicity. Erratic fluctuations in brightness were detected (Kristian et al., 1967) also in Cyg X-2 along with changes in radial velocity of several hundred km s^{-1} (Burbidge et al., 1967; Kraft and Demoulin, 1967; Kraft and Miller, 1969), but again no certain evidence of binary motion could be demonstrated. The main evidence for the duplicity of Cyg X-2 remained its composite spectrum: a late F or G-type star is accompanied by the emission lines of He II (later λ4650 C III–N III [Bopp and Vanden Bout, 1972]), rather like the spectrum of binary system Nova Per (1901). These results thus led to an epoch in which the mass transfer binary model for galactic compact X-ray sources was decidedly moot, although Wilson (1970) was able to give a binary picture for Cyg X-2 that explained all the observations at least qualitatively. This state of affairs continued until the remarkable properties of the X-ray source Cen X-3 were discovered by UHURU (Giacconi et al., 1971; Schreier, et al., 1972).

Cen X-3 contains an X-ray pulsar with period 4.8 s which is eclipsed every 2.087 days by an unseen (i.e., X-ray dark) companion. The eclipse lasts for a time D equal to 23 % of the period, and the reduced X-ray flux seen during the eclipse is presumed to arise as a result of scattering in a circumstellar cloud. The orbital velocity of the pulsar was derived from its smooth oscillatory period change, (Schreier et al., 1972), and the spectroscopic mass function was found to be

$$\frac{m_2^3 \sin^3 i}{(m_1 + m_2)^2} = 15 \, m_\odot, \tag{2}$$

where m_1 and m_2 are the masses of the pulsar and of the unseen companion, respectively, and i is the inclination of the orbit (angle between the line of sight and the imaginary plane of the sky).*

Application of the mass-transfer model enables one to place limits on the masses

* In this paper, the quantity m_1 will always refer to the more massive component when the stars were on the main sequence.

of the objects. This comes about because the unseen component must fill its lobe of the inner zero-velocity surface if it is to be a source of matter for accretion by the pulsar. The radius R of the unseen component, defined in the usual way as the cube root of the product of the three lobe dimensions, is then given in units of the separation a as a function of the mass ratio $m_2 m_1$ (cf., e.g., Kopal, 1959; Paczynski, 1971a). At the same time, if we assume that the radius of the pulsar is small compared with the companion, (which seems justified by the large amplitude of the pulsar variation), we have that R/a is a simple trigonometric function of D, the eclipse duration, and i. Thus for any assumed value of i, the mass ratio can be obtained, which, when combined with the mass function, leads to masses for the two components. Treatment of this problem by van den Heuvel and Heise (1972) leads to upper mass limits $m_1 = 0.70 \, m_\odot$ (the pulsar) and $m_2 = 17 \, m_\odot$ (unseen companion) when $i = 90°$. As i decreases, the upper limit decreases, and the value of the mass ratio is quite sensitive to the duration D. It is also sensitive in the same way to the treatment of the geometry, since it is not quite correct to assume that the occulting disk has an eclipse dimension equal to the radius of the star as defined above. A more nearly accurate discussion of this point was made by Wilson (1972), who concluded that the upper limit to the mass of the pulsar must be near $0.23 \, m_\odot$, rather than the $0.70 \, m_\odot$ derived by van den Heuvel and Heise.

These considerations make it clear that optical identification for Cen X-3 is urgently required to assist in establishing the correct physical parameters for the system. Information on optical candidates changes almost daily, but at the time of writing, the candidate put forward by Kristian *et al.* (1972), viz., LR Cen, seems somewhat doubtful. According to Hiltner (1972), the period of this optical eclipsing binary is 2.095 ± 0.001 days, as determined in 1972 from photoelectric observations, a value too far from the X-ray figure of 2.08712 ± 0.00004 days. This confirms the earlier conclusion by Elliot and Liller (1972) based on photographic photometry. It is fortunate if this is the case: the phasing of the optical observations of LR Cen and the X-ray data indicate that the X-ray source is in front of the dark companion at principal X-ray eclipse of Cen X-3, a result requiring an extraordinarily complex (impossible?) model for the system. It remains to be seen if the 11th magnitude emission line star, WRA 795, suggested by Margon (1972) as a candidate, is in fact the same as Cen X-3.

Quite aside from the question of the correct physical parameters for Cen X-3, the binary star model for compact X-ray sources must now be taken seriously, and the question raised: how do systems come into existence in which a 'biologically old' collapsed star is accompanied by a presumably 'biologically young' and massive object which overflows its lobe of the zero-velocity surface? A considerable amount of work has been done in recent years, principally by Kippenhahn and his associates, on these questions, and it is the relevant parts of these papers that I now wish to treat briefly. A more extensive review of the evolution of mass-exchange binary systems has been given in an excellent article by Paczynski (1971a) to which the reader is referred for details.

2. Evolution of Mass-Exchange Binaries: Some Examples

As is well known, stars undergoing evolutionary processes in the deep interior can experience drastic changes in radius, by factors of 10 or 100 or more. If such a star is a member of a binary system in which the initial separation is several stellar radii, it will encounter its lobe of the inner zero velocity surface, and its subsequent evolution will be characterized by loss of mass through the inner Lagrangian point L_1. This, however, is not the only mechanism by which mass loss through L_1 can be generated. It is possible for close binaries to radiate enough energy in the form of gravitational waves (Kraft *et al.*, 1962; Kraft, 1966; Faulkner, 1971) that the period is significantly reduced in a time shorter than the nuclear lifetime. In this case, the lobe around the larger star will shrink, eventually cutting down into the star itself. This mechanism for driving mass loss through L_1 has been considered by Faulkner (1971) in connection with the U Gem stars. We return to this point later, but now consider only those binaries undergoing mass transfer as a result of some kind of exhaustion of nuclear fuel.

To fix ideas, consider the evolution of a single star with $m = 5\ m_\odot$. In Figure 1, we reproduce Iben's (1967) well-known HR diagram, i.e., plot of the evolutionary path in a temperature-luminosity array. Of major interest here are the events associated with points 4 and 7. After an initial period of core hydrogen burning, a thick hydrogen-burning shell is established at point 4, and the star rapidly moves to the

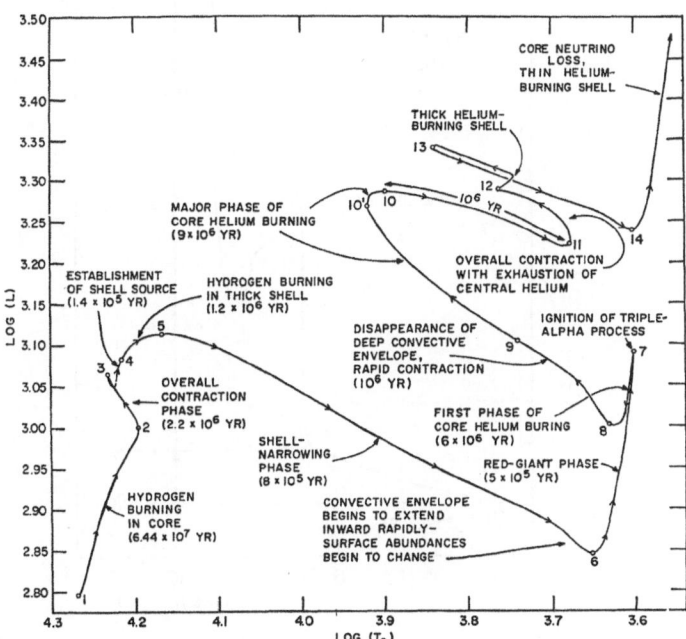

Fig. 1. Evolution in the HR diagram of a star of mass 5 $m_\odot$ (after Iben, 1967).

right increasing its radius as the exhausted core continues to contract. At point 7, the radius begins to decrease in response to the onset of core He-burning, even though after point 7, the energy production of the hydrogen burning shell exceeds that of the He-burning core. Very crudely we can say that core contraction is accompanied by an increase in the radius, part of the increased energy generation being taken up in the gravitational potential energy of the star. On the other hand, ignition of nuclear fuel in the center with corresponding expansion of the core burning region seems to be accompanied by a contraction of the stellar envelope.

That these ideas are likely to be relevant in connection with real binary stars is demonstrated in Figure 2, which is taken from the work of Paczynski (1971a). Consider a binary with primary mass $m_1 = 5\,m_\odot$ and with mass ratio $m_1/m_2 = 2/1$. From Kepler's third law, we have $a^3/P^2 = m_1 + m_2$ in appropriate units, a known number. If r is the mean radius of the lobe corresponding to the primary, then r/a is a known function $F\,(m_2/m_1)$. From Iben's evolutionary track, the radius R of the primary star is obtained at every point on the track, for example at the critical points 1, 4, and 7. If we set $R = r$ at these critical points, then the separation a is found from F, and critical limiting periods indicated in Figure 2, are found from Kepler's third law. If, for example, the actual period of the real binary is less than 1.5 days (corresponding to point 4 of the track), then the primary in the system will

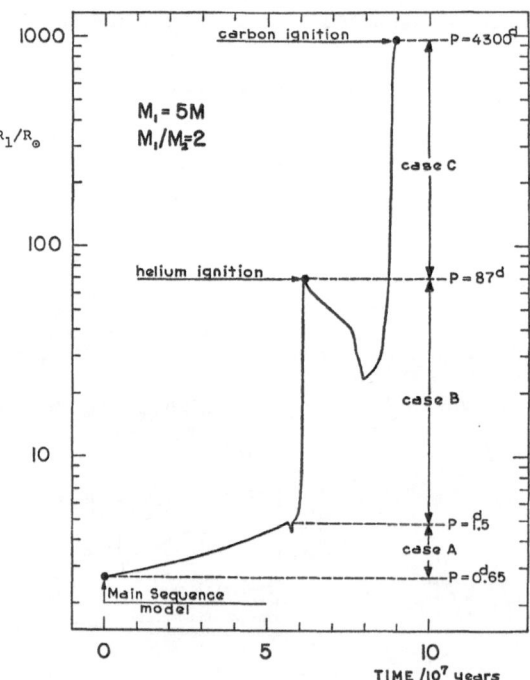

Fig. 2.   Case A, Case B, Case C evolution for a binary system with $m_1 = 5\,m_\odot$ and $m_2 = 2.5\,m_\odot$ (after Paczynski, 1971a, Figure 1).

encounter its zero-velocity surface before hydrogen exhaustion in the core. Following Kippenhahn and Weigert (1967), this is known as 'Case A' binary evolution. If the period of the binary, on the other hand, is more than 1.5 days but less than 87 days, the surface will be encountered during the stage of shell hydrogen burning ('Case B'), and so forth. These periods are quite characteristic of spectroscopic binaries; indeed, about 40% of all binaries have periods less than 100 days (Kuiper, 1935). This estimate of frequency would not be greatly changed in dealing with evolving stars of mass less than 5 $m_\odot$, or in dealing with mass ratios different from 2/1. Thus we may expect mass transfer to be a quite common experience in the evolution of spectroscopic binaries.

We now consider examples of mass transfer binaries in which, after mass loss through L_1, the primary evolves into a degenerate configuration on a time-scale short compared to its main sequence lifetime. Calculations have been carried out by many investigators during the past five years, among them Kippenhahn, Weigert, Giannone, Refsdal, Lauterborn, Barburo, Ziolkowski, Kriz, and Paczynski (see Paczynski [1971a] for a comprehensive list of references). All calculations assume the conservation of total mass and total orbital angular momentum. We chose two representative examples here, one in which the primary is a massive main sequence star ($m \gtrsim 3 \, m_\odot$) of the kind that would now be evolving off the main sequence in the Pleiades or α Per clusters, and one in which the primary is of low mass ($m \lesssim 3 \, m_\odot$), similar to F-type stars near the turnoff of the old galactic cluster NGC 752. According to Paczynski (1971a), a qualitative difference in evolution occurs depending on whether the mass of the primary is greater than, or less than about $3m_\odot$. For Case A evolution, an originally massive primary will still be burning hydrogen slowly in the core even after the secondary has begun its expansion, having already acquired the mass ejected by the primary. In this case, the primary and secondary simply exchange roles. In Case B evolution, however, the primary evolves into an object of small radius, regardless of its initial mass. In addition, for primaries of small mass, the same ultimate state is reached independent of whether the evolution proceeds as in Case A or Case B. We therefore choose our examples entirely from Case B.

EXAMPLE 1

$m_1 = 9.0 \, m_\odot$, $m_2 = 3.1 \, m_\odot$, $P = 4.85$ days (Kippenhahn and Weigert, 1967). The critical surface is reached *after* hydrogen exhaustion in the core, and the mass-loss proceeds on the Kelvin time-scale of only 4×10^4 y, about 10^{-2} of the main sequence life-time. At the end of the stage of rapid mass transfer, the primary has a mass of 2.0 $m_\odot$, and has transferred most of its mass to the secondary, which moves up the main sequence to a position corresponding to a star undergoing core hydrogen burning at 10.1 $m_\odot$. The old primary fills its lobe of the zero-velocity surface burning hydrogen in a shell source. However, before the old secondary is able to exhaust a sufficient supply of hydrogen to move off the main sequence, the primary ignites He in the core, contracts away from the surface, and becomes a star near the He-burning main sequence. This sequence of events is depicted in the HR diagram in Figure 3. With angular momentum conserved, the period is somewhat lengthened, viz., 12.7 days. According to Kippenhahn and Weigert (1967), the first stage of rapid mass transfer can be identified with objects like β Lyr, and the final state is similar to numerous Wolf-Rayet binaries. We note that the two final objects have about the same luminosity even though the masses are very different: this is because the more massive component burns hydrogen while the less massive burns helium. The nuclear lifetime of the less massive component will therefore be significantly shorter than its companion.

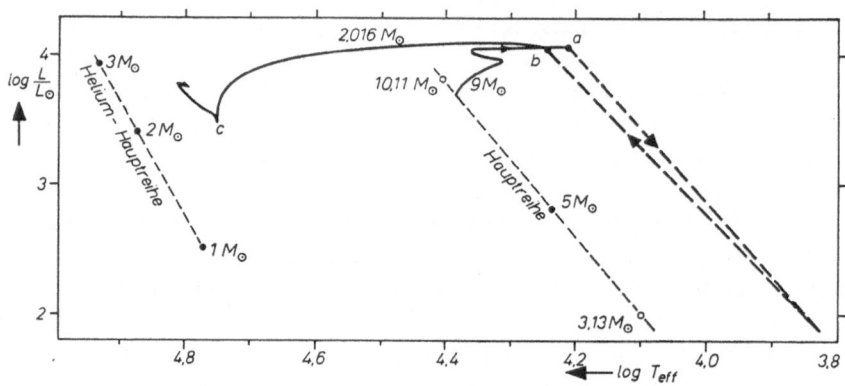

Fig. 3.   Evolution in the HR diagram of a system with $m_1 = 9.0\, m_\odot$, $m_2 = 3.1 m_\odot$, and $P = 4.85$ days (after Kippenhahn and Weigert, 1967, Figure 4).

How the more highly evolved star reaches its final state of exhaustion, and therefore degeneracy, has not been calculated in detail, and several branching scenarios are possible, depending on the mass of the original primary. In the system considered here, the helium burner has a mass of $2.0\, m_\odot$. According to Paczynski (1971b), helium stars with masses between 1.0 and about 3 $m_\odot$ evolve into the red giant region for a second time. The helium star in our system may therefore transfer additional matter to its main sequence companion and settle down eventually as a helium or carbon white dwarf. If its final mass were (say) $1.0\, m_\odot$, the final period would be 77 days. Subsequent evolution would find the 11 solar mass new primary eventually evolving to fill its critical zero-velocity surface, and matter would be expected to accrete onto its white-dwarf companion, a picture conceivably identifiable with an X-ray source. It is easy to see, however, that if too much matter is transferred in this penultimate stage so that the mass ratio becomes very large, the period will become quite long and thus establishment of conditions suitable for the generation of an X-ray source may be delayed to Case C or may not take place at all.

If the mass of the original primary had been somewhat smaller, the mass of the resulting helium star would have been too small for 'red-giant'-type evolution to take place. As an example, consider a system with initial masses $m_1 = 4.0\, m_\odot$, $m_2 = 2.7\, m_\odot$ and initial period 3.26 days. According to Ziolkowski (1970), in the end the primary becomes a helium star of mass $0.56\, m_\odot$, the secondary becomes a hydrogen-burning main sequence star of mass 6.11 $m_\odot$, and the orbital period is 100 days. The helium star ultimately becomes a white dwarf without passing through the red-giant stage (Paczynski, 1971b). Subsequent evolution of this object could produce an X-ray source, although again the period is rather uncomfortably long.

A third branching of possibilities supposes the existence of a primary sufficiently massive that its pure helium descendant has mass $\gtrsim 3\ m_\odot$ and cannot further lose mass by ejection in the second red giant stage. Such an object will pass rapidly through He and C burning stages, ignite still heavier elements, and will presumably become a supernova, leaving behind a neutron star or black hole. Whether a binary system dissolves or becomes an accelerated system of longer period depends on whether more or less than half the total mass is ejected, respectively (Blaauw, 1961; Boersma, 1961; Gott, 1972). Since in the present case it is the less-massive component that blows up, the system will in every case remain bound.

This possibility was considered in some detail for the case of Cen X-3 by van den Heuvel and Heise (1972). Their semi-empirical calculation is shown schematically in Figure 4. They begin with a close binary in which $m_1 = 16\ m_\odot$, $m_2 = 3\ m_\odot$ and the initial period is 3 days. Using the evolutionary tracks of Paczynski, they find that the descendant object has masses 15 $m_\odot$, 4$m_\odot$, and the period is 1.53 days. The less-massive component is a pure helium burning star, and the more massive object is still on the hydrogen-burning main sequence. The mass transfer occurred on a time scale equal to 1/300th of the original main sequence lifetime. After about 1/4th of this nuclear time, the helium star explodes. Van den Heuvel and Heise assume, as an example, that a mass of 3.5 $m_\odot$ is ejected to infinity, leaving behind a neutron star of mass 0.5 $m_\odot$. If the ejection occurs in a spherically symmetrical way, the whole system is accelerated by 30 km s^{-1} owing to recoil, and the binary period is increased to 2.17 days (van den Heuvel, 1968). When the 15 $m_\odot$ star eventually fills its critical surface owing to core hydrogen exhaustion, matter will flow onto the surface of the neutron star and an X-ray source similar to Cen X-3 could conceivably result.

Van den Heuvel and Heise interpret the X-ray pulsations as if the object were a classical pulsar: the 4.8 s oscillation is presumed due to the rotation of a neutron star with an embedded dipole magnetic field. If the oscillation is driven instead by the pulsations of a degenerate star (Blumenthal et al., 1972), the period is too long to admit of a neutron star (cf. Thorne and Ipser, 1968), and a white dwarf is required. But, as we have seen in the earlier discussion, consideration of angular momentum makes it difficult to reconcile the large mass-ratio inferred for Cen X-3 with the rather short orbital period unless matter is removed in a supernova outburst, a process probably resulting in a neutron star. However, before we are driven to embrace uncritically the neutron star model we should remember that all calculations of mass-transfer in binaries assume conservation of orbital angular momentum and total mass. If, during some stage of mass-transfer, matter is also lost through one of the outer Lagrangian points, the period could be significantly decreased (Kuiper, 1941; Kruszewski, 1966). If matter were thus lost from the system by the original

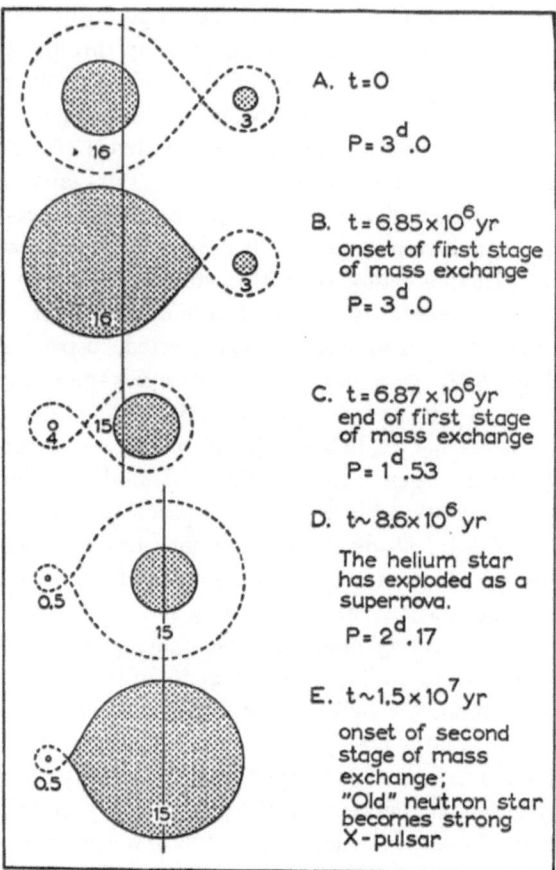

Fig. 4. Possible evolution of a massive binary into an X-ray source like Cen X-3 (after van den Heuvel and Heise, 1972).

secondary, the binary could possibly have a short period in the end, the more evolved component having arrived safely in the white dwarf stage without a supernova explosion. Detailed calculations of this possibility have not, of course, been made.

EXAMPLE 2

$m_1 = 2.0$ $m_\odot$, $m_2 = 1.0$ $m_\odot$, $P = 1.15$ days (Kippenhahn *et al.*, 1967). In this typical low-mass case, the primary encounters the critical surface after the onset of shell hydrogen burning in 5.7×10^8 yr, and loses all but 0.26 $m_\odot$ to the secondary in only 6.9×10^6 yr (see Figures 5 and 6). At this point the original primary fills its critical lobe, forming a semi-detached system of mass-ratio $\frac{1.0}{1}$ and period 24.1 days which lasts for about 10^7 yr. The so-called 'subgiant components of eclipsing binary systems' are of this kind (Plavec, 1968); a typical object is DN Ori (Smak, 1964), in which $P = 12.97$ days, and in which the A2 and giant F5 components have masses of 2.65 and

0.18 $m_\odot$, respectively. The subgiant component develops a degenerate core which naturally stops contracting, and never succeeds in igniting helium. In the end we have a main sequence A-type star of mass 2.74 $m_\odot$ accompanied by a low mass white dwarf with the same orbital period found above, 24.1 days. The velocities of the stars in their orbits are about 10 and 100 km s^{-1}; only the slower moving component will be seen in the spectrum of the combined light. Considering the errors of radial velocity determinations for A-type stars, we think it likely that many systems of this kind will be missed.

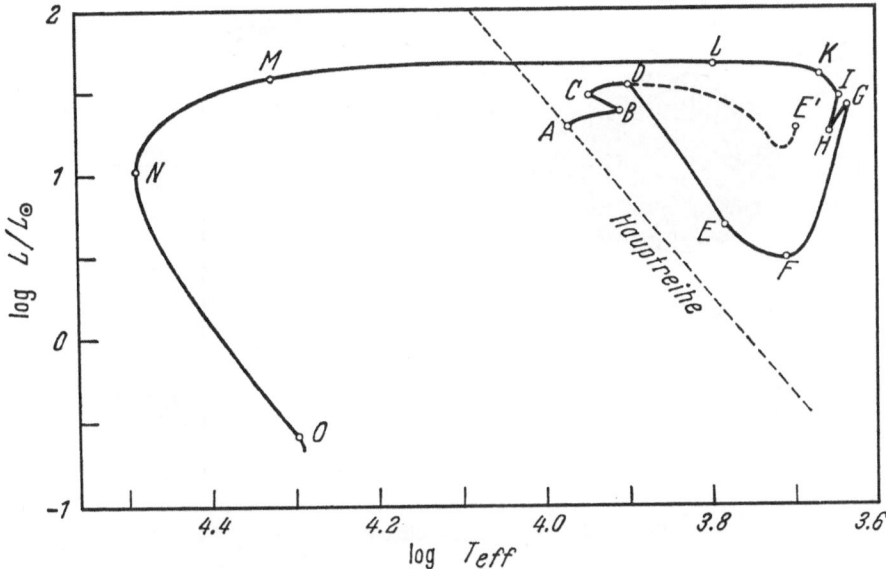

Fig. 5. Evolution in the HR diagram of the primary in a binary with $m_1 = 2.0\ m_\odot$, $m_2 = 1.0\ m_\odot$, and $P = 1.15$ days (after Kippenhahn *et al.*, 1967), At point D, Case B evolution begins. At K, the star becomes a 'subgiant', and at O, a white dwarf. The period at K and O is 24.1 days.

The subsequent evolution of the system will lead to a swelling-up of the new primary, spilling of mass through L_1, and capture of this material by the white dwarf, i.e., a possible X-ray source. The accretion heating of the white dwarf may make it visible in the combined optical spectrum. However, if the mass transfer is fast, the radial velocity variations of the degenerate component may well be masked by motions of gas streams within its lobe of the zero-velocity surface, or by streams of matter ejected from the outer Lagrangian points.

3. Some Boundary Conditions on Possible Mass-Transfer Models

In making choices between the various cases considered in the previous Section, three additional observational facts about galactic compact X-ray sources must be kept in mind. First, although many are located in the galactic plane at random longitudes,

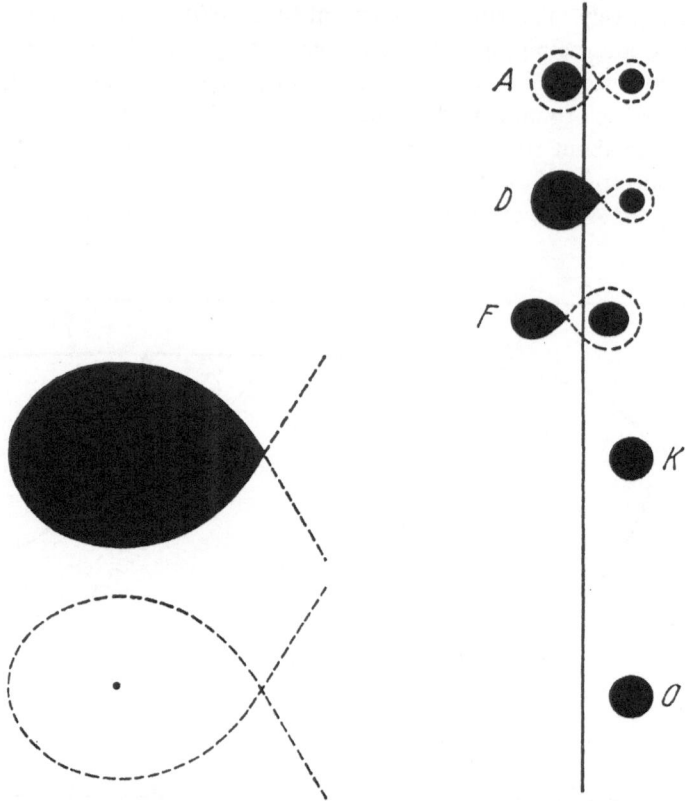

Fig. 6.   Scaled dimensions of the binary considered in Figure 5.

a quite significant fraction are associated with the region of the galactic nucleus (Giacconi *et al.*, 1972). Unless these are objects of an entirely unexpected sort, many are presumably mass transfer binaries in the stellar population associated with the galactic nuclear region, viz., an old metal-rich population similar to the stars in galactic clusters like M 67 and NGC 752 (Arp, 1965; Van den Bergh, 1972) or globular clusters like M 71 (Arp and Hartwick, 1971) and NGC 6352 (Hartwick and Hesser, 1972). This means we are dealing with binary systems in which the primaries are of relatively low mass, i.e., in the range 1 to 2 $m_\odot$ (cf. also Salpeter, 1972). This suggests that at least some of the X-ray sources are like the system of Example 2. Sco X-1 and Cyg X-2 could fall into this category. Optically, we see only the X-ray source in the former but we see both components in the latter. If the radial velocity variations due to orbital motion are spoiled by gas streams in the case of the X-ray component, our only hope is to detect the velocity variations of the G-type component in Cyg X-2. Unfortunately, this may be nearly impossible if it is very much more massive than its companion, as the above models suggest.

A second type of boundary condition is derived from the fact that U Gem stars and old novae consist of a late-type star that overflows its zero-velocity surface and transfers matter to its degenerate companion. *Why then are not old novae and U Gem stars detected as sources of hard X-rays?* The Prendergast-Burbidge model gives a clue to understanding this anomaly. It will be recalled that the model requires a mass-transfer rate of 2×10^{19} gm s^{-1} to give the required X-ray luminosity. In Figure 7, we plot the rate of mass-loss by hydrogen shell-burning primaries in mass-transfer systems as a function of the mass of the primary. Only models by Kippenhahn and his associates were used for a variety of mass ratios and periods; the result would not be affected by choosing models of other investigators. In a steady state, the ejected matter is, of course, accreted by the secondary. We assume that these rates

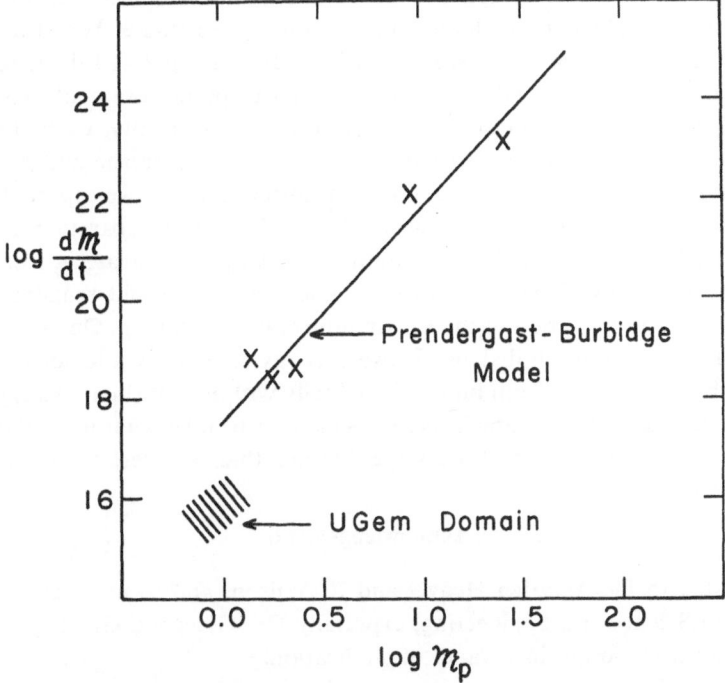

Fig. 7. The mass transfer rate of primaries in Case A or Case B evolution as a function of primary mass.

can be taken to apply to the case when the secondary is a degenerate object since the driving mechanism has a time-scale depending only on the Kelvin time of the primary. We see from the Figure that the rate required by the Prendergast-Burbidge model is comfortably in the range of ejection rates computed for evolving primaries. Thus the required accretion rate can be supplied by stars undergoing evolution to the right in the HR diagram for the first time. This does not seem to us to be accidental.

If the mass transfer rate were reduced by one or two orders of magnitude, the

X-rays would not be produced by the model (Prendergast, private communication). This is because the inner part of the disk does not become sufficiently hot and the outer part is too opaque. But mass-transfer in U Gem stars and many old novae is not driven by hydrogen exhaustion in the primary component, but rather by the emission of gravitational waves (Faulkner, 1971). This is because the masses are too small for significant nuclear burning to have taken place. The corresponding transfer rate is therefore of order 10^{16} gm s^{-1} (Faulkner, 1971), too small to produce X-ray emission. That these objects could be responsible for the discrete sources needed to explain the soft X-ray background (Gorenstein and Tucker, 1972)is unlikely, since their numbers per unit volume of space are too small by a factor of 10^4.

Our third and final boundary condition has to do with the number of X-ray sources expected per unit volume of space if all are derived from mass transfer binaries. Consider for example stars in the low mass group of Section 2. We assume no loss of angular momentum or mass. We start with stars having $P < 3$ days, so that the period of the evolved descendant is short enough to permit eventual mass transfer under Case B. From the luminosity function in the solar vicinity, we find that there are about 2×10^{-5} such binaries per pc^3 (Kuiper, 1935; Anderson and Kraft, 1972) and from the ratio of main sequence to mass-transfer lifetimes, we expect about 10^{-7} X-ray sources per pc^3, or one about every 100 pc. This mean separation is too small by a factor of 5 to 10, but one can easily think of ways to increase it. For example, perhaps the white dwarf becomes unstable under accretion and remains an X-ray source for only a tiny fraction of the Kelvin time of its companion. On the other hand, we could increase the mean distance between X-ray sources by a factor of 2 or 3 by assuming they are derived from binaries originally with B-type (high mass) primaries. At the present state of our knowledge, it seems best to take comfort in the fact that the predicted numbers are at least larger, rather than smaller, than the numbers required by observation.

Acknowledgements

I am indebted to Drs van den Heuvel and R. Wilson, and to members of the staff at American Science and Engineering, expecially Drs Giacconi, Gursky, and Tucker for communicating results in advance of publication.

Appendix

Cyg X-1 has not been discussed here even though it has been identified with the 9th magnitude mass-transfer spectroscopic binary HDE 226868 (Webster and Murdin, 1972; Bolton, 1972). The most certain evidence that this star is, in fact, the X-ray source, stems from the position of the variable radio source given by Hjellming and Wade (1971). If, however, variable radio emission is a normal feature of mass transfer binaries that are not necessarily X-ray emitters (cf. Hjellming et al., 1972), then we may not be justified in assuming that HDE 226868 is identical with Cyg X-1. For this reason, discussion of the properties of HDE 226868 was not included here.

References

Anderson, K. and Kraft, R.: 1972, *Astrophys. J.* **172**, 631.
Arp, H.: 1965, *Astrophys. J.* **141**, 43.
Arp, H. and Hartwick, F.: 1971, *Astrophys. J.* **167**, 499.
Blaauw, A.: 1961, *Bull. Astron. Inst. Netherlands* **15**, 265.
Blumenthal, G., Cavaliere, A., Rose, W., and Tucker, W.: 1972, *Astrophys. J.* **173**, 213.
Boersma, J.: 1961, *Bull. Astron. Inst. Netherlands* **15**, 291.
Bolton, C.: 1972, *Nature* **235**, 271.
Bopp, B. and Vanden Bout, P.: 1972, *Publ. Astron. Soc. Pac.* **84**, 68.
Burbidge, E., Lynds, C., and Stockton, A.: 1967, *Astrophys. J.* **150**, L95.
Cameron, A. and Mock, M.: 1967, *Nature* **215**, 464.
Chodil, G., Jopson, R., Mark, H., Seward, F., Swift, C.: 1965, *Phys. Rev. Letters* **15**, 605.
Crawford, J. and Kraft, R.: 1956, *Astrophys. J.* **123**, 44.
Elliot, J. and Liller, W.: 1972, *IAU Circ.* No. 2395.
Faulkner, J.: 1971, *Astrophys. J.* **170**, L99.
Giacconi, R., Gursky, H., Kellogg, E., Schreier, E., and Tananbaum, H.: 1971, *Astrophys. J.* **167**, L67.
Giacconi, R., Gursky, H., Kellogg, E., Murray, S., Schreier, E., and Tananbaum, H.: 1972, *Astrophys. J.*, (in press).
Gorenstein, P. and Tucker, W.: 1972, *Astrophys. J.*, (in press).
Gott, R.: 1972, *Astrophys. J.* **173**, 227.
Grader, R., Hill, R., Seward, F., and Toor, A.: 1966, *Science* **152**, 1499.
Hartwick, F. and Hesser, J.: 1972, *Bull. Am. Astron. Soc.* **4**, Part 1, p. 241.
Hiltner, W.: 1972, *IAU Circ.*, No. 2398.
Hiltner, W., Mook, D., and Lynds, C.: 1970, *Ann. Rev. Astron. Astrophys.* **8**, 139.
Hjellming, R. and Wade, C.: 1971, *Astrophys. J.* **168**, L21.
Hjellming, R., Wade, C., and Webster, E.: 1972, *Nature* **236**, 43.
Iben, I.: 1967, *Ann. Rev. Astron. Astrophys.* **5**, 571.
Kippenhahn, R. and Weigert, A.: 1967, *Z. Astrophys.* **65**, 251.
Kippenhahn, R., Kohl, K., and Weigert, A.: 1967, *Z. Astrophys.* **66**, 58.
Kopal, Z.: 1959, *Close Binary Systems*, John Wiley and Sons, New York.
Kraft, R.: 1962, *Astrophys. J.* **135**, 408.
Kraft, R.: 1964, *Astrophys. J.* **139**, 457.
Kraft, R.: 1966, *Trans. IAU* **12B**, 519.
Kraft, R. and Demoulin, M.: 1967, *Astrophys. J.* **150**, L183.
Kraft, R. and Miller, J. 1969, *Astrophys. J.* **155**, L159.
Kraft, R., Mathews, J., and Greenstein, J.: 1962, *Astrophys. J.* **136**, 312.
Kristian, J., Brucato, R., and Westphal, J.: 1972, (in press).
Kristian, J., Sandage, A., and Westphal, J.: 1967, *Astrophys. J.* **150**, L99.
Kruszewski, A.: 1966, in Z. Kopal (ed.), *Advances in Astron. and Astrophys.* Vol. 4, Academic Press, Inc., New York, p. 233.
Krzeminski, W. and Smak, J.: 1971, *Acta Astron.* **21**, 133.
Kuiper, G.: 1935, *Publ. Astron. Soc. Pac.* **47**, 121.
Kuiper, G.: 1941, *Astrophys. J.* **93**, 133.
Margon, B.: 1972, *IAU Circ.*, No. 2398.
Paczynski, B.: 1971a, *Ann. Rev. Astron. Astrophys.* **9**, 183.
Paczynski, B.: 1971b, *Acta Astron.* **21**, 1.
Peterson, L. and Jacobson, A.: 1966, *Astrophys. J.* **145**, 962.
Plavec, M.: 1968, in Z. Kopal (ed.), *Advances in Astron. and Astrophys.* Vol. 6, Academic Press, Inc. New York, p. 201.
Prendergast, K. and Burbidge, G.: 1968 *Astrophys. J.* **151**, L83.
Salpeter, E.: 1972, 'Remarks before the Princeton Symposium on X-ray Sources', Jan. 15, 1972.
Sandage, A., Osmer, P., Giacconi, R., Gorenstein, P., Gursky, H., Waters, J., Bradt, H., Garmire, G., Sreekantan, B., Oda, M., Osawa, K., and Jugaku, J.: 1966, *Astrophys. J.* **146**, 316.
Schreier, E., Levinson, R., Gursky, H., Kellogg, E., Tananbaum, H., and Giacconi, R.: 1972, *Astrophys. J.* **172**, L79.

Shklovskii, I.: 1967, *Astrophys. J.* **148**, L1.
Smak, J.: 1964, *Publ. Astron. Soc. Pac.* **76**, 210.
Thorne, K. and Ipser, J.: 1968, *Astrophys. J.* **152**, L71.
Van den Bergh, S.: 1972, *Publ. Astron. Soc. Pac.* **84**, 306.
Van den Heuvel, E.: 1968, *Bull. Astron. Inst. Netherlands* **19**, 432.
Van den Heuvel, E. and Heise, J.: 1972, *Nature*, (in press).
Webster, B. and Murdin, P.: 1972, *Nature* **235**, 37.
Westphal, J., Sandage, A., and Kristian, J.: 1968, *Astrophys. J.* **154**, 139.
Wilson, R.: 1970, in L. Gratton (ed.), 'Non-Solar X- and Gamma-Ray Astronomy', *IAU Symp.* **37**, 242.
Wilson, R.: 1972, *Astrophys. J. Letters* **174**, L27.
Ziolkowski, J.: 1970, *Acta Astron.* **20**, 213.

5. HARD COSMIC X-RAY SOURCES

L. E. PETERSON

Dept. of Physics, University of California, San Diego, La Jolla, Calif. 92037, U.S.A.

Abstract. A review of the observational status of X-ray sources detected in the 20 ≃ 500 keV range is presented. Of the approximately 115 sources listed in the March 1972 edition of the UHURU 2–6 keV sky survey catalog, about 15 sources have been studied in hard X-rays. Most of the data have been obtained from balloons, although the OSO-3, and more recently the OSO-7, have contributed. With the exception of CEN A, the SMC, and possibly M-87, all the sources detected at higher energies are galactic and heavily concentrated in the galactic plane. The Crab Nebula has been measured to about 500 keV in continuous emission and a component at the ≃ 33 ms pulsar period comprising about 20% of the total emission has been detected to ∼10 MeV. Objects such as SCO-1 and CYG-2 are characterized by an exponential spectrum, which varies over a 10 min. time scale about a factor of two, and a flatter spectrum extending to above 40 keV which exhibits independent variability. Objects such as CYG-1 and possibly CYG-3 have a multi-component power law spectrum extending to over 100 keV, and may vary many factors over a period of weeks. Other sources generally not yet identified with optical or radio candidates, located in the Galactic Center and the Centaurus/Crux region also show considerable variability, and in one case may have been detected to nearly 500 keV. Only upper limits at about 2×10^{-4} photon $(cm^2 \, s \, keV)^{-1}$ in the 20–50 keV range exist for most supernova remnants and extragalactic sources.

1. Introduction

This is a brief review of the observational knowledge of cosmic X-ray sources in the twenty to several hundred keV range. This work is not intended to be complete, but will summarize recent results, indicate future directions, and update a similar review given at a preceeding IAU Symposium (Peterson, 1970). We will also include data obtained by UCSD on the OSO-7 satellite and on recent balloon flights which have not yet appeared in the open literature. In particular, observations of hard X-rays will be interpreted in terms of the catalog of sources generated from the UHURU data (Giaconni *et al.*, 1972).

The majority of X-ray emitters observed by rockets and by the UHURU satellite have a soft spectrum, therefore, the relative sensitivity must be increased at higher energies to detect the same number of sources. Since sensitivity is a more difficult problem at these energies (Peterson *et al.*, 1971), the number of observed sources is less. Of the 115 sources listed in the UHURU 2–6 keV catalog, only about 20 have been detected above 15 keV. The sensitivity level with detectors available at the present time is in the range of 1 to 2×10^{-4} photons $(cm^2 \, s \, keV)^{-1}$ in the 20–50 keV range. The most useful work in this range has been obtained by extending the spectrum, which must be a signature of the radiating mechanism, and studying the time variability which relates to models based on compact sources or binary configurations. Discovery and identification is best accomplished at lower energies.

Only two objects, the Crab Nebula (Peterson and Jacobson, 1970) and a source near the Galactic Center (Johnson *et al.*, 1972) have a measured spectrum extending significantly beyond 300 keV. Several cosmic sources, GX 1+4, GX 304−1, and

Bradt and Giacconi (eds.), X- and Gamma-Ray Astronomy, 51–73. All Rights Reserved.
Copyright © 1973 by the IAU.

GX 301–2, all of which are variable, have been discovered first from balloon observations. All confirmed measurements thus far have been limited to continuum type of spectra, although γ-ray lines have been searched for to a sensitivity level of about 10^{-3} photons $(cm^2 s)^{-1}$, and Haymes (Johnson *et al.*, 1972) has reported a barely significant measurement of a line near 0.511 MeV from the Galactic Center. Two extra-galactic sources, CEN A or NGC 5128 (Lampton *et al.*, 1972) and the Small Magellanic Cloud (Price *et al.*, 1971) has been positively detected at energies greater than 20 keV. A number of results have also appeared on the Virgo Cluster, or possibly M 87, which indicate a detectable spectrum beyond 50 keV.

2. Instrumentation

Before discussing measurements of various sources and their physical significance, it is useful to indicate some concepts of detection technique. Detectors used in the 20–500 keV range usually consist of scintillation counters with active or semiactive anti-coincidence collimators (Peterson, 1970; Peterson *et al.*, 1971), although Glass (1969) has used proportional counters to obtain very large areas. Scintillation telescopes have thus far been less than a few hundred cm^2 with apertures not smaller than a few degrees. Table I indicates instruments used to obtain certain significant observations in the >20 keV range, and includes reference citations in this paper.

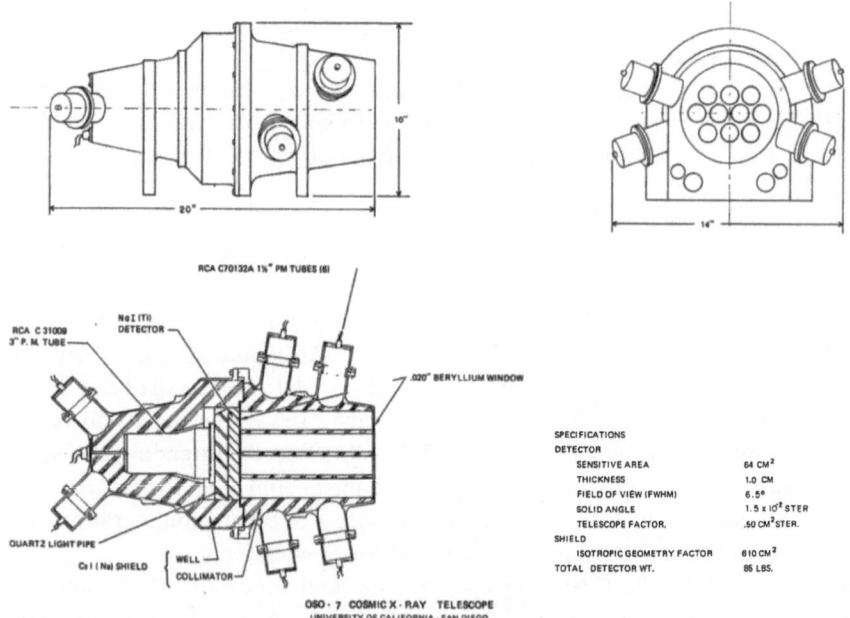

Fig. 1. The telescope used by UCSD on the OSO-7 satellite consists of a combination of scintillation counter elements providing detection and active anti-coincidence collimation features. This detector is similar to those used by others in the 20–500 keV energy range.

TABLE I

Detectors and observations – typical results

Institution and dates	Detector type	Energy range (keV)	Detector geometry Area (cm²)	Thickness (cm)	FWHM (degrees)	Vehicle and obs. dates	Published results	References cited
GSFC 1965–1967	Cs I(T*l*) Plastic anti and graded collimator	~20–100	20	0.2	~20	*Balloons* 18 Dec. 1966 15 Dec. 1966 15 July 1967	Galactic Center SCO-1 TAU-1 Diffuse Comp.	Boldt et al. (1968) Riegler (1969)
MIT 1965–1969	Na I(T*l*) with plastic anti and graded collimator	18 ~ 105 keV	358	0.1	13 × 13	*Balloons* 15 Feb. 1967 15 Oct. 1967 24 Oct. 1967 20 Mar. 1969 16 Apr. 1969	CYG-1/CYG-3 TAU-1 CEN-2 GX 304-1 SCO-1 GX 301-2 M-87	Lewin et al. (1970b) Lewin et al. (1970b) Lewin et al. (1968a) Lewin et al. (1971b) McClintock et al. (1969)
1966–1968	Na I(T*l*) with Na I(T*l*) anti shield and Pb collimator	23–97	56.3	0.1	8.4	*Balloons* 16 May 1967 25 May 1967 27 June 1967	CYG-1 SCO-1 Gal. Center	Overbeck and Tananbaum (1968)
1969–Present	Na I(T*l*) with Na I(T*l*) anti shield Pb collimator	18–50 keV	45	0.1	1.5 × 13	*Balloons* 15 Oct. 1970	GX 301-2 GX 304-1 GX 1+4 GX 3+1 $335° \leq l^\pi \leq 350°$	McClintock et al. (1971 Lewin et al. (1971b) McClintock et al. (1972)

Institution and dates	Detector type	Energy range (keV)	Area (cm²)	Thickness (cm)	FWHM (degrees)	Vehicle and obs. dates	Published results	References cited
UCSD 1967–1968	Na I(T*l*) with Cs I(T*l*) collimator	7.7–190	9.4	0.5	25	*OSO-III* 7 Mar. 1967 to 8 June 1968 and intermittently thereafter	Diffuse Comp.	Schwartz et al. (1970) Schwartz (1970) Hudson et al. (1970) Pelling (1971)
							SCO-1	Peterson (1970)
							CYG region M87 Norma Region CEN Region	Schwartz et al. (1971)
RICE 1967–Present	Na I(T*l*) with Na I(T*l*) and Plastic anti collimator	31–544	~18	5	~18	*Balloon* 29 Aug. 1967 23 Apr. 1968	CYG-1	Haymes and Harnden (1970) Johnson et al. (1972)
		23–930				4 Jun. 1969 25 Nov. 1970	Gal. Center VELA Pulsar	Harnden et al. (1972) Haymes et al. (1972)
TATA inst. 1968–1971	Na I(T*l*) with passive graded collimator	~20–150 keV	97	0.4	18.6	*Balloon* 28 Apr. 1968	SCO-1	Agrawal et al. (1969)
						27 Dec. 1968	CYG-1	Agrawal et al. (1971b)

Table 1 (continued)

UCSD

	Detector	Energy				Date	Source	Reference
1970–Present	Na I(Tl) with graded collimator and Plastic anti	~20–150 keV	80.5	0.4	12.6	16 Apr. 1969 5 Apr. 1971	CYG1-1, -2, -3 Gal. Center TAU-1 TWX-1 SGO-1 CYG	Agrawal et al. (1972a)
1968–Present	Na I(Tl) with Cs I(Tl) phoswich and Cs I(Na) honeycomb collimator	15–500 keV	~34.2	1.0	5.9	*Balloon* 9 June 1968 15 Apr. 1969 10 June 1969 17 July 1969 9 Sept. 1970	CP 1919 CYG-1 CYG-2 CYG-3 CYG-1	Matteson (1971) Peterson (1972) (this paper) Laros et al. (1972)
						25 Sept. 1970	M87	
						20 Oct. 1970	M31 3C273	
1971	Na I(Tl) with Cs I(Na) collimator	~7–550 keV	64	1.0	6.5	*OSO-7* 29 Sept. 1971 to present	Gal. Plane VELA	Peterson (1972) (this paper)

Most observations have been obtained from balloons or satellites because of the short exposure during a rocket flight. Although atmospheric cut off at depths of 2–3 gm/cm^2 usually limits measurements to above 20 keV, MIT (McClintock *et al.*, 1972) has succeeded in using exceedingly large balloons to carry instruments to a depth less than 1.5 gm/cm^2. Recently, significant satellite observations have been obtained by UCSD on OSO-3 (Schwartz *et al.*, 1970; Schwartz, 1970), by Frost (1969) on the OSO-5 and Brini on the OSO-6. The UCSD experiment on OSO-7 will add further knowledge on cosmic X-ray sources.

The telescope configuration on the OSO-7, shown in Figure 1, is described here since it is also typical of those used recently on balloons. The sensitive element consists of a 1 cm thick Na I(T*l*) crystal inside a massive Cs I(Na) anticoincidence shield. Holes drilled in the Cs I(Na) collimator define the aperture. This detector has an effective area of 64 cm^2, a FWHM response of 6.5°, and operates over the 7.0–550 keV range. Instruments located in the rotating wheel of an OSO usually point radially outward, and therefore scan great circles with the nominal two-second rotation period. In the UCSD instrument, each event is analyzed into 128 energy-loss channels and transmitted as an event signature which also contains satellite clock time to 0.625 ms. This time may be used to obtain the X-ray direction, knowing the instantaneous satellite rotation vector, or as event timing for pulsar analysis.

3. Source Location and Identification

Here we discuss hard X-ray sources by regions on the celestial sphere, indicating relative source strengths, variability, and identification. In particular, we attempt to identify sources measured from balloon or satellite work with entries in the UHURU catalog (Giacconi *et al.*, 1972). Source identification is difficult because of variability in strength and spectral shape, and because of confusion due to close spacing.

A. GALACTIC PLANE SCAN

Here we present preliminary results obtained from the UCSD instrument on the OSO-7, which was launched September 29, 1971. These results were obtained in December 1971 when the OSO-7 was scanning along the galactic equator, and represent about five 90-min orbits of quick-look data for an exposure per source of about 50 s. The sources, locations and scan range are shown on the map in Figure 2, where pre-UHURU source names are used. The counting rate as a function of galactic longitude is shown in Figure 3 for the energy ranges 7.0–26 keV and >26 keV. Also shown is an equivalent UHURU scan (Giacconi *et al.*, 1971) obtained approximately one year earlier. Sources which appear consistently in both UHURU and OSO scans are indicated.

The most striking feature is the softness of the X-ray sources. Although the 7–26 keV range shows nearly as many sources as in the UHURU scan, the >26 keV shows only two sources, CYG-1 and the Crab Nebula. The OSO-7 scan does not show the Galactic Center on this date since it coincided with the Sun, and a solar cell inhibits

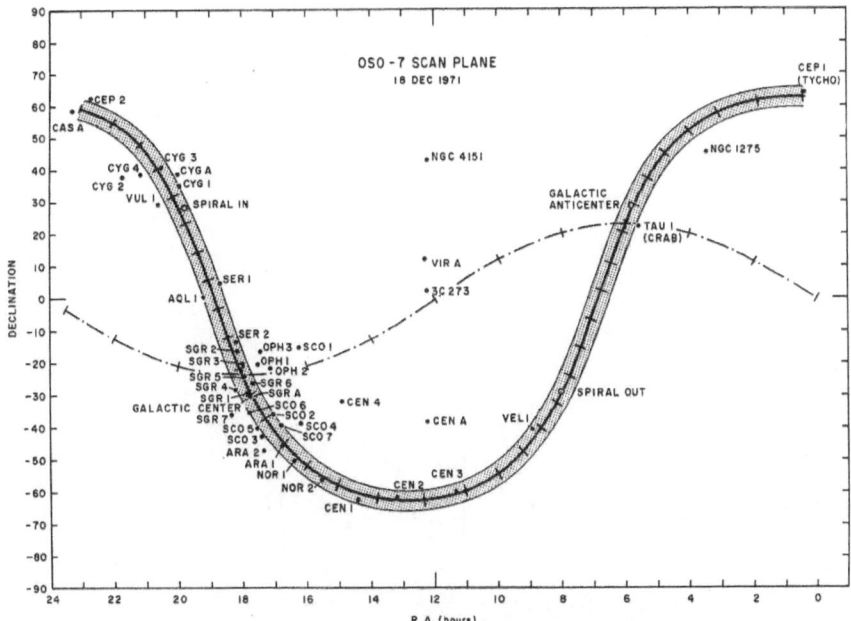

Fig. 2. This map of the celestial sphere shows most of the X-ray sources detected previous to the UHURU. The scan path along the galactic plane of the OSO-7 on 18 December 1971 is also shown.

readout to prevent excessive telemetry deadtime due to solar X-rays. No correction in either of these scans has been made for the aperture response; in particular, the Crab Nebula was nearly out of the OSO-7 aperture. It is not immediately clear that differences between UHURU and 7–26 keV OSO-7 scans occur because of a soft spectrum, the aperture response, or are due to time variations in the one year interval.

The relatively strong source 2U0613+09, at approximately 200° galactic longitude was not observed in the OSO-7 data. The source VEL-1 (2U0900-40) is observable in the OSO-7 data, and additional results on the Vela Region will be indicated later in this paper. The pulsating source CEN-3 (Schreier *et al.*, 1972) apparently has a significantly softer energy spectrum than the Crab Nebula. CEN-2, the first source observed to have a nova-like behavior, is apparently not indicated in either the UHURU catalog or the OSO data. The source as identified CEN-1 in Figure 3 may actually be GX 304-1, which was discovered by Lewin *et al.* (1971a) and is indicated as 2U1258-61 in the AS & E catalog. Furthermore, the OSO-7 data clearly indicates more than one source in this region. The sources in the Lupus/Norma complex are not also well resolved in the OSO results, however, the rather strong source CIR X-1 at $l^{II}=322°$ (2U1516-56) either has an exceedingly soft spectrum or was unobservable one year later in hard X-rays. In general, sources in the Lupus/Norma region have been reported as being variable both by the UHURU and by balloon (McClintock *et al.*, 1972) and satellite results (Schwartz *et al.*, 1971). The source ARA-1 (2U1641-45 ?) reported as

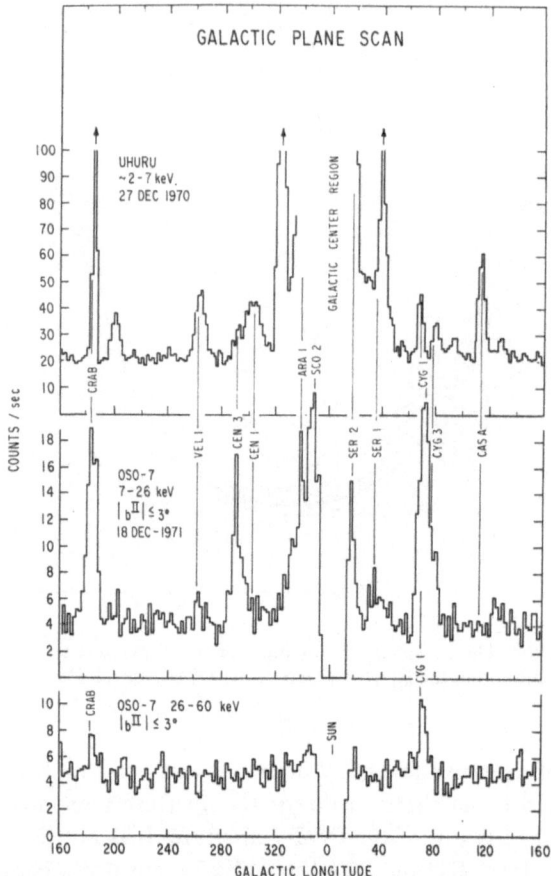

Fig. 3. Sources identified in the OSO-7 galactic plane scan compared with those seen at lower energies one year earlier by the UHURU. An obvious feature
is the softness of the typical X-ray source spectrum.

variable in the UHURU catalog, is apparently soft, although it may be confused with
other strong variable sources within a few degrees galactic longitude. A source within
20° of the Galactic Center, possibly SCO-2 or SCO-6, may just show in the >26 keV
data. Galactic Center sources, as already indicated, have been gated out by the OSO-7;
they will be discussed in a later section.

Northward from the Galactic Center, SER-2 (2U1813-14) also shows strongly.
The strong source identified as SER-1 was nearly out of the aperture, and therefore
the structure near $l^{II} \cong 30°$ may be associated with other sources in this region. CYG-1
and CYG-3 are clearly resolved in the 7–26 keV range, however, at the higher energies
only CYG-1 is well above background. According to Gorenstein *et al.* (1970), the
supernova remnant CAS A has a steep spectrum, and may therefore not be expected
to show in the OSO-7 data.

B. GALACTIC CENTER REGION

Earlier observations of the Galactic Center are somewhat confused by the wide aperture telescopes (Peterson, 1970; Boldt *et al.*, 1968; Riegler, 1969; Agrawal *et al.*, 1971a; Overbeck and Tananbaum, 1968) and the apparent high variability of the sources. Recent observations by Lewin *et al.* (1971a) over the 18–50 keV range have indicated the presence of a strong source GX 1 + 4, and possible fluxes from GX 3 + 1 and GX 5 − 1. The spectrum of GX 9 + 9 has also been measured by the Tata Institute (Agrawal *et al.*, 1971a). Sources at these locations are identified in the UHURU catalog, and most have also been indicated in earlier rocket surveys by MIT (Bradt *et al.*, 1971), as shown in Figure 4. Although possibly not completely resolved, GX 1 + 4 is

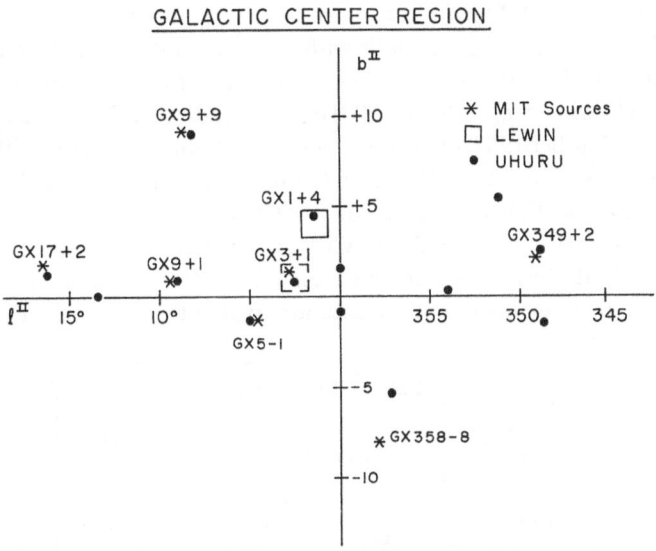

Fig. 4. Source locations in the galactic center region as identified by MIT and UHURU surveys. Only the source GX 1 +4, and possibly GX 3 +1 have been isolated in >20 keV X-rays, although numerous results obtained with wide angle telescopes on the total region have been reported.

reported by Lewin as having a hard spectrum corresponding to a temperature of 28 ± 12 keV, or a power law number spectrum with an index of − 2.4 ± 0.7. Agrawal *et al.* (1971a) report a similar spectrum from the galactic center region, even though the sources were not resolved. Haymes (Johnson *et al.*, 1972) has also measured this region with a wide aperture detector and determined a hard spectrum with a power law index of 2.37 ± 0.05 from 23–930 keV. Although this was tentatively associated with GX 5 − 1, most likely it should be identified with GX 1 + 4 since the power law index agrees with Lewin. According to the UHURU catalog, there is a diffuse source (2U1743-29) of about 2° extent associated with the Galactic Center which balloon measurements have not yet been able to resolve. Haymes has also identified a γ-ray line structure at approximately 0.5 MeV with an intensity of $1.8 \pm 0.5 \times 10^{-3}$ photons $(cm^2\ s)^{-1}$ associated with the Center region. Although there are qualifications to this observation, verification would add a new and exciting element to γ-ray astronomy.

C. CENTAURUS/CRUX

This region is characterized by outstanding examples of source variability and flaring (Lewin et al., 1971b; McClintock et al., 1971; Schwartz et al., 1971). CEN-2, which was first observed to have a nova-like behavior in soft X-rays in early 1967 (Chodil et al., 1968), was thought to have been observed in a series of balloon flights and by OSO-III in hard X-rays. There is no source identified in the UHURU catalog at the improved galactic longitude of $310.2 \pm 1°$ given by Francey (1971). It is not obvious that CEN-1, CEN-4 and CEN-5 were ever observed from balloons.

During 1969 and 1970, Lewin (McClintock et al., 1971) has reported two new sources, GX 304−1 and GX 301−2, both with high variability. The former is certainly identified with 2U1258-61; it had been observed earlier and mistakenly thought to be CEN2 (Lewin et al., 1970b, 1971b). The factor of 7 decrease reported for CEN2 between October 1967 and March 1969 is probably associated with GX 304−1. GX 301−2 is identified with 2U-1223-62. Figure 5, adapted from Schwartz et al. (1971) shows the location of sources in the Centaurus region, compared with the UHURU catalog. CEN-1, CEN-2 and CEN-5 apparently have not been verified by the UHURU. Little has been published on the spectra of these sources at higher energies, although some lower energy data are available (Hill et al., 1972).

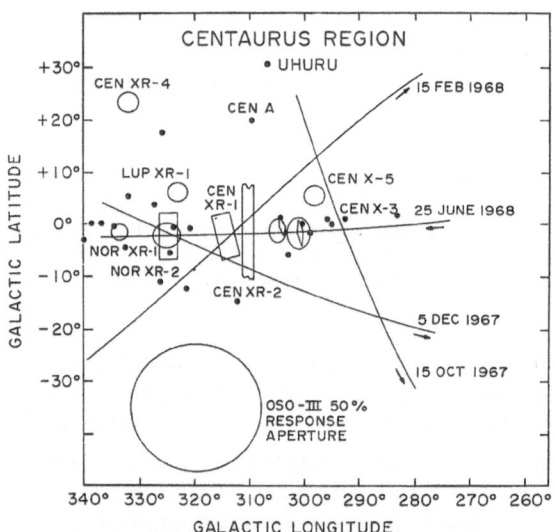

Fig. 5.   The Centaurus region showing known X-ray sources. The ellipses and diamonds are variable sources observed by the MIT balloon group. The scan lines are those of the OSO-III. The sources GX 304-1 and GX 301−2 have been observed to exhibit considerable variability in hard X-rays. The redetermined position of CEN-2, shown here, implies hard X-rays earlier associated with it are due to other sources.

CEN 3, the remarkable modulated pulsating source (Schreier *et al.*, 1972) believed associated with a rotating neutron star as a binary companion to a larger star, has not yet been observed from balloons, although it is evident from the OSO-7 data that the spectrum cuts off rather steeply above 26 keV.

D. CYGNUS SOURCES

Early balloon and satellite observations failed to resolve the Cygnus sources so that interpretation in terms of variability and spectrum may be unclear. CYG-1, assumed to be the strongest source in this region has a hard spectrum (Haymes and Harnden, 1970; Webber and Reinert, 1970) and has been observed many times from balloons. In particular, the work of Overbeck and Tananbaum (1968) showed this source to be highly variable in hard X-rays. These sources are resolved in more recent observations, so unambiguous knowledge of their spectrum and variability is now available.

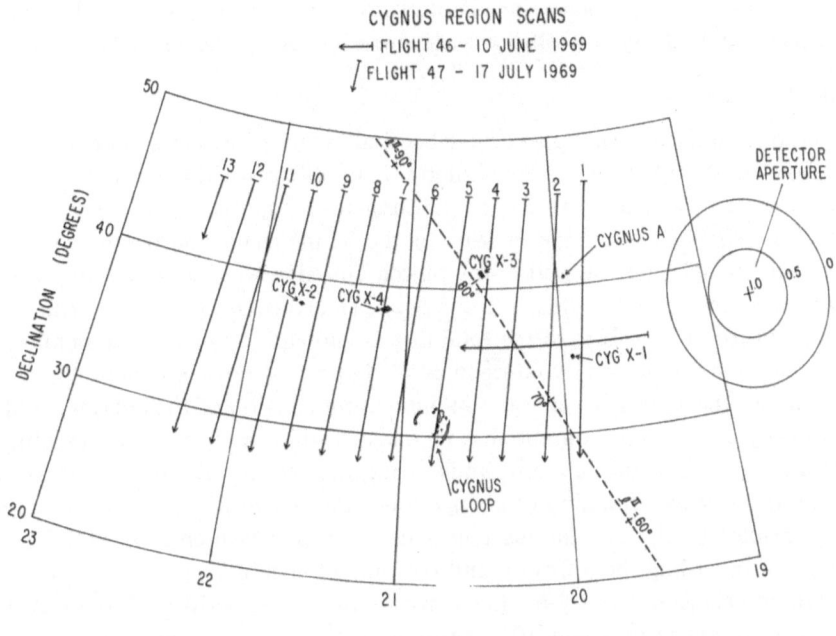

Fig. 6. This shows the location of identified X-ray sources and other interesting objects in the Cygnus region. The scan paths are those of Matteson during his high resolution balloon survey of the region. Only CYG-1, -2, and -3 have been detected >20 keV.

Figure 6, taken from the thesis of Matteson (1971) shows the location of the principal sources, CYG-1, CYG-2 and CYG-3. Apparently, CYG-4 has never been observed from balloons; furthermore, it is not indicated in the UHURU catalog. Only upper limits exist for the Cygnus Loop and the radio source CYG A (Matteson, 1971).

E. SUMMARY

From the spatial distribution of hard X-ray sources, inferences can be made about the nature of the emitting objects. Except for a paucity of sources in Orion and in the Galactic anti-center direction, these tend to cluster in a manner which delineates spiral arms. These sources are therefore probably not of extreme Pop. I, but may be old Pop. I, and perhaps even some Pop. II objects distributed toward the Galactic center. This suggests X-ray 'stars' are not associated with the formation of normal stars, but are a part of stellar evolution not yet understood. The spatial distribution of similar X-ray emitting objects provides information relative to their place in stellar evolution and their distance. At present, these sources are not well enough understood to permit these investigations.

4. Source Types

This is a more complete discussion of the various sources in terms of their characteristic spectra and the type of objects with which they have been identified.

A. THE CRAB NEBULA

Observations on this unique source have been summarized by Peterson and Jacobson (1970). The number spectrum is represented as a power law with an index of -2.25 from about 1–500 keV, as shown in Figure 7. Long term time variations have not been noted. Following the discovery in 1969 of the 33 ms radio pulsar located near the center of the nebula, optical and X-ray pulsed emission were soon measured. Pulsed X-ray measurements now extend to 3.5 MeV, (Kurfess, 1971; Orwig, *et al.*, 1971), and account for at least 20% of the total flux as shown in Figure 7. A compilation of data on the pulsed and steady components (Figure 1 of Fazio's review, this volume p. 303) shows that at higher energies an increasing fraction of the energy is pulsed.

The steady emission presumably has its origin within the 1' or 2' size identified by rocket observations with the amorphous nebulosity. No information is yet available on the source size as a function of energy. Since the synchroton process is thought to be responsible for the continuous component, such measurements would provide information regarding the diffusion and life-time of energetic electrons in the nebula itself. Observations to detect γ-ray line emission have been performed, but only upper limits at an intensity of about 10^{-3} photons $(cm^2\ s)^{-1}$ are available (Peterson and Jacobson, 1970). Based on the Cf^{254} hypothesis for the explanation of the 55 day-light curve of type I supernovae, intensities of about 3×10^{-4} photon $(cm^2\ s)^{-1}$ are expected for γ-rays due to Am^{241} at 62 keV, Cf^{251} at 180 keV and Cf^{249} at 390 keV. Although the status of observational data on the Cf^{254} hypothesis has not changed since last reviewed, the entire idea seems less likely in view of the association of the central pulsar with a collapsed star.

B. SUPERNOVA REMNANTS

No other observed X-ray source or supernova remnant has all the characteristics of

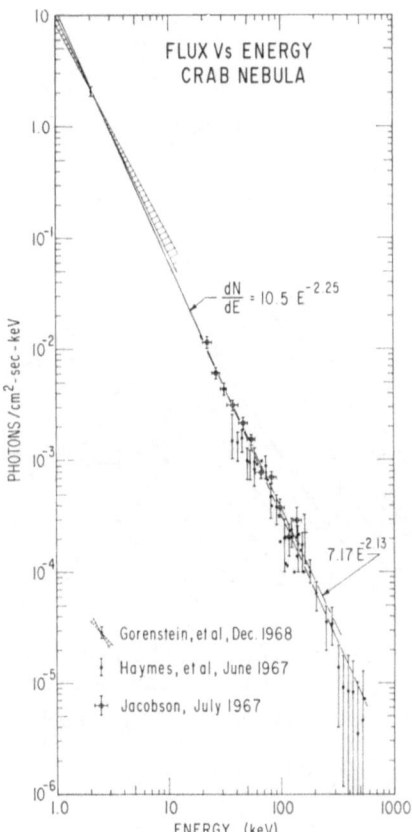

Fig. 7. The spectrum of the Crab nebula determined between 1 and 500 keV, assembled from rocket and balloon results. Only certain selected data are shown here, though all the published data were taken into account (see Gorenstein *et al.*, 1970, and Pounds [this volume] for references). The power law spectrum is thought to be indicative of the synchroton process. About 20 % of the emission in this spectrum is pulsed.

the Crab Nebula with the possible exception of VELA X; other supernova remnants have not yet been observed above 10 keV despite many attempts. Such objects, observed in the 2–10 keV range (Gorenstein *et al.*, 1970) have a spectrum characteristic of a hot gas not exceeding 10 or 20×10^6 K and a relatively low luminosity of about 10^{36} ergs s^{-1}. In particular, CAS A, Tycho, and the Cygnus Loop (Matteson, 1971) have limits $\sim 6 \times 10^{-4}$ photons (cm^{2}s keV)$^{-1}$ in the 20–50 keV range. A hard X-ray component, if it exists in these objects, must have a luminosity less than a few percent of the total.

Figure 8 shows the Vela region which contains the Vela supernova remnant, radio pulsar PSR 0833-45, and an UHURU source identified with this object. Harnden *et al.* (1972) have reported a source in this region, and have found marginal evidence for pulsed X-ray emission in the 23–80 keV range with a period 150 ns less than the

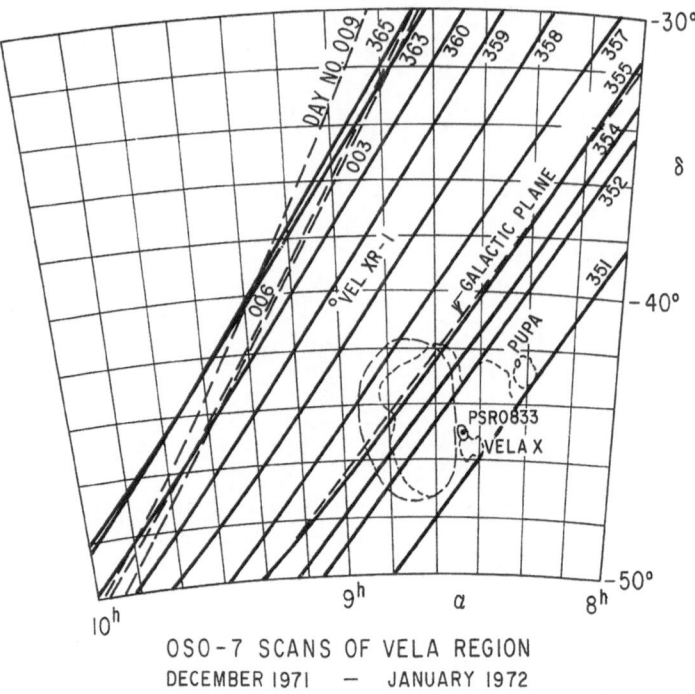

OSO-7 SCANS OF VELA REGION
DECEMBER 1971 — JANUARY 1972

Fig. 8. This shows locations of the X-ray source VEL-1, and objects associated with the Vela X
supernova remnant, together with scan paths of the OSO-7 X-ray instrument in
December 1971/January 1972.

nominal 89^+ ms. Except for the pulsed feature, this emission could be associated with
the variable X-ray source VEL-1. The OSO-7 scanned this region from December 17,
1971 through January 10, 1972 and failed to detect X-rays from the location of VEL X
at a 2σ limit of 4×10^{-2} keV $(cm^2 \; s \; keV)^{-1}$ in the 7–26 keV range.

C. SCO-1 TYPE SOURCES

SCO-1 is characterized by an exponential spectrum whose equivalent temperature is
on the order of 50×10^6 K and is identified with a flickering blue star-like object. The
X-ray spectrum in the 1–10 keV range varies a factor of 2, apparently at random,
although a recent analysis (Angel *et al.*, 1971) has indicated oscillation in the 1–10 Hz
range persisting for a minute. Early observations indicated the spectrum above
40 keV also contained an additional variable component. This has now been verified to
about 120 keV (Agrawal *et al.*, 1969; Lewin *et al.*, 1970a; Haymes *et al.*, 1972). Simultaneous optical observations (Toor *et al.*, 1970; Pelling, 1971) have shown a cou-
pling to the lower energy X-rays, and therefore indicated a common region. Radio and
optical variations show no such correlation (Lampton *et al.*, 1971). Soft X-ray and
optical observations may be interpreted in terms of a hot gas model which becomes
optically thick in the UV. The hard X-ray and radio fluxes are inconsistent with this

simple model and may require an additional source region. Perhaps simultaneous observations of these components are also appropriate, since both have demonstrated time variability.

In addition to the random variations, SCO-1 also exhibits a flare-like phenomenon which extends into the hard X-ray region (Lewin *et al.*, 1968a) and is associated with an optical increase (Hudson *et al.*, 1970; Pelling, 1971). The bursts have about a ten-minute duration, a factor of two increase in X-ray intensity and a spectrum similar to that of the quiescent state.

Because of its intensity, SCO-1 has been the most studied source of this type. However, CYG-2 has also been tentatively identified with a star-like flickering object (Kraft and Demoulin, 1967). The compilation of data shown in Figure 9 taken from Matteson's thesis (1971) indicates the exponential spectrum believed associated with the 50×10^6 K gas and the flat, high energy component. Both features of the spectrum in this source may also show variability.

D. CYG-1 TYPE SOURCES

CYG-1 is characterized by a spectrum which extends to several hundred keV (Haymes and Harnden, 1970); rapid, possibly quasi-periodic variations in soft X-rays (Terrell,

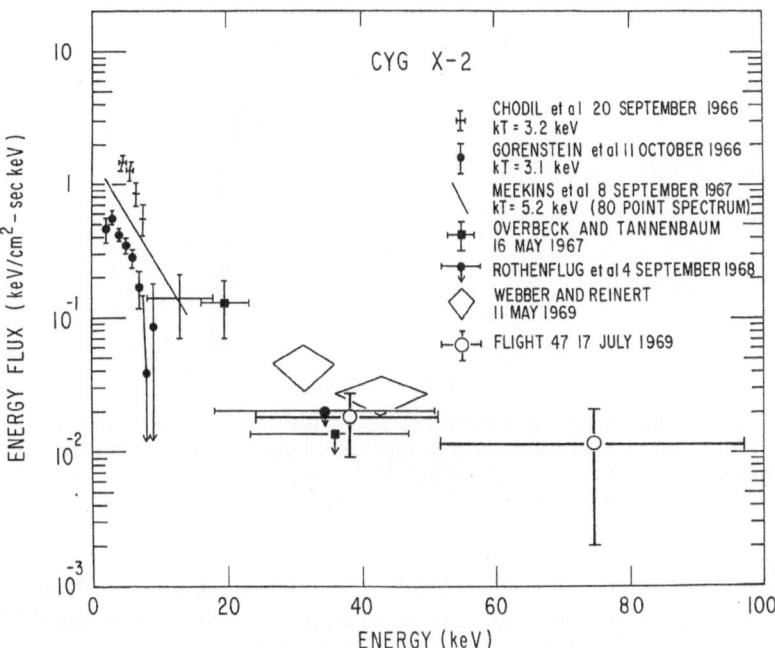

Fig. 9. Spectral data on CYG-2, taken from Matteson's thesis (1971), where the references may be found. At energies below 15 keV a thermal spectrum with factor two intensity and temperature variations is observed, while higher energies show a hard spectral component which is also variable. A similar high energy feature is seen in the spectrum of SCO-1.

1972); and large changes in hard X-ray intensity (Overbeck and Tananbaum, 1968). Recent observations have indicated that the hard X-ray spectrum of CYG-1 is characterized by 'high' and 'low' states. Figure 10 shows the results obtained by Matteson (1971) together with other spectral data. A flight on 10 June 1969 indicated a hard

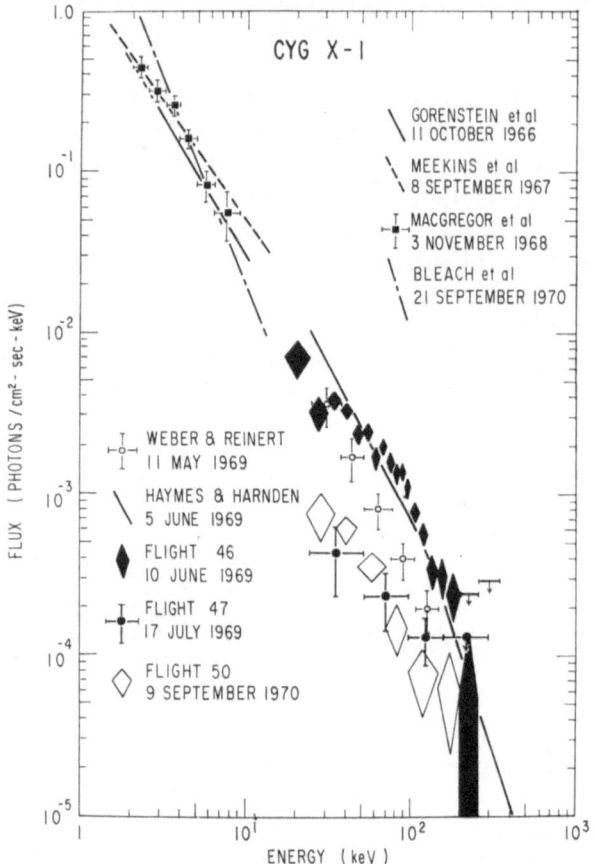

Fig. 10. A compilation of rocket and balloon spectral data on CYG-1, taken from Matteson (1971). During a series of balloon observations, May to June 1969, large variations in intensity and complex spectral changes occurred at energies above 20 keV. The intensity on 17 July 1969 is the lowest ever measured from CYG-1 at energies above 20 keV.

spectrum, in agreement with other observations which were made during the preceeding month. Less than one month later, an additional observation found CYG-1 in a much weaker state, as did an observation by USCD about a year later. Reference citations shown in Figures 9, 10, and 11, may be obtained from Matteson (1971).

The source has now been tentatively identified with a 5.6 day binary (Bolton, 1972; Webster and Murdin, 1972). Unlike CEN-3, 5.6 day variations possibly associated with an occultation of the X-ray emitting neutron star by large central star have not been observed in the soft X-ray region by UHURU. The OSO-7 observed CYG-1 between 16 and 22 December 1971 when the source was generally in a 'high' state. Although large changes in hard X-rays were observed, no turn-off occurred at the

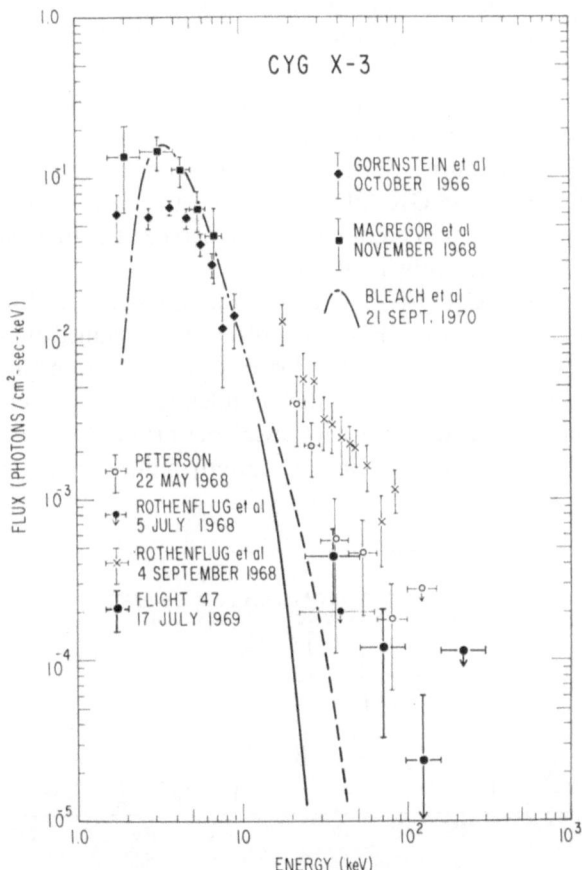

Fig. 11. CYG-3 has a remarkable similarity to CYG-1 as this collection of data (Matteson, 1971) shows. The flux at 100 keV is sometimes that of CYG-1 with a factor of ten variations occurring at energies above 20 keV. The relative lack of X-rays at energies below 4 keV may be due to absorption in a column density of more than 10^{23} atoms/cm^2. Extrapolation of fits to the low energy spectrum indicated by the solid and broken lines cannot account for the high energy spectra.

time predicted from the optical observations. Furthermore, no evidence has been found for periodic fluctuations in the 160 ms to 1.2 s range by Agrawal *et al.* (1971b). Matteson (1971) has also searched for longer period variations and found none.

CYG-3, whose composite spectrum is shown in Figure 11 (Matteson, 1971), has many characteristics similar to CYG-1. At present no optical candidate exists for

CYG-3, nor is there convincing evidence for binary modulation. This source does how-ever exhibit absorption at lower energies (Bleach *et al.*, 1971; Gorenstein *et al.*, 1967).

E. NOVA-LIKE AND VARIABLE SOURCES

These sources, such as CEN-2 (Chodil *et al.*, 1968), are characterized by a sudden increase in soft X-rays and a decay on a time scale typically that of galactic novae, 30 days.

In view of the re-determination of the position of CEN 2 (Francey, 1971) and the consequent re-evaluation of previous balloon observations, (Lewin *et al.*, 1971b) it is not clear that these sources are observable in hard X-rays. It is clear that some sources, such as GX 1 + 4 and GX 301 − 1 have appeared after not being seen in earlier surveys of the same region, and continue to be observed. The time behaviour of these sources in soft X-rays is presently unknown.

Some sources such as GX 304 − 1 have general variability. VEL-1, identified with 2U0900-40, may also be in this class. This object was scanned by the OSO-7 from December 17, 1971 to January 10, 1972. Observed time variations shown in Figure 12 are obtained from about four orbits per day of quick-look data. Each point represents an average of one hour of data for each available orbit. Despite clear evidence of variability, there is no obvious periodicity or pattern in these preliminary observations.

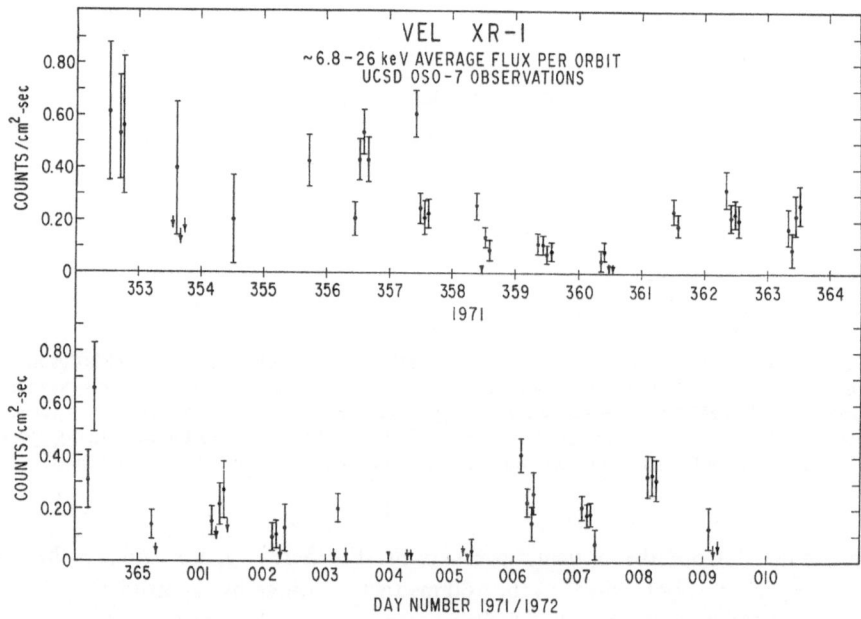

Fig. 12.   Averages of the 7–26 keV X-ray flux from VEL-1 over selected one hour intervals during a
two week period show considerable variability but no obvious periodicities
in the ∼ 1 day to 10 day regime.

5. Extra Galactic Sources

CEN A is the only extra-galactic source whose measured spectrum clearly extends to high energies. The total spectrum, taken from the work of Lampton *et al.* (1972) is shown in Figure 13, and is represented by a power law whose index is between −1.45 and −2.0 over the 1–180 keV range. This object has been compared with other extra-galactic emitters detected by UHURU (Kellogg *et al.*, 1971).

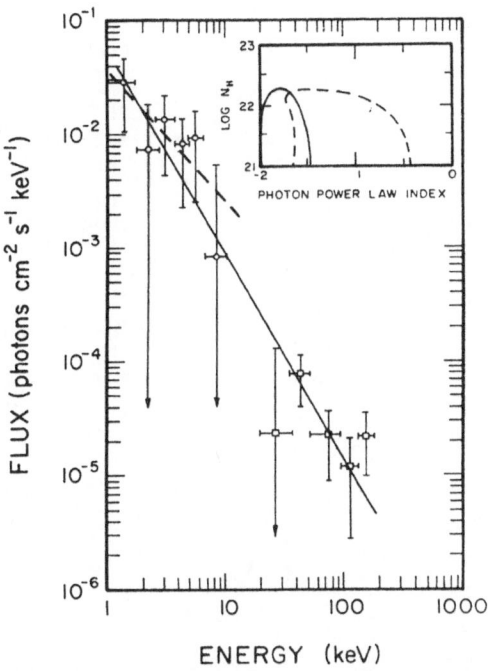

Fig. 13. Spectrum of NGC 5128 extending over the 1–180 keV range. Errors are ±1σ statistical confidence. The broken line shows the best power-law fit to rocket data only, while the solid line is the best power-law fit to all data. The inset shows contours of constant probability for the fits. (Lampton *et al.*, 1972).

Price *et al.* (1971) have detected the SMC to about 50 keV and determined an energy spectrum with an index $\alpha = 0.15 \pm 0.20$. This flux seems to disagree with an upper limit reported by Lewin *et al.* (1968b). Since the hard source noted by the UHURU (Leong *et al.*, 1971) in the Small Magellanic Cloud is apparently compact and variable, these observations may be consistent.

Data from the Virgo region is usually reported as relevant to M87, even though telescope apertures do not separate M 87 from other objects in Virgo. Although Haymes (Fishman *et al.*, 1970) has reported detectable fluxes on several occasions, other workers (Peterson, 1970; McClintock *et al.*, 1969) have indicated substantially lower upper limits. Figure 14 shows a summary of recent measurements and upper limits taken from Laros *et al.* (1972), reporting recent observations by UCSD on extragalactic emitters. UCSD may have detected a flux from the Virgo region at the 2σ confidence level. If all observations are assumed correct, a two component model is required, one of which produces a soft spectrum associated with the radio galaxy M 87, and the other a hard component due to at least one other compact and variable object.

The Andromeda Nebula (M31), 3C273, and the Large Magellanic Cloud have also been searched for hard X-ray emissions. The most recent upper limit obtained by

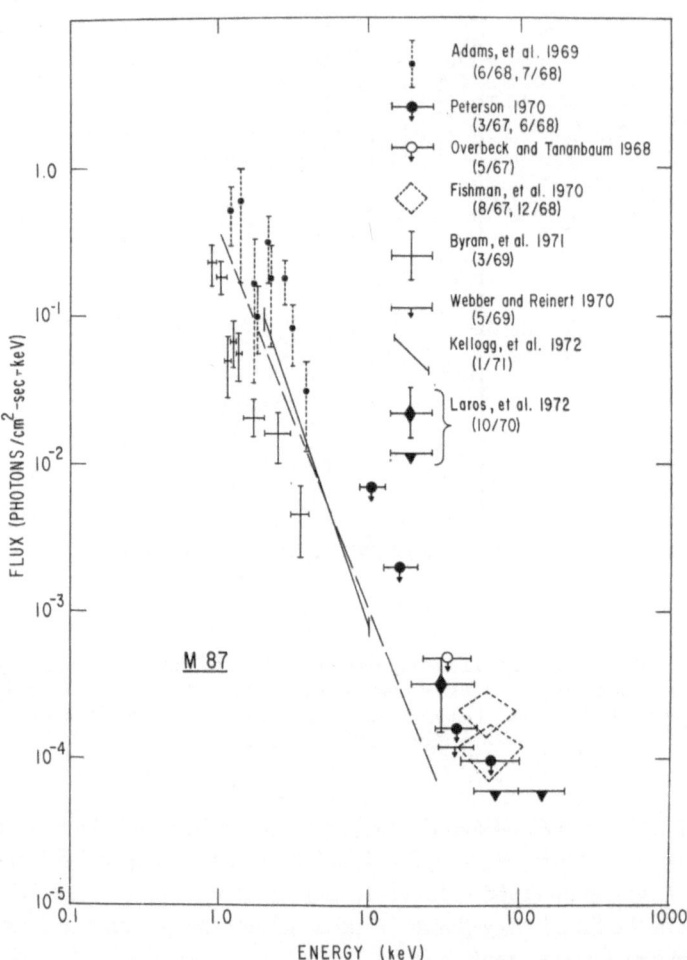

Fig. 14. Selected X-ray observations of M87 (Laros *et al.*, 1972) showing observation dates. Mc Clintock *et al.* (1969) also has an upper limit in the 35–52 keV range of 1×10^{-4} photons $(cm^2 \; s \; keV)^{-1}$ at the 1σ confidence level. A power law average of all the data in the 1–10 keV (dashed line) is shown along with the single most sensitive measurement (solid line) and the extreme measurements which have been taken as evidence for variability.

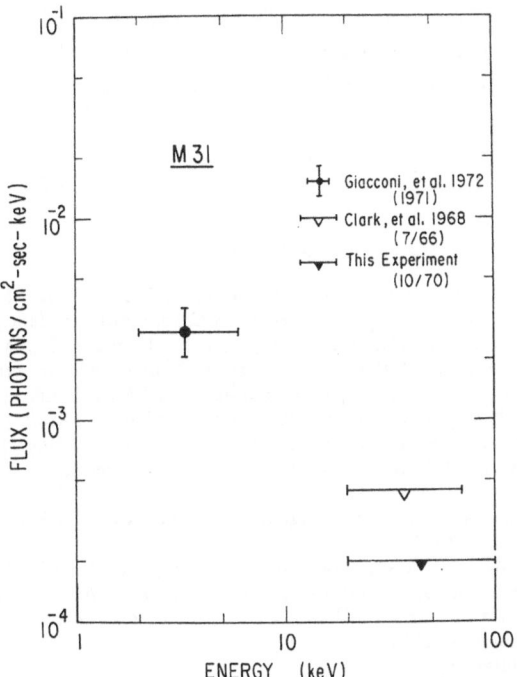

Fig. 15. X-ray observations of M31, including recent upper limits by UCSD. The positive result obtained by the UHURU satellite has yet to be confirmed by other experiments.

UCSD on M31 is shown in Figure 15. A comparable limit has also been obtained for 3C 273. A 2σ upper limit of $\sim 6 \times 10^{-4}$ photons $(cm^2 \ s \ keV)^{-1}$ over the 24–52 keV range was also obtained by Matteson (1971) for CYG A.

6. Conclusion

This paper summarizes the present status of X-ray observations in the $20 \simeq 500$ keV range on galactic and extra-galactic sources. Because of limitations of sensitivity and observing time, results have not been as extensive as those in the range where the UHURU has produced a wealth of new discoveries. Hard X-ray emission is a certain indicator of non-thermal particle populations, and must, therefore, be associated with energetic processes in extreme astrophysical situations. In the immediate future, the OSO-7 will provide new results. By 1975 the HEAO-A will carry instruments to provide a further increment in sensitivity and observation time.

Acknowledgements

The author ackowledges useful discussions with various investigators regarding the contents of this paper. Dr Mel Ulmer in particular has provided extensive discussion and criticism of the manuscript. Many of my associates, Drs Baity, Laros, Matteson, Pelling, Ulmer and Wheaton have also contributed results which appear in this paper. This research was supported by NASA under Contracts NGR 05-005-003 and NAS 5-11080.

Note added in proof. Further analysis has discovered an 8.7 ± 0.2 d periodicity for Vela X-1 X-rays.

References

Agrawal, P. C., Biswas, S., Gokhale, G. S., Iyengar, V. S., Kunte, P. K., Manchanda, R. K., and Sreekantan, B. V.: 1969, *Proc. 11th Int. Conference on Cosmic Rays*, Budapest, Hungary.

Agrawal, P. C., Biswas, S., Gokhale, G. S., Iyengar, V. S., Kunte, P. K., Manchanda, R. K., and Sreekantan, B. V.: 1971a, *Proc. 12th Int. Conference on Cosmic Rays*, Hobart (Australia), 1, 20.

Agrawal, P. C., Gokhale, G. S., Iyengar, V. S., Kunte, P. K., Manchanda, R. K., and Sreekantan, B. V.: 1971b, *Proc. 12th Int. Conference on Cosmic Rays*, Hobart (Australia), 1, 26.

Angel, J. R. P., Kestenbaum, H., and Novick, R.: 1971, *Astrophys. J. Letters* 169, L57.

Bleach, R. D., Boldt, E. A., Holt, S. S., Schwartz, D. A., and Serlemitsos, P. J.: 1972, *Astrophys. J.* 171, 51.

Boldt, E., Doong, H., Serlemitsos, P., and Reigler, G. R.: 1968, *Can. J. Phys.* 46, S444.

Bolton, C. T.: 1972, *Nature* 235, 271.

Bradt, H., Burnett, B., Mayer, W., Rappaport, S., and Schnopper, H.: 1971, *Nature* 229, 96.

Chodil, G., Mark, H., Rodrigues, R., and Swift, C. D.: 1968, *Astrophys. J. Letters* 152, L45.

Fishman, G. J., Harnden, F. R., Jr., and Haymes, R. C.: 1970, in L. Gratton (ed.), 'Non-Solar X- and Gamma-Ray Astronomy', *IAU Symp.* 37, 116.

Francey, R. J.: 1971, *Nature* 229, 229.

Frost, K. J.: 1969, *Astrophys. J. Letters* 158, L159.

Giacconi, R., Kellogg, E., Gorenstein, P., Gursky, H., and Tananbaum, H.: 1971, *Astrophys. J. Letters* 165, L27.

Giacconi, R., Murray, S., Gursky, H., Kellogg, E., Schreier, E., and Tananbaum, H.: 1972, ASE-2919, American Science & Engineering, Cambridge, Mass. (submitted to *Astrophys. J.*).

Glass, I. S.: 1969, *Astrophys. J.* 157, 215.

Gorenstein, P., Giacconi, R., and Gursky, H.: 1967, *Astrophys. J. Letters* 150, L85.

Gorenstein, P., Kellogg, E. M., and Gursky, H.: 1970, *Astrophys. J.* 160, 199.

Harnden, F. R., Jr., Johnson, W. N., III, and Haymes, R. C.: 1972, *Astrophys. J. Letters* 172, L91.

Haymes, R. C. and Harnden, F. R., Jr.: 1970, *Astrophys. J.* 159, 1111.

Haymes, R. C., Harnden, F. R., Jr., Johnson, W. N., III, Prichard, H. M., and Bosch, H. E.: 1972, *Astrophys. J. Letters* 172, L47.

Hill, R. W., Burginyon, G., Grader, R. J., Palmieri, T. M., Seward, F. D., and Stoering, J. P.: 1972, *Astrophys. J.* 171, 519.

Hudson, H. S., Peterson, L. E., and Schwartz, D. A.: 1970, *Astrophys. J. Letters* 159, L51.

Jacobson, A S.: 1968, Ph. D. Thesis, UCSD.

Johnson, W. N., III., Harnden, F. R. Jr., and Haymes, R. C.: 1972, *Astrophys. J. Letters* 172, L1.

Kellogg, E., Gursky, H., Leong, C., Schreier, E., Tananbaum, H., and Giacconi, R.: 1971, *Astrophys. J. Letters* 165, L49.

Kinzer, R. L., Noggle, R. C., Seeman, N., and Shore, G. H.: 1971, *Nature* 229, 187.

Kraft, R. P. and Demoulin, M.: 1967, *Astrophys. J. Letters* 150, L183.

Kurfess, J. D.: 1971, *Astrophys. J. Letters* 168, L39.

Lampton, M., Bowyer, S., Welch, J., and Grasdalen, G.: 1971, *Astrophys. J. Letters* 164, L61.

Lampton, M., Margon, B., Bowyer, S., Mahoney, W., and Anderson, K.: 1972, *Astrophys. J. Letters* 171, L45.

Laros, J. G., Matteson, J. L., and Pelling, R. M.: 1972, UCSD (submitted for publication to *Astrophys. J.*)

Leong, C. Kellogg, E., Gursky, H., Tananbaum, H., and Giacconi, R.: 1971, *Astrophys. J. Letters* 170, L67.

Lewin, W. H. G., Clark, G. W., and Smith, W. B.: 1968a, *Astrophys. J. Letters* 152, L55.

Lewin, W. H. G., Clark, G. W., and Smith, W. B.: 1968b, *Nature* **220**, 249.

Lewin, W. H. G., McClintock, J. W., Ryckman, S. G., Glass, I. S., and Smith, W. B.: 1970a, *Astrophys. J. Letters* **162**, L109.

Lewin, W. H. G., McClintock, J. E., and Smith, W. B.: 1970b, *Astrophys. J. Letters* **159**, L193.

Lewin, W. H. G., Ricker, G. R., and McClintock, J. E.: 1971a, *Astrophys. J. Letters* **169**, L17.

Lewin, W. H. G., McClintock, J. E., Ryckman, S. C., and Smith, W. B.: 1971b, *Astrophys. J. Letters* **166**, L69.

Matteson, J. L.: 1971, 'An X-ray Survey of the Cygnus Region in the 20–300 keV Energy Range', Thesis, UCSD.

McClintock, J. E., Lewin, W. H. G., Sullivan, R. J., and Clark, G. W.: 1969, *Nature* **223**, 162.

McClintock, J. E., Ricker, G. R., and Lewin, W. H. G.: 1971, *Astrophys. J. Letters* **166**, L73.

McClintock, J. E., Ricker, G. R., Ryckman, S. G., and Lewin, W. H. G.: 1972, *Astrophys. J. Letters* **173**, L57.

Orwig, L. E., Chupp, E. L., and Forrest, D. J.: 1971, *Nature* **231**, 171.

Overbeck, J. W. and Tananbaum, H. D.: 1968, *Astrophys. J.* **153**, 899.

Pelling, R. M.: 1971, 'A Study of Simultaneous Optical and X-ray Observations of Scorpius X-1,' Thesis, UCSD.

Peterson, L. E.: 1970, in L. Gratton (ed.), 'Non-Solar X- and Gamma-Ray Astronomy', *IAU Symp.* **37**, 59.

Peterson, L. E. and Jacobson, A. S.: 1970, *Publ. Astron. Soc. Pac.* **82**, 412.

Peterson, L. E., Pelling, R. M., and Matteson, J. L.: 1971, UCSD-SP-71-04, SPARMO Symp., Seattle, Wash. (to be published in *Space Sci. Rev.*).

Price, R. E., Groves, D. J., Rodrigues, R. M., Seward, F. D., Swift, C. D., and Toor, A.: 1971, *Astrophys. J. Letters* **168**, L7.

Riegler, G. R.: 1969, 'An X-Ray Survey of the Sky From Balloon Altitudes', Thesis, University of Maryland.

Schreier, E., Levinson, R., Gursky, H., Kellogg, E., Tananbaum, H., and Giacconi, R.: 1972, *Astrophys. J. Letters* **172**, L79.

Schwartz, D. A.: 1970, *Astrophys. J.* **162**, 439.

Schwartz, D. A., Hudson, H. S., and Peterson, L. E.: 1970, *Astrophys. J.* **162**, 431.

Schwartz, D. A., Peterson, L. E., and Hudson, H. S.: 1971, Preprint X-661-71-445, Goddard Space Flight Center, Maryland (to be published in *Astrophys. J.*).

Terrell, N. J., Jr.: 1972, *Astrophys. J. Letters* **174**, L35.

Toor, A., Seward, F. D., Cathey, L. R., and Kunkel, W. E.: 1970, *Astrophys. J.* **160**, 209.

Webber, W. R. and Reinert, C. P.: 1970, *Astrophys. J.* **162**, 883.

Webster, B. L. and Murdin, P.: 1972, *Nature* **235**, 37.

6. SIMULTANEOUS X-RAY, OPTICAL AND RADIO
OBSERVATIONS OF GALACTIC X-RAY SOURCES

W. A. HILTNER

University of Michigan, Ann Arbor, Mich., U.S.A.

Abstract. Simultaneous observations of the X-ray source Sco X-1 are reviewed and discussed. Several conclusions can be drawn from the observations. (a) There is no correlation of the radio intensities with either the optical or X-ray intensities. (b) The X-ray intensity is high and variable only when the blue optical magnitude is brighter than 12.6–12.7. (c) There is often correlation of X-ray and optical flares, but the relationship is not one to one.

Attention is called to the paucity of simultaneous observations of other identified X-ray sources.

In order to make simultaneous X-ray, optical and radio observations it is obvious that the X-ray source must be identified in other spectral regions of interest. This requirement has emerged as one of the principal obstacles for more extensive observations in spectral regions other than the X-ray. However, as will become apparent, it is not the only one.

The number of galactic X-ray sources, excluding the old novae, that have reliable identifications in other spectral regions are very few. In the optical region Sco X-1, Tau X-1 (the Crab Nebula) and Cyg X-2 are identified with confidence. Cyg X-1 and 2U 0900−40 may be associated with optical spectroscopic binaries (Bolton, 1972; Webster and Murdin, 1972; and Hiltner and Osmer, 1972), the secondary components of which may possibly be the X-ray object. Both sources are variable in X-ray intensity (Giaconni *et al.*, 1972) and both spectroscopic binaries have been reported as probable variables in the optical region (Kukarkin and Soluyanov, 1972; and Hiltner *et al.*, 1972). The two X-ray sources for which observations at other wavelengths should prove most exciting are Cen X-3 and Her X-1. Both are eclipsing binaries (Schreier *et al.*, 1972) as observed in the X-ray region. Neither has been identified in the optical or radio. Their identification should receive the highest priority.

The need for simultaneous observations at different spectral regions implies that the sources are variable in more than one region and non-periodic. It now seems relatively safe to assume that most, if not all, X-ray sources are variable. The three galactic sources listed above with reliable optical and radio identifications all are variable in intensity at both X-ray and optical wavelengths and two are known to be variable at radio frequencies. Simultaneous observations can therefore possibly assist in developing a model for the X-ray stars.

Only Sco X-1 has been observed simultaneously at two or more spectral regions, and these observations will be discussed in some detail, after a few comments on Cyg X-2. Kristian *et al.* (1967) showed that Cyg X-2 was variable in optical intensity and spectroscopic studies (Kraft and Demoulin, 1967; and Kraft and Miller, 1969) of the optical counterpart have been published. However, there have been no simultaneous observations reported in the literature. The neglect of this object by optical astronomers is no doubt a consequence of its faintness. Its B magnitude is 15.2, a factor of 5 to 10 fainter than the optical counterpart of Sco X-1. No attempt will be made to discuss Tau X-1, the Crab Nebula, but I refer the reader to the many reviews and extensive literature on this object. Furthermore, the variable optical component is periodic.

Our discussion by necessity will therefore be centered around Sco X-1 of which there are now sufficient simultaneous observations for analysis. Before discussing the simultaneous observations it is useful to review the independent observations because of their much greater abundance.

At the time of its optical identification in 1966 Sandage *et al.* (1966) showed that the optical counterpart was highly variable in light intensity. Subsequent extensive observations (Hiltner and Mook, 1967, 1970; Westphal *et al.*, 1968) showed irregular light variations with the following characteristics: (1) There are often night to night

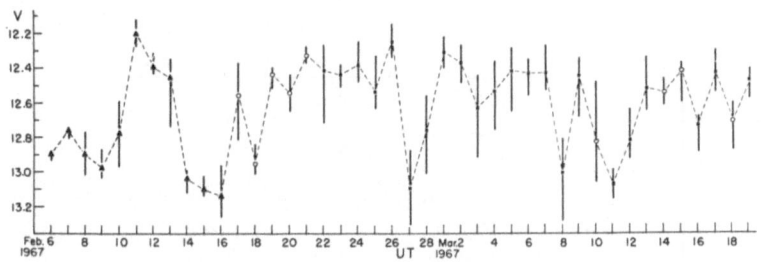

Fig. 1.   Average visual magnitude of Sco X-1 for 42 consecutive nights. (Mook, 1967). Bars through the data points represent the total range in *V* over which Sco X-1 varied on that night.

variations of a magnitude, (2) smooth variations of one-half magnitude in an hour or two, (3) flares with an amplitude of about 0.2 mag and a rise time as short as 90 s, and (4) a flickering with an amplitude of about 0.02 mag on a time scale of a minute, more or less. The first three types of variations are shown in Figures 1 and 2 and the last in Figure 3. It has been shown by Westphal *et al.* (1968) that the flickering is superimposed on all the other types of variations listed above. No periodicity has been established although transient oscillations have been reported after the object has flared (see Figure 3).

Rocket flights have shown that the X-ray intensity of Sco X-1 in the 2 to 10 keV range is variable, perhaps by a factor of two. The first X-ray flare was observed in 1967 by Lewin *et al.* (1968). During this balloon flight the intensity of hard X-rays was observed to increase (Figure 4) by a factor of four in ten minutes. The high X-ray intensity then decayed with a characteristic time of about thirty minutes. Further balloon flights by Lewin *et al.* (1970) confirm these earlier observations of large variations in X-ray intensity. Changes in X-ray intensity on a shorter time scale have been observed by Evans *et al.* (1970) as shown in Figure 5.

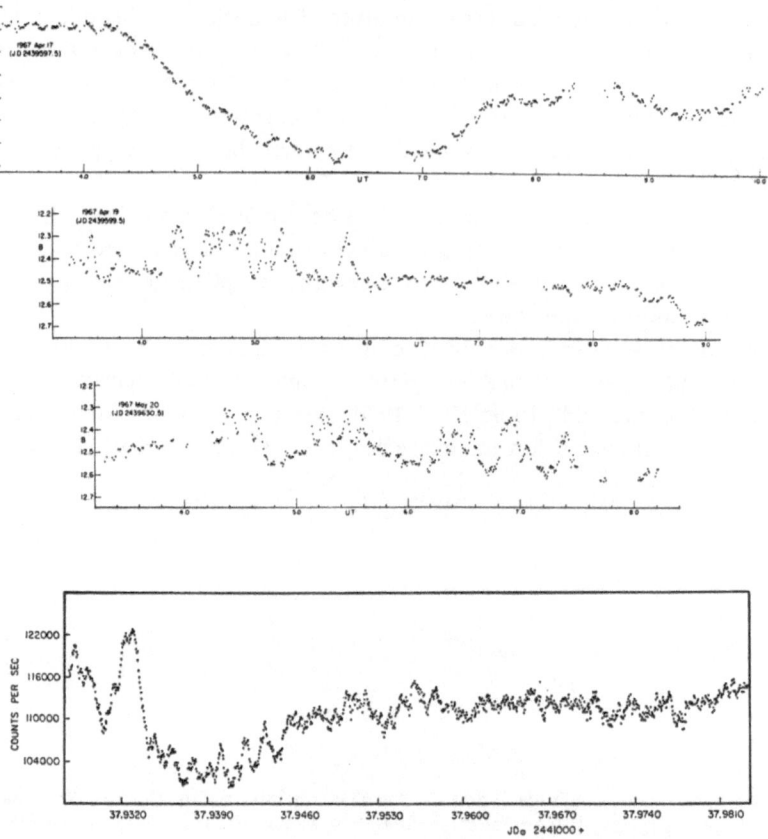

Fig. 3.   Portion of the light curve of Sco X-1 on 1971 March 27 (Robinson and Warner, 1972). Each
point represents an integration for 4 s. Note the 'oscillations' that persisted for about 15 min
and then faded.

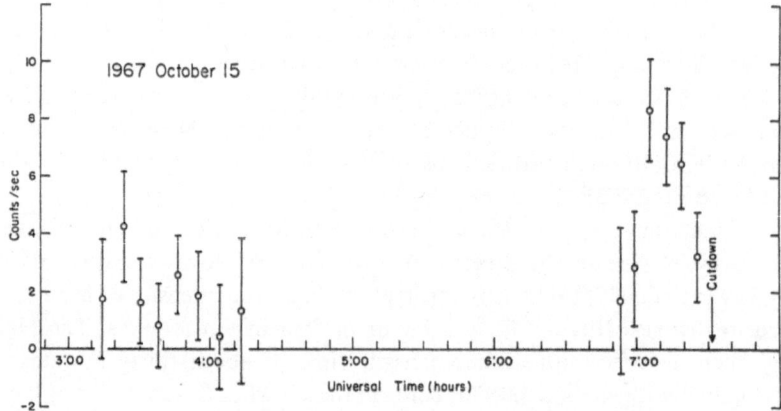

Fig. 4.   Intensity of X-rays from Sco X-1 in the 20–30 keV range as a function of universal time on
1967 October 15 (Lewin *et al.*, 1968).

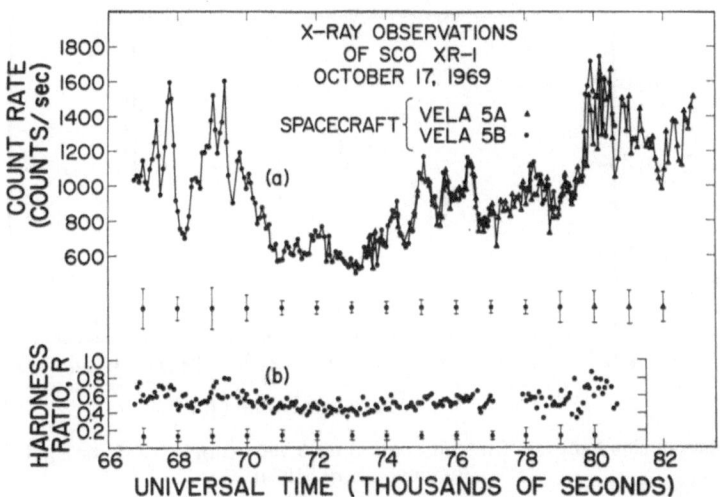

Fig. 5. X-ray observations of Sco X-1 on 1969 October 17. (a) Observed 3–12 keV counting rate versus time and (b) spectral hardness ratio versus time. Typical error bars are shown below each curve (Evans *et al.*, 1970).

Radio observations by Hjellming and Wade (1971) have shown that Sco X-1 consists of a highly variable source centered on the optical counterpart and two satellite sources on opposite sides of the central source and probably constant in intensity. This configuration has been confirmed by Braes and Wiley (1972). Further observations by Wade and Hjellming (1972) showed from about 300 h of observing that major flaring of the central source was in progress 15% of the time, dormant (<0.003 flux units, the limit of detection) 20%, constant at about 0.01 flux units 5% and for the remaining 60% the object was relatively weak (0.01 to 0.03 flux units) but erratically and continuously variable.

The need for simultaneous X-ray, optical and radio observations is obvious for Sco X-1 is highly variable in all wavelengths and non-periodic.

The earliest attempt to make simultaneous observations has been reported by Burginyon *et al.* (1970). Eight rocket flights were made while the optical counterpart of Sco X-1 was monitored for blue magnitude. The eight X-ray observations were made in the B magnitude interval from 13.2 to 12.4. Some correlation of the X-ray intensity and the B magnitude can be seen in Figure 6, but the uncertainties in the observations are only a little smaller than the effect.

The OSO-3 and Vela satellites gave the earliest opportunity to obtain more detailed correlations of X-ray and optical intensities of Sco X-1. Hudson *et al.* (1970) observed two X-ray flares in Sco X-1 for which there was also optical coverage. These flares, one of which is shown in Figure 7 (June 3), correlated nicely at the two spectral region. The X-ray enhancement amounted to about a factor of two over the quiescent emission in the 7.7 to 12.5 keV energy range.

These correlations have been confirmed by further observations by Pelling (1972)

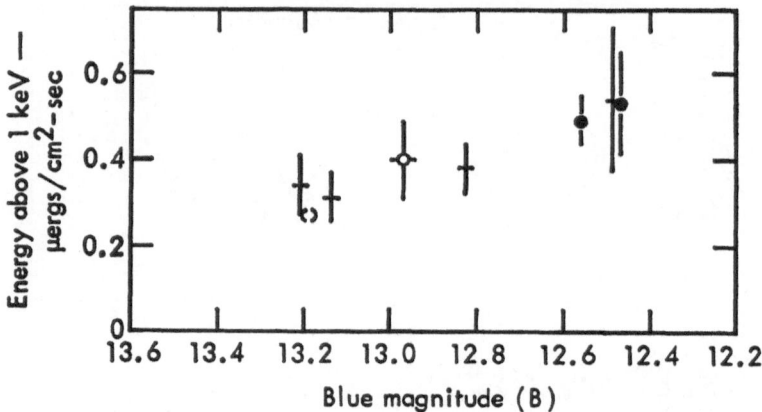

Fig. 6. Results of simultaneous X-ray and optical observations of Sco X-1 for eight rocket flights (Burginyon *et al.*, 1970).

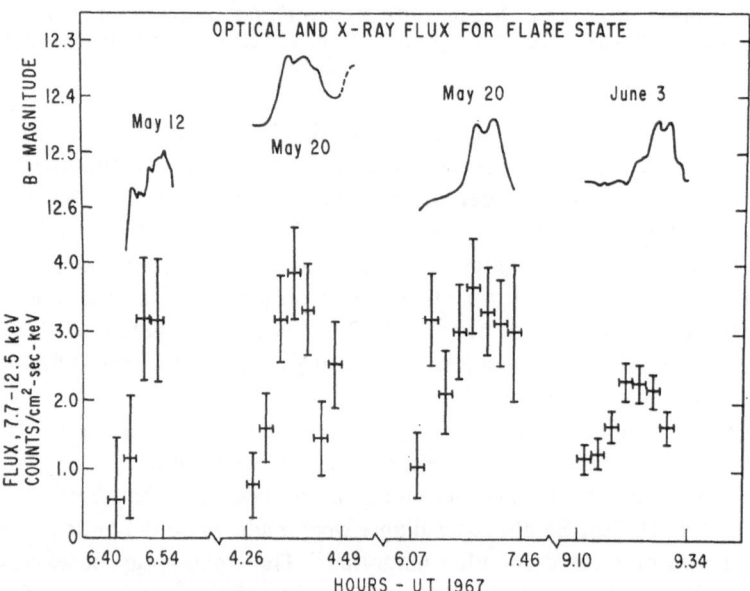

Fig. 7. Simultaneous optical and X-ray data for four flare episodes (Pelling, 1972). The night of June 3 was reported by Hudson *et al.* (1970).

who reported three additional X-ray flares observed by OSO-3 that were correlated with optical flares. They are also shown in Figure 7. A correlation of optical and X-ray flares was also confirmed by Evans *et al.* (1970). Their observations are shown in Figure 8. The X-ray intensity increase is significantly greater than that of the optical, a factor of two versus about 20%.

The correlation outside of flaring may be very different. The OSO-3 observations

(Pelling, 1972) are shown in Figure 9. The two cases shown in this figure differ only in assumed aspect of the satellite, the extremes, so that the true X-ray flux probably falls somewhere between. In Figure 10 are shown the Vela satellite observations obtained during the World Wide Watch of Sco X-1 in 1970. (Details will be published elsewhere.)

In both figures the data suggest a general decrease in X-ray flux as the object brightens optically until the magical magnitude 12.6–12.7 is reached – the magnitude from which the object may (but not necessarily) flare in the optical range. This weak anti-correlation can also be seen in Figure 8. Further observations are needed to confirm or reject the reality of this slight anti-correlation.

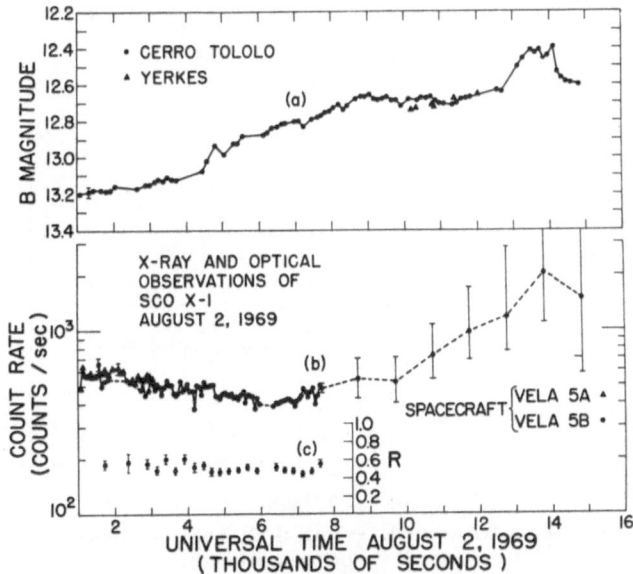

Fig. 8.   Simultaneous X-ray and optical observations of Sco X-1. (a) B magnitude versus time, (b) 3–12 keV X-ray counting rate observed by the Vela 5 satellites and (c) ratio of counting rate in the two energy channels 6–12 and 3–6 keV. After 8000 s the error bars represent upper and lower limits of the counting rate (Evans *et al.*, 1970).

Kastenbaum *et al.* (1971) have looked for correlation of X-ray and optical flickering. Their results from a rocket flight are shown in part in Figure 11. The two power spectra for the two spectral regions show a strong maximum at a period of 20 s. The correlation of the oscillations persisted for about two minutes.

The X-ray satellite SAS-A (UHURU), was programmed so that during the period March 24 to 28, 1971, Sco X-1 was observed by the satellite while optical and radio observations were made from ground-based observatories. The optical and radio coverage was for a longer period. The optical observations were made at Cerro Tololo Inter-American Observatory, Leiden Southern Station, the University of Michigan and the University of Texas and the radio observations were made at National Radio

Astronomy Observatory and Westerbork*. The observations from March 22 to April 2, 1971 are shown in Figure 12. During these twelve days there was flare activity in both the optical and radio observations as well as periods of quiescence. The optical

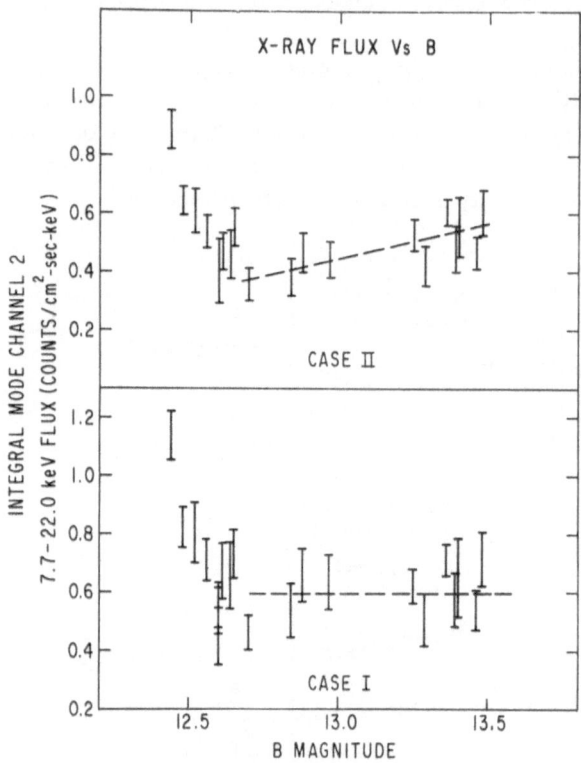

Fig. 9.   X-ray intensity versus B magnitude (Pelling, 1972). The two curves differ in assumed satellite aspect; the opposite extremes.

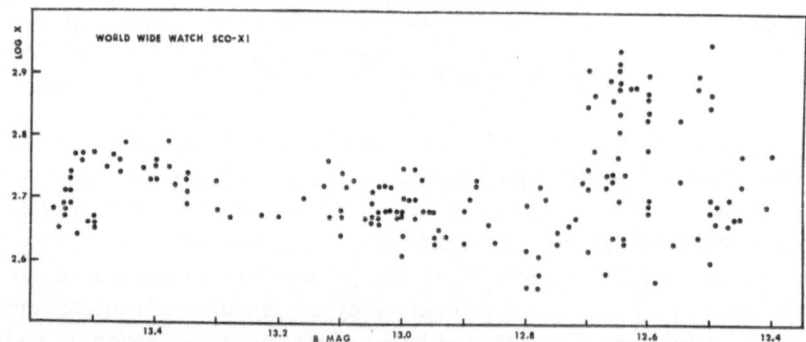

Fig. 10.   X-ray intensity versus B magnitude from the 'World Wide Watch' of Sco X-1 in 1970. Details to be published elsewhere.

* Participating observers included H. Bradt, L. Braes, W. Hiltner, R. Hjellming, W. Kunkel, G. Miley, J. Pel, J. Thomas, P. Vanden Bout, C. Wade and B. Warner.

intensity was always low when there was radio flaring and on occasion both spectral regions were faint simultaneously. During optical flaring the radio emission was in the 'dormant' or 'relatively weak' state. Judging from these observations apparently there is no obvious correlation between the optical and radio intensities.

Figure 13 shows all the available observations, X-ray, optical and radio, for the period March 24–28, 1971. An inspection of the figure suggests the following. (1) When the object is faint optically it is also faint in the X-ray region. (2) When it is bright at optical wavelengths it is bright and active in X-ray. (3) No correlation between X-ray and radio is apparent.

A close inspection of the figure on the night of March 26 will reveal a rapid increase in optical brightness followed an hour later by a rapid increase in X-ray intensity. One must not conclude that this represents a related phenomenon but rather only the

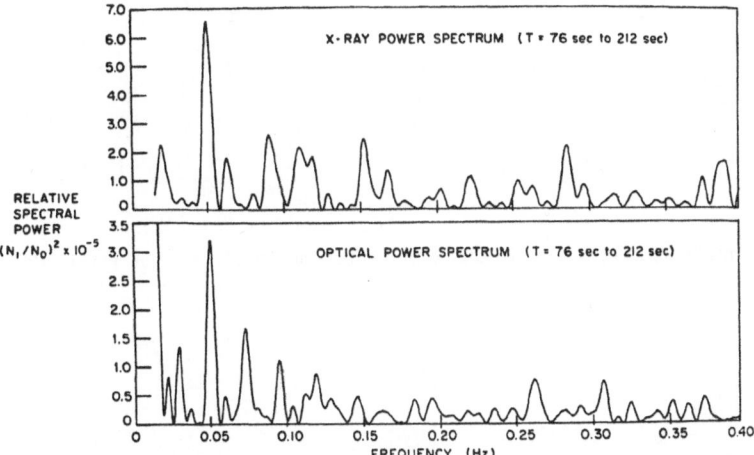

Fig. 11. The power spectrum of Sco X-1 in the X-ray and optical spectral regions for a duration of 136 s (Kestenbaum *et al.*, 1971). Note the large increase in power at both wavelengths at 0.05 Hz (20 s period).

absence of optical coverage during the rise in X-ray intensity. Note that Sco X-1 had just reached the critical magnitude from which it may flare when the optical coverage was discontinued and the X-ray intensity was still low.

Figures 14 and 15 show the nights of October 27 and 28 respectively in more detail. These figures clearly show that the optical and X-ray flares are related, but not one to one. The decrease in X-ray intensity near 10.5 UT on March 27 with no change in optical intensity is very striking while other changes are nicely correlated. Note that on these two nights Sco X-1 was relatively faint and quiet in the radio region.

Figure 16 shows the late October to early November observations of Sco X-1 made by OSO-7 along with radio observations made at the National Radio Astronomy Observatory. These observations give beautiful confirmation that the radio and X-ray intensities are unrelated.

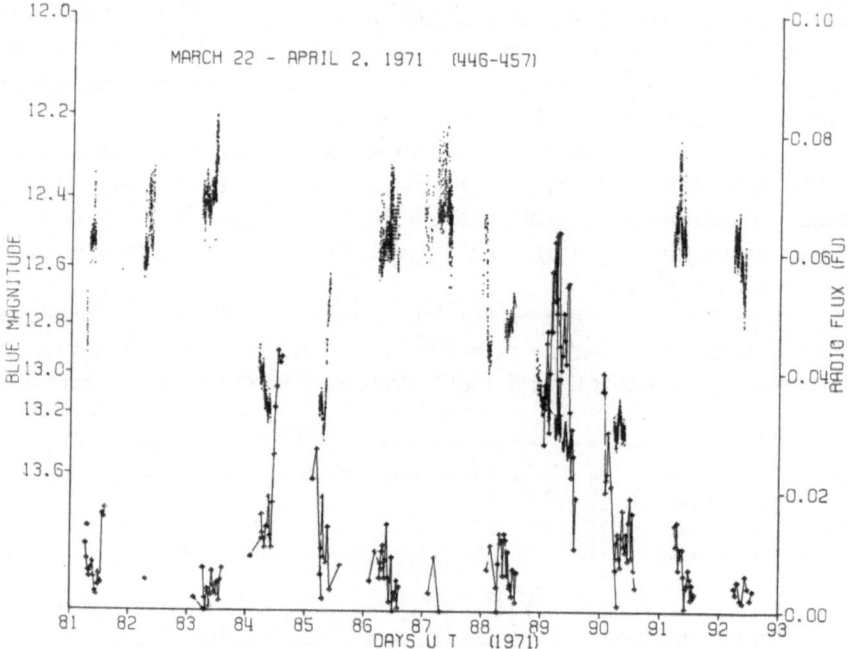

Fig. 12.   Simultaneous optical and radio observations of Sco X-1 for the period 1971 March 22
to April 2.

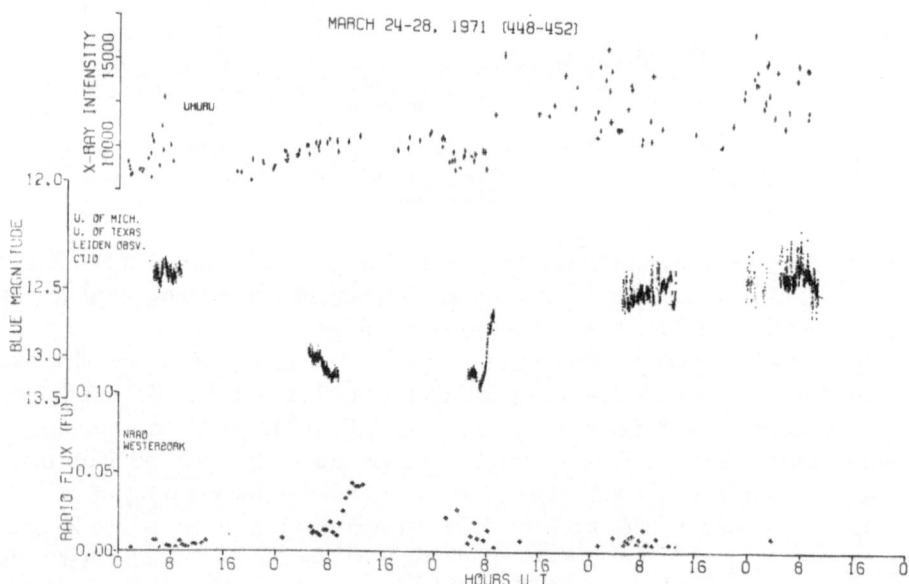

Fig. 13.   Simultaneous X-ray, optical and radio observations of Sco X-1 from 1971 March 24
to March 28.

In summary, the data now available strongly support the following conclusions:

(a) There is no correlation of the radio intensities with either the optical or X-ray intensities. Thus far, the data show that whenever Sco X-1 has flared in the radio region it has been faint in the X-ray and optical spectral regions and vice versa.

(b) The X-ray intensity is high only when the blue optical magnitude is brighter

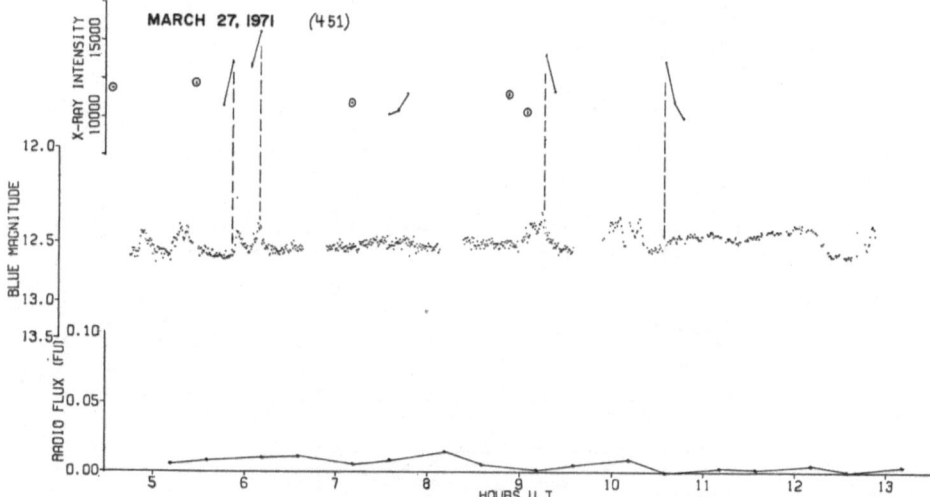

Fig. 14. Simultaneous X-ray, optical and radio observations of Sco X-1 for 1971 March 27. The vertical lines were drawn to aid in visual analysis.

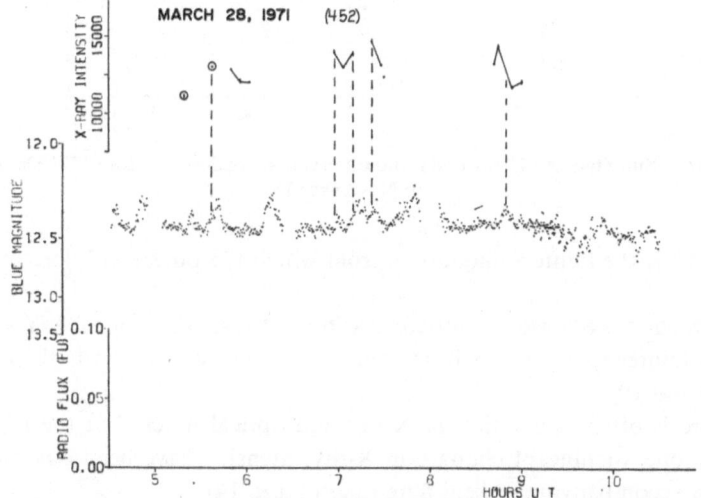

Fig. 15. Simultaneous X-ray, optical and radio observations of Sco X-1 for the night of 1971 March 28. The vertical lines are for visual aids. Preliminary estimates of the UHURU aspect errors yield an uncertainty in X-ray intensity of about 5%. This is comparable to the smallest of the changes in intensity seen in Figures 14 and 15.

Sco X-1

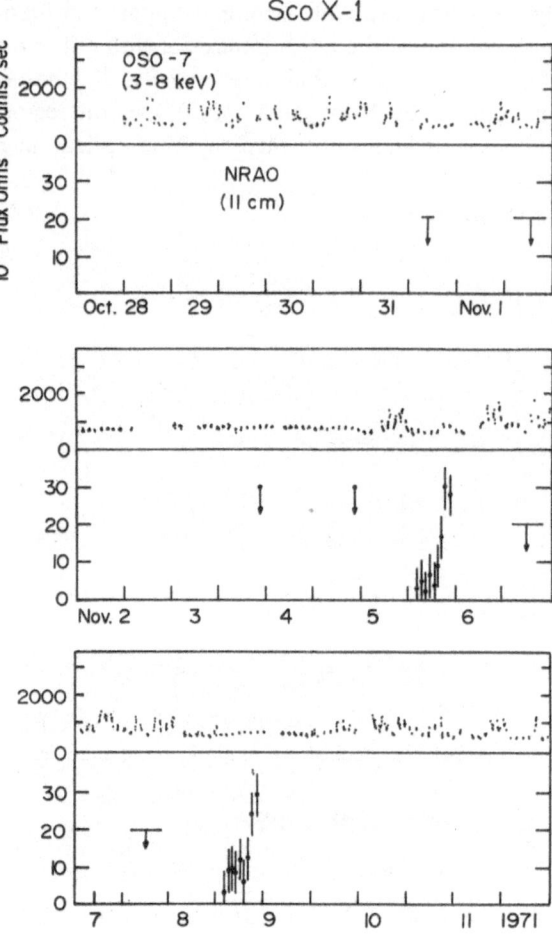

Fig. 16.   Simultaneous X-ray and radio observations of Sco X-1 from 1971 October 28
to November 11.

than 12.6–12.7, the faintest magnitude from which the object will flare in the optical
region.

(c) There may be a weak anti-correlation between the optical magnitude when
Sco X-1 is fainter than 12.7 in B and the X-ray intensity, which is then always low
(Figures 9 and 10).

(d) There is often correlation of X-ray and optical flares, but the relationship is
not one to one. Significant changes in X-ray intensity have been observed to occur
without an accompanying optical flare (e.g. Figure 14).

Further observations are always needed to more firmly establish the correlation of
the X-ray and optical intensities. The absence of simultaneous radio and/or radio and
optical flares needs further confirmation. Simultaneous X-ray and optical observations

with good time resolution are needed in order to more firmly establish the correlation of flares in these two spectral regions. Then, simultaneous observation must be extended to other X-ray sources that can only lead to a better theoretical appreciation of these objects.

References

Bolton, C.: 1972, *Nature* **235**, 271.

Braes, L. L. E. and Miley, G. K.: 1972, *Astron. Astrophys.*, in press.

Burginyon, G. A., Grader, R. J., Hill, R. W., Price, R. E., Rodrigues, R., Seward, F. D., Swift, C. D., Hiltner, W. A., and Mannery, E. J.: 1970, *Astrophys. J.* **161**, 987.

Evans, W. D., Belian, R. D., Conner, J. P., Strong, I. B., Hiltner, W. A., and Kunkel, W. E.: 1970, *Astrophys. J. (Letters)* **162**, L115.

Giacconi, R., Murray, S., Gursky, H., Kellogg, E., Schreier, E., and Tananbaum, H.: 1972, *Astrophys. J.*, in press.

Hiltner, W. A. and Mook, D. E.: 1967, *Astrophys. J.* **150**, 851.

Hiltner, W. A. and Mook, D. E.: 1970, *Astron. Astrophys.* **8**, 1.

Hiltner, W. A. and Osmer, P.: 1972, IAU Circ. No. 2398.

Hiltner, W. A., Werner, J., and Osmer, P.: 1972, *Astrophys. J. Letters*, in press.

Hjellming, R. M. and Wade, C. M.: 1971, *Astrophys. J. Letters* **164**, L1.

Hudson, H. S., Peterson, L. E., and Schwartz, D. A.: 1970, *Astrophys. J. Letters* **159**, L51.

Kestenbaum, H., Angel, J. R. P., Novick, R., and Cocke, W. J.: 1971, *Astrophys. J. Letters* **169**, L49.

Kraft, R. P. and Demoulin, M. H.: 1967, *Astrophys. J. Letters* **150**, L183.

Kraft, R. P. and Miller, J.: 1969, *Astrophys. J. Letters* **155**, L159.

Kristian, J., Sandage, A., and Westphal, J.: 1967, *Astrophys. J. Letters* **150**, L99.

Kukarkin, B. V. and Soluyanov: 1972, IAU Circ. No. 2395.

Lewin, W. H. G., Clark, G. W., and Smith, W. B.: 1968, *Astrophys. J. Letters* **152**, L55.

Lewin, W. H. G., McClintock, J. E., Ryckman, S. G., Glass, I. S., and Smith, W. B.: 1970, *Astrophys. J. Letters* **162**, L109.

Mook, D. E.: 1967, *Astrophys. J. Letters* **150**, L25.

Pelling, R. M.: 1972, thesis.

Robinson, E. L. and Warner, B.: 1972, preprint.

Sandage, A. R., Osmer, P., Giacconi, R., Gorenstein, P., Gursky, H., Waters, J., Bradt, H., Garmire, G., Sreekantan, B., Oda, M., Osawa, K., and Jugaku, J.: 1966, *Astrophys. J.* **146**, 316.

Schreier, E., Levinson, R., Gursky, H., Kellogg, E., Tananbaum, H., and Giacconi, R.: 1972, *Astrophys. J. Letters* **172**, L79.

Wade, C. M. and Hjellming, R. M.: 1972, *Astrophys. J.*, in press.

Webster, B. L. and Murdin, P.: 1972, *Nature* **235**, 37.

Westphal, J., Sandage, A., and Kristian, J.: 1968, *Astrophys. J.* **154**, 139.

7. RADIO OBSERVATIONS OF X-RAY SOURCES

L. L. E. BRAES and G. K. MILEY

Leiden Observatory, The Netherlands

Abstract. Observational work on radio emission from galactic X-ray sources is reviewed. Some problems associated with the detection of these weak radio sources are mentioned and our present knowledge of radio identifications is summarized. The radio properties of the most studied objects are discussed in detail.

1. Introduction

The complementary nature of different branches of astronomy has rarely been more apparent than in the study of galactic X-ray sources. Here we shall review the radio work on these sources, excluding objects identified with supernova remnants. Also we concentrate on the observational aspects, leaving the theoretical implications to Dr Hjellming (this volume, p. 98).

2. Observational Techniques

The first thing to bear in mind is the weakness of the radio emission we are dealing with. For example, the radio flux from Sco X-1, the brightest object in the X-ray sky, is more than four orders of magnitude weaker than that of Cas A, the brightest radio source. In this flux range, present single-dish radio telescopes are not noise limited, but are chiefly troubled by their inability to separate the many confusing sources in their beam. Instruments which use interferometric techniques to synthesize large apertures combine high sensitivity with high resolving power, and it is therefore no coincidence that almost all the observations we shall describe were made with these aperture-synthesis arrays.

A typical aperture-synthesis observation at wavelength λ (see, e.g., Swenson, 1969) consists of measuring the complex fringe visibility outputs of n interferometers with baselines increasing in length by a factor d out to a maximum of $D = nd$. If a radio source is continually tracked over a range of hour angles, the Earth's rotation provides a two-dimensional array of projected baselines. A Fourier transform of the visibility data then yields a map of the radio brightness distribution with an east-west angular resolution $\approx \lambda/D$, over an area of sky determined by the beam of the primary antennas.

We shall be concerned mainly with contributions made by the National Radio Astronomy Observatory's (NRAO) three-element interferometer at Green Bank, U.S.A., operating at frequencies of 2695 and 8085 MHz, and the synthesis radio telescope at Westerbork (WSRT), the Netherlands, operating at 1415 MHz. Photographs of these arrays are shown in Figures 1 and 2. For a description of the characteristics of the instruments we refer to Hogg *et al.* (1969), Baars and Hooghoudt (1972), and Casse and Muller (1972); the relevant instrumental parameters are listed in Table I.

Because of its longer maximum baseline and higher observing frequencies, the

Bradt and Giacconi (eds.), X- and Gamma-Ray Astronomy, 86–97 All Rights Reserved.
Copyright © 1973 by the IAU.

Fig. 1. The NRAO three-element interferometer.

TABLE I

Instrumental parameters of the NRAO and Westerbork
synthesis telescopes

	NRAO interferometer		WSRT
Number of telescope elements	3		12
Diameter of elements (m)	25		25
Number of simultaneous baselines	3		20
Maximum length (m)	2700		1600
Orientation (azimuth in degrees)	242		90
Operating frequencies (MHz)	8085	2695	1415
wavelengths (cm)	3.7	11.1	21.2
Bandwidth (MHz)	2×30	2×30	1×4
Resolution in right ascension (s″)	3	8	22
Sensitivity[a] (10^{-29} W m^{-2} Hz^{-1})	4	4	4
Half-power beamwidth (degrees)	0.1	0.3	0.6

[a] 3 rms noise after 12 hr observation.

Fig. 2.   The Westerbork synthesis radio telescope (Copyright 'Aerophoto Eelde').

resolving power of the NRAO interferometer is superior to that of the WSRT. The broader bandwidth and lower-noise front ends of the NRAO instrument compensate in sensitivity for the fewer telescope elements.

A problem inherent in using the aperture-synthesis technique is that of confusion by diffraction grating rings. The response of a synthesis array is usually described in terms of a synthesized beam, i.e. the calculated response of the instrument to a point source observed with the same baseline coverage. In general, this beam will have diffraction grating rings due to the discrete nature of the baseline coverage, and these rings will surround the source at intervals whose separation on the synthesis map is inversely proportional to d. For many complex regions the limiting factor is not the sensitivity or the resolution, but the magnitude of confusion from grating rings. Here the WSRT has two immediate advantages. First, the large number of simultaneous baselines results in much fewer rings. Secondly, because of the east-west orientation of the telescope elements the rings are well defined ellipses, rather than complicated spiral patterns as produced by the skew baseline of the NRAO interferometer.

The highly variable nature of the radio emission from X-ray sources also poses somewhat of a problem because a typical measurement is a synthesis of several hours observation. A variation in the flux of a source during this period causes a smearing of the response on the synthesis map. The variability may be analysed in small time intervals, but a limitation on the time resolution is, of course, set by the signal to noise ratio.

3. Radio Identification of X-Ray Sources

Location of a radio source within current X-ray error boxes is not sufficient for a secure identification, for the probability of a chance coincidence is in general not negligible. It is only when a source also exhibits unusually large and rapid variations that an identification is regarded as definite. This additional criterion is imposed because of the unique variability behaviour of the radio object coincident with the Sco X-1 X-ray star.

So far, the X-ray sky has been searched in only a preliminary way at radio wavelengths. The availability of aperture-synthesis instruments has restricted searches to declinations north of $\sim -40°$, and only the brightest sources with positions known to better than a few minutes of arc have been observed.

Table II summarizes the present status of the results. It contains all the published NRAO and Westerbork identification data together with some unpublished results of Westerbork measurements. Note that the quoted flux densities do not refer to simultaneous observations.

Of the 14 sources in the table, five now have well established radio counterparts. Two others have radio sources within the X-ray error boxes, but these have not yet been found to be variable. Upper limits are available for the radio emission of the remaining seven objects.

We should also mention the result of a recent radio search for Her X-1 (2U 1705+ +34) conducted by the MIT group; with the NRAO interferometer four sources

TABLE II
NRAO and Westerbork observations of X-ray sources

Source		Radio position		Flux density (10^{-29} W m^{-2} Hz^{-1})			References[a]
		$\alpha(1950)$	$\delta(1950)$	1415 MHz	2695 MHz	8085 MHz	
	2U 0114+63	–	–	<10	–	–	1
	2U 0352+30	–	–	< 5	–	–	2
Sco X-1	2U 1617−15	$16^h17^m01^s7 \pm 0^s6$	−15°32'17" ± 8"	20	7	3	1
		16 17 04.47 ±0.06	−15 31 15.8 ± 4.5	< 5 − 63	< 5 − 260	<5 − 70	1, 3, 4, 5, 6
		16 17 06.8 ±0.4	−15 30 14 ± 5	23	12	3	
GX 340+0	2U 1641−45	–	–	–	< 5	< 5	7
GX 349+2	2U 1702−36	–	–	–	< 5	< 5	7
GX 3+1	2U 1744−26	–	–	< 5	< 5	–	1, 8
GX 5−1	2U 1757−25	17 58 03.5 ±0.3	−25 04 53 ±12	10	< 5	–	8, 9
GX 9+1	2U 1758−20	17 58 34.0 ±0.3	−20 30 58 ± 5	< 5	< 5 − 8	< 5	1, 7, 8
GX 13+1	2U 1811−17	–	–	<15	–	–	1
GX 17+2	2U 1813−14	18 13 10.8 ±0.2	−14 03 13 ± 3	< 5 − 11	< 5 − 22	–	1, 6, 8
	2U 1954+31	19 55 00.2 ±0.2	+31 47 23 ± 3	33	–	–	1
Cyg X-1	2U 1956+35	19 56 28.87 ±0.02	+35 03 55.0 ± 0.3	< 5 − 21	< 5 − 15	<5 − 17	6, 8, 10, 11, 12, 13
Cyg X-3	2U 2030+40	20 30 37.63 ±0.03	+40 47 12.5 ± 0.5	23 − 230	120 − 321	123 − 515	14, 15
Cyg X-2	2U 2142+38	–	–	< 2	< 5	–	1, 8, 16

[a]
1. Braes and Miley, unpublished
2. Braes and Miley, 1972a
3. Hjellming and Wade, 1971a
4. Braes and Miley, 1971b
5. Wade and Hjellming, 1971
6. Braes and Miley, 1971c
7. Zaumen et al., 1972
8. Hjellming and Wade, 1971b
9. Braes et al., 1972
10. Braes and Miley, 1971a
11. Hjellming et al., 1971
12. Wade and Hjellming, 1972
13. Braes and Miley, 1972b
14. Braes and Miley, 1972c
15. Hjellming et al., 1972
16. Braes and Brouw, 1971

were detected which they consider as candidates for the radio counterpart of this pulsing X-ray source (Doxsey *et al.*, 1972).

4. Remarks on Individual Objects

In this section we shall successively discuss in some detail the results for Sco X-1, the Cygnus sources, and the sources in the galactic centre region.

A. SCORPIUS X-1

Sco X-1 was the first X-ray star to be detected at radio wavelengths (Andrew and Purton, 1968). Variation in its radio flux was first reported by Ables (1969).

Perhaps the most intriguing feature is its triple radio structure, first revealed by NRAO observations (Hjellming and Wade, 1971a) and confirmed later by Westerbork measurements (Braes and Miley, 1971b). Sco X-1 appears remarkably like a miniature quasar, with a central highly variable component coinciding in position to within 1″ with the X-ray star, and two companion sources located symmetrically on either side of it, 1.′18 to the north-east and 1.′24 to the south-west.

The outer components have non-thermal spectra, with spectral indices ∼ −1 above 1415 MHz. From a comparison with the 408 MHz result obtained by Jauncey (1971),

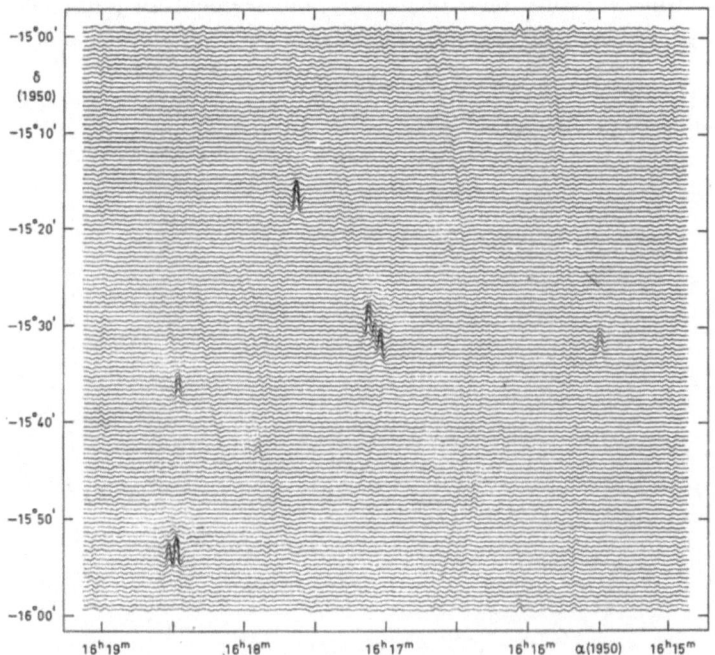

Fig. 3. Westerbork 1415 MHz map of the Sco X-1 field, presented as a series of intensity profiles in right ascension. The map combines 99 hr of observations to give an rms noise of about 0.5×10^{-29} W m^{-2} Hz^{-1}.

at least one appears to become optically thick below this frequency. Because these
properties are typical of normal extragalactic sources, the question arises whether the
companions are physically related to the X-ray star, or whether they are just background objects. As Sco X-1 is situated in an uncrowded region well away from the
galactic plane, the probability of a chance coincidence is small. This can be seen from
Figure 3, showing the 1415 MHz map of the surrounding $1° \times 1°$ region. Additional
evidence for a relationship comes from recent, unpublished, NRAO and Westerbork
results indicating that the north-east companion is variable.

If the two sources are relativistic plasmoids ejected from the star, it is conceivable
that their proper motions should be measurable. Assuming a distance of 300 pc, a
transverse velocity of 0.01 c would result in an apparent expansion of $2''$ per year.
Such a measurement is difficult, however, because of distortion by sidelobes from the
variable central component and by the elongated beam due to the low declination.
Present results are inconclusive.

The central component has been studied extensively (Wade and Hjellming, 1971;
Braes and Miley, 1971c). Its spectrum is non-thermal with an index typically ~ -0.5
between 2695 and 8085 MHz. The few simultaneous NRAO and Westerbork measurements suggest that it becomes optically thick between 1415 and 2695 MHz.

Figure 4 illustrates how the activity of the central source can vary from day to day;
in 24 hr its mean flux density increased from 7 to 44×10^{-29} W m^{-2} Hz^{-1}. It also
varies erratically on a time scale of minutes, but searches for pulsar-like behaviour
have so far proved negative (see, e.g., Taylor *et al.*, 1972). Lampton *et al.* (1971) failed

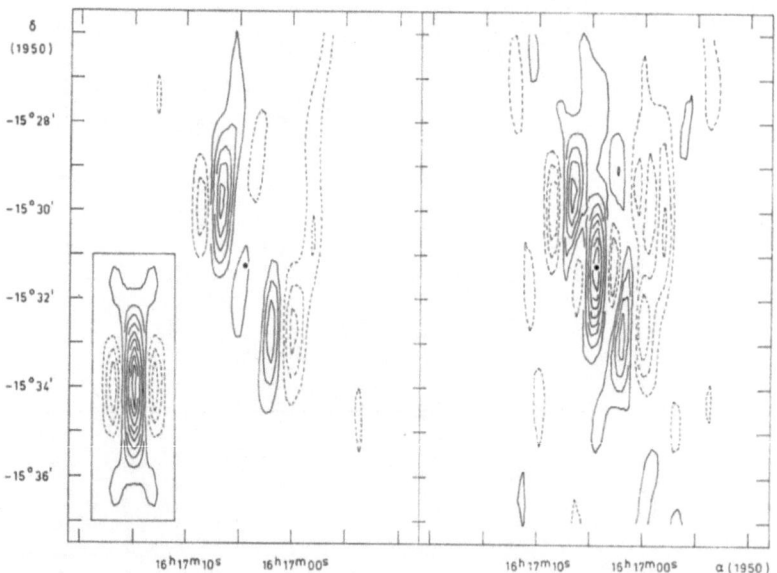

Fig. 4.   1415 MHz maps of Sco X-1 on March 29 (left) and March 30, 1971 (right). The contours
give levels of equal brightness, the dashed ones representing negative intensities. For comparison,
the antenna pattern is shown inset. The black dot marks the position of the X-ray star.

to find any correlation between the optical and radio variations, and Dr Hiltner (this volume, p. 74) just pointed out that though considerably more data now exist, evidence for a direct correlation is still lacking.

B. CYGNUS X-1

Like that of Sco X-1, the radio counterpart of Cyg X-1 is strongly variable, but in a very different way. It appeared suddenly within a three-weeks period, just over a year ago (Braes and Miley, 1971a; Hjellming and Wade, 1971b). Previously, the Cyg X-1 field had been observed three times at NRAO and once at Westerbork without success. Figure 5 shows that apart from a slight decrease in flux after its initial

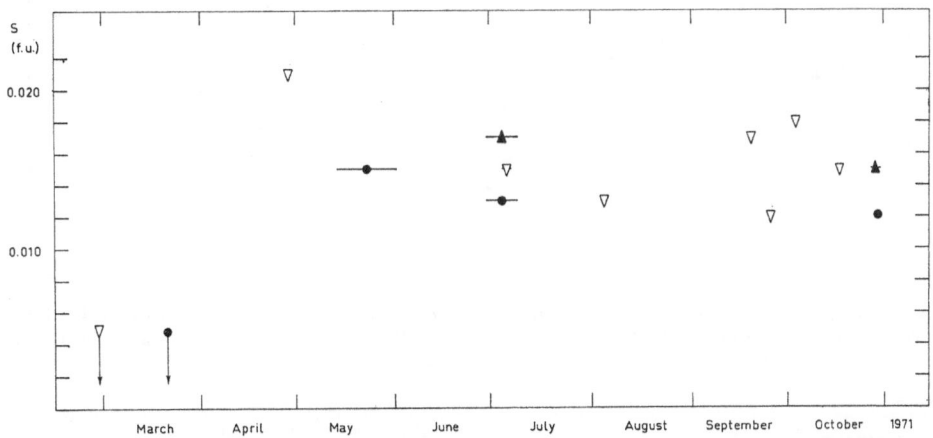

Fig. 5. The flux density of Cyg X-1 as a function of time. Included are Westerbork measurements at 1415 MHz (▽) and NRAO data at 2695 (●) and 8085 MHz (▲). Where measurements have been combined this is indicated by horizontal bars. The rms error in an individual point is about 0.004 fu $= 4 \times 10^{-29}$ W m^{-2} Hz^{-1}.

appearance, there is little evidence for subsequent radio variability. Also no pulsar-like behaviour has been found.

A compilation of data at four frequencies from Westerbork, NRAO, and Algonquin Radio Observatory, indicates that the Cyg X-1 radio source has a peculiar spectrum characteristic of the active extragalactic sources, with a minimum between 1415 and 8085 MHz (Hjellming et al., 1971).

Figure 6 shows how the optical identification of Cyg X-1 was made possible by the interplay of X-ray and radio techniques. The detection of a radio counterpart narrowed down the positional uncertainty and led to the identification with the B0 Ib star HDE 226868. Subsequent refinements of the early radio positions shown in Figure 6 left no doubt regarding the positional coincidence with this spectroscopic binary (Wade and Hjellming, 1972; Braes and Miley, 1972b).

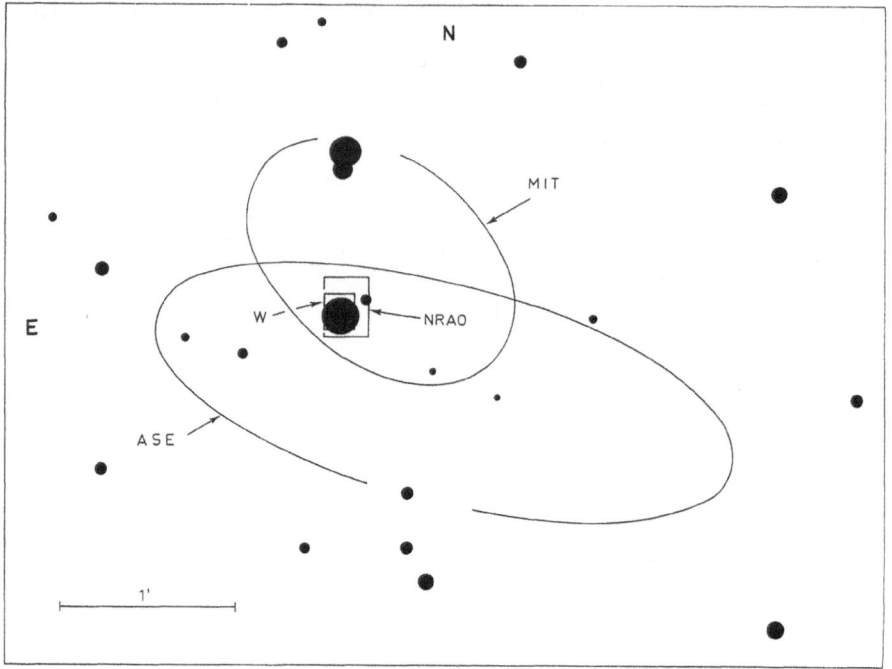

Fig. 6. Identification of Cyg X-1 with HDE 226868. Superimposed on the star field are the UHURU (private communication) and MIT (Rappaport *et al.*, 1971) X-ray error ellipses and the (3 rms) radio error boxes.

C. CYGNUS X-2

No radio emission has yet been detected from the star generally assumed to be the optical counterpart of Cyg X-2. There is a point source nearby, but earlier (Westerbork) indications of variability have not been confirmed (Braes and Brouw, 1971; Hjellming and Wade, 1971b).

D. CYGNUS X-3

Recently, Westerbork observations have led to the detection of a highly variable radio source near the X-ray position of Cyg X-3 (Braes and Miley, 1972c). Over a period of eight months its 1415 MHz flux density was observed to vary by an order of magnitude, and also on a time scale of hours it appears to vary appreciably. As Dr Hjellming will relate (this volume, p. 98), this extreme variability has been confirmed by subsequent NRAO measurements.

The Cyg X-3 radio source is remarkably strong; the ratio of its radio to X-ray emission is about 50 times greater than for Cyg X-1 and exceeds that for Sco X-1 by more than two orders of magnitude.

Cyg X-3 is located in the crowded Cyg X radio complex, and Figure 7 shows the many sources in the field. This map is a complete synthesis resulting from a combination

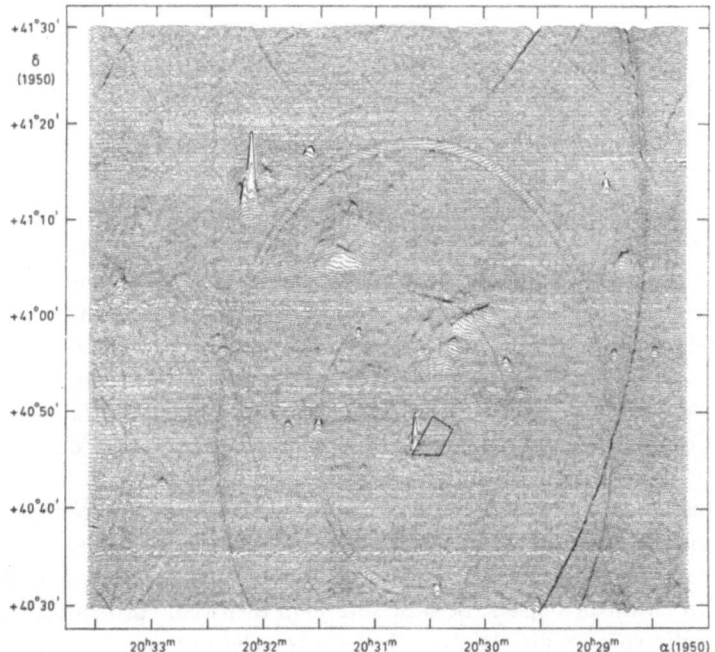

Fig. 7. Westerbork complete-synthesis map of the 1415 MHz brightness distribution in a $1° \times 1°$ region surrounding Cyg X-3. The quadrilateral marks the 90% confidence limits of the 2ASE X-ray position for Cyg X-3. Noise in this figure amounts to about 0.7×10^{-29} W m^{-2} Hz^{-1}.

of data from four separate days. The variable source stands out due to the surrounding grating rings, caused by the changes in flux level during the period covered by the observations.

An attempt at optical identification has met with no success, but this is not surprising since the whole region is obscured by the Great Cygnus Rift.

E. GALACTIC CENTRE SOURCES

If these objects are indeed at the distance of the galactic centre, one would expect their apparent radio strengths to be relatively low and difficult to measure. Radio identification problems are aggravated by the low declination and increased confusion.

At present, only GX 9+1 and GX 17+2 have been positively detected (Hjellming and Wade, 1971b; Braes and Miley, 1971c; Zaumen et al., 1972); both sources exhibit short-term variability of the Sco X-1 type. There is also a radio source located within the GX 5−1 X-ray error box (Braes et al., 1972), but definite identification must await the detection of variability. As can be seen from Table II, NRAO and Westerbork searches at the positions of GX 3+1, 13+1, 340+0, and 349+2 have been negative. Cruder upper limits at 4.6 cm were previously set by Andrew et al. (1970).

No optical counterparts have yet been found for any of the galactic centre sources.

5. Concluding Remarks

From this review it is clear that the radio observations are as yet in their infancy; we only have limited measurements of some of the strongest X-ray sources. Even so, we can already distinguish two classes, namely those that exhibit short-term variability (e.g. Sco X-1) and those that do not (Cyg X-1). The failure to find any direct correlations between the radio behaviour and the optical and X-ray characteristics implies that the radio emission originates from a different region of the objects.

It is important that simultaneous observations over a wide range of frequencies are carried out to illucidate the spectral properties of the sources, and because of their unpredictable behaviour and large dispersion in luminosity repeated monitoring is needed. It is interesting to note that observations of Cyg X-1 over an eight-month period would not have revealed it to be variable, and therefore failure to measure variability does not rule out a possible radio identification. Neither does the complete absence of radio emission mean that an X-ray source will never be detectable.

We can anticipate that within the next few years as more and better X-ray positions become available, the number of secure radio identifications will increase manyfold. Also, intercontinental interferometry may be used to explore the structural variations of some of the stronger objects. Because several of their radio properties are so typical of extragalactic sources and because they are so relatively nearby, it is tempting to speculate that radio observations of galactic X-ray sources may contribute to our knowledge of violent events taking place far outside our galaxy.

Note added in proof: An unprecedented radio outburst was observed in Cygnus X-3 during early September 1972 (IAU Circ. No. 2440). Its spectral behaviour resembled that of an expanding synchrotron source with a maximum flux density of 22×10^{-26} W m^{-2} Hz^{-1} at 10630 MHz. This giant flare will be discussed in *Nature* during October 1972.

References

Ables, J. G.: 1969, *Astrophys. J.* **155**, L27.
Andrew, B. H. and Purton, C. R.: 1968, *Nature* **218**, 855.
Andrew, B. H., Purton, C. R., Rappaport, S., Bradt, H., and Schnopper, H. W.: 1970, *Astrophys. J.* **161**, L173.
Baars, J. W. M. and Hooghoudt, B. G.: 1972, *Astron. Astrophys.*, in preparation.
Braes, L. L. E. and Brouw, W. N.: 1971, *Astron. Astrophys.* **12**, 320.
Braes, L. L. E. and Miley, G. K.: 1971a, *Nature* **232**, 246.
Braes, L. L. E. and Miley, G. K.: 1971b, *Astron. Astrophys.* **14**, 160.
Braes, L. L. E. and Miley, G. K.: 1971c, *Veröff. Remeis-Sternw. Bamberg* **IX**, No. 100, 173 (IAU Coll., No. 15, 'New Directions and New Frontiers in Variable Star Research').
Braes, L. L. E. and Miley, G. K.: 1972a, *Nature* **235**, 273.
Braes, L. L. E. and Miley, G. K.: 1972b, *Nature Phys. Sci.* **235**, 147.
Braes, L. L. E. and Miley, G. K.: 1972c, *Nature* **237**, 506.
Braes, L. L. E., Miley, G. K., and Schoenmaker, A. A.: 1972, *Nature* **236**, 392.
Casse, J. L. and Muller, C. A.: 1972, *Astron. Astrophys.*, in preparation.
Doxsey, R., Murthy, G. T., Rappaport, S., Zaumen, W., and Spencer, J.: 1972, *Astrophys. J.* **176**, L15.

Hjellming, R. M. and Wade, C. M.: 1971a, *Astrophys. J.* **164**, L1.
Hjellming, R. M. and Wade, C. M.: 1971b, *Astrophys. J.* **168**, L21.
Hjellming, R. M., Wade, C. M., Hughes, V. A., and Woodsworth, A.: 1971, *Nature* **234**, 138.
Hjellming, R. M., Hermann, M., and Webster, E.: 1972, *Nature* **237**, 507.
Hogg, D. E., Macdonald, G. H., Conway, R. G., and Wade, C. M.: 1969, *Astron. J.* **74**, 1206.
Jauncey, D. L.: 1971, *Nature Phys. Sci.* **230**, 200.
Lampton, M., Bowyer, S., Welch, J., and Grasdalen, G.: 1971, *Astrophys. J.* **164**, L61.
Rappaport, S., Zaumen, W., and Doxsey, R.: 1971, *Astrophys. J.* **168**, L17.
Swenson, G. W.: 1969, *Ann. Rev. Astron. Astrophys.* **7**, 353.
Taylor, J. H., Huguenin, G. R., and Hirsch, R. M.: 1972, *Astrophys. J.* **172**, L17.
Wade, C. M. and Hjellming, R. M.: 1971, *Astrophys. J.* **170**, 523.
Wade, C. M. and Hjellming, R. M.: 1972, *Nature* **235**, 271.
Zaumen, W., Murthy, G. T., Rappaport, S., Hjellming, R. M., and Wade, C. M.: 1972, *Nature* **235**, 378.

8. RADIO COUNTERPARTS OF X-RAY SOURCES AND X-RAY COUNTERPARTS OF RADIO STARS

R. M. HJELLMING

National Radio Astronomy Observatory*, Green Bank, West Virginia, U.S.A.

Abstract. Some of the implications of the data on radio counterparts of X-ray sources are discussed. Data on the radio counterpart of Cyg X-3 is compared with the data on the radio star β Persei (Algol), to show that similar radio emission mechanisms are involved. The data for such radio stars not yet known as X-ray sources imply that such objects should be transient X-ray sources.

The principal data concerning the radio counterparts of X-ray sources have just been reviewed by Braes and Miley (this volume, p. 86). This paper will discuss some of the implications of the radio data and a class of radio stars which should be transient X-ray sources, although they have not yet been detected as such. Up until the recent detection of Cyg X-3 (Braes and Miley, 1972b; Hjellming *et al.*, 1972b), the latter conclusion was largely theoretical; however, the newly observed radio properties of Cyg X-3 make it a virtual certainty that the radio binary stars (like β Persei and β Lyrae) and at least some of the X-ray stars involve the same phenomena at radio wavelengths.

Let us begin by referring to Table I, which summarizes the properties of the known radio counterparts of compact galactic X-ray sources. The radio variability and flux

TABLE I
X-ray stars with known radio counterparts

| Name | Radio characteristics | | Flux density (f.u.) | | | Optical identification |
	Spectrum	Factor of variability	Min.	Max.	Freq.	
Sco X-1	Mostly non-thermal some thermal	At least 60	<0.004	0.25	2695	Peculiar blue star variable in emission lines and optical continuum
GX17+2	Non-thermal	At least 2	<0.005	0.02	2695	?
GX9+1	Non-thermal	At least 1.5	<0.005	0.008	2695	?
Cyg X-1	Non-thermal and thermal	Off then on, by at least factor of 3, steady since	<0.005	0.015 0.021	8085 1415	BO Ib spectroscopic binary peculiar emission lines
Cyg X-3	Mostly thermal some non-thermal	At least 10	0.02 0.1 0.1	0.23 0.32 0.52	1415 2695 8085	

* The National Radio Astronomy Observatory is operated by Associated Universities, Inc. under contract with the U.S. National Science Foundation.

density levels are well known, except for the newest and strongest entry, Cyg X-3. The first three sources are similar in that they are dominated by variable non-thermal radio emission; we can say that this is characteristic of radio counterparts of the Sco X-1 type.

1. Sco X-1

Only Sco X-1, of the sources dominated by non-thermal radio emission, is strong enough to give much information. Back in the days when there was one measurement (Andrew and Purton, 1968) of a radio flux at one frequency at one time and another measurement (Ables, 1969) at another frequency at another time, there were a number of theoretical papers interpreting the radio spectrum of Sco X-1. Since it has been proven to be a triple radio source (Hjellming and Wade, 1971a; Wade and Hjellming, 1971b; Braes and Miley, 1971a), with a central component sitting on the star which is presumably the X-ray source, that is variable by at least a factor of 60, there have been, to my knowledge, no theoretical papers interpreting Sco X-1. To the radio astronomer the potentially most interesting aspect of Sco X-1 is the double source surrounding the X-ray star and the central component, as shown in Figure 1. This is,

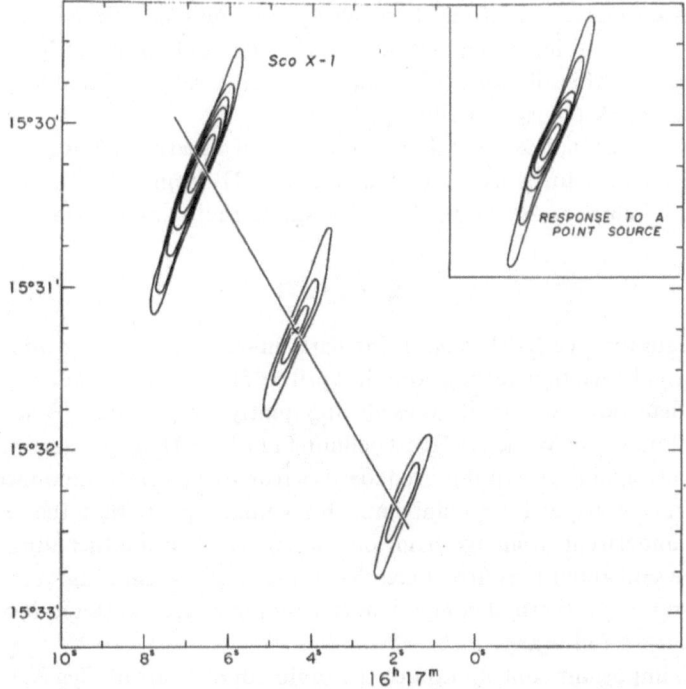

Fig. 1. A map of the Sco X-1 field derived from 2695 MHz observations with the NRAO interferometer during February, March, and April 1971. The position of the X-ray star is shown by a ×. The diagonal line through the component sources is at position angle 29°. The coordinate grid is for epoch 1950.0. The synthesized beam image of a point source is shown in the upper right hand corner.

of course, because of the common but yet largely unexplained presence of double radio sources symmetrically placed about interesting galaxies. Evidence is accumulating, but is not yet completely conclusive, that the NE companion has a variable radio flux density; if and when the evidence becomes conclusive it will be a virtual certainty that the radio companions are associated with Sco X-1. Because of the non-thermal spectra (Hjellming and Wade, 1971a; Wade and Hjellming, 1971b; Braes and Miley, 1971a) of the companions and the association with the X-ray star, it is hard to avoid the conclusion that the most likely explanation will identify the companions with complexes of coupled plasma and magnetic fields which are blasted with cosmic ray electrons from the X-ray star. One cannot say whether the two companions represent the only regions with the magnetic fields needed to produce a radio source, or whether a more extensive shell is being bombarded by two highly beamed streams of cosmic ray electrons. The resolution of these questions may have enormous significance for the physics of double radio sources.

Very little can be said about the nature of the variable central component. Because of its dominantly non-thermal nature, one tends to think first of a variably supplied synchotron spectrum. Unfortunately, the spectrum flattening proceeds while the flare is still clearly on the rise. One must conclude that the relative supply of higher and lower energy cosmic ray electrons alters during the course of a flare in the sense of the relative proportion at higher energies increasing. The only other suggestion of reproducible behavior while flaring is the suggestion of a repeated step-like behavior at both frequencies (Wade and Hjellming, 1971b).

There is also clear indication of the rare presence of a thermal component in Sco X-1. Two successive days in February 1970 (Wade and Hjellming, 1971b), just before the great flare of February 26, 1970, showed a stable radio source with a perfectly flat spectrum.

2. Cyg X-1

The radio data for Cyg X-1 has been singularly uninteresting in its own right – except for the unusual fact that during March-April 1971, the radio source jumped from below the detection limit to its present apparently steady state (Braes and Miley, 1971b; Hjellming and Wade, 1971b; Hjellming et al., 1971; Braes and Miley, 1972a). The enormous significance of this event derives from the recently announced UHURU result (see review paper by Tananbaum, this volume, p. 9) that the time-averaged X-ray flux underwent a change from one steady level to another during this time. Little can be said about the current Cyg X-1 radio source except that it may have both a thermal and a non-thermal component (Hjellming et al., 1971) and it has remained steady now for a full year.

The most important contribution of the radio counterpart of Cyg X-1 is its role in the identification of Cyg X-1 with a spectroscopic binary with a 9th magnitude BOIb component.

Since short time scale variability of a radio source near an X-ray source has been adopted as a necessary criterion for telling a radio counterpart of an X-ray source from the very common and relatively steady background sources, clear identification of a Cyg X-1 type radio counterpart will always be difficult. We have no idea how common discontinuities in radio and X-ray properties might be. Cases where a steady radio source sits within an X-ray error box will allow identification of a radio counterpart of the Cyg X-1 type only if it is identifiable with an obviously peculiar star. However, one must wrestle with the dilemma of whether there are radio stars which behave the same way but are not X-ray sources, as we will now discuss.

3. Radio Binary Stars

There are three radio binary stars that are not yet known to be X-ray sources: β Persei (Algol), β Lyrae (Wade and Hjellming, 1972; Hjellming et al., 1972a), and Antares B (Wade and Hjellming, 1971a; Hjellming and Wade, 1971c). They are all variable on time scales of hours, as are the Sco X-1 type radio counterparts, but their dominant characteristic is that they appear like variable thermal bremsstrahlung sources when flaring. The recently detected radio counterpart to Cyg X-3 (Braes and Miley, 1972b; Hjellming et al., 1972b) shows radio characteristics virtually identical to the radio binary stars, particularly β Persei (Algol).

4. β Persei (Algol)

Because it is the strongest of the radio binary stars, β Persei (Algol) can be taken as the prototype of the class. It is noteworthy that the discovery of the Cyg X-3 radio counterpart, which is very similar to β Persei, occurred a week after a paper was written (Hjellming, 1972) arguing that the simplest interpretation of the β Persei radio data required it to be a transient X-ray source. Let us therefore summarize the arguments (Hjellming, 1972) that lead to this result.

With the exception of a couple of hours on January 22–23, 1972, the radio spectrum of β Persei has always appeared thermal. Initially it was believed (Wade and Hjellming, 1972; Hjellming et al., 1972a) that a thermal interpretation was unlikely; but now there are compelling reasons for just the opposite view. The radio flux density for a high temperature thermal source can be written (Hjellming, 1972) as

$$S_\nu \cong \frac{2kT_e}{\lambda^2} \, \Omega_s [1 - \exp{(-0.28 \, ET_e^{-1.45} \nu_{\text{GHz}}^{-2.05})}] \tag{1}$$

where T_e = electron temperature, Ω_s = solid angle subtended by the emitting source, E = emission measure, k = the Boltzmann constant, ν = the frequency, and λ = wavelength. Measurement of S_ν at two frequencies, when the argument in the exponential (optical depth) is large enough, allows empirical determination of $T_e\Omega_s$ and $ET_e^{-1.45}$. Knowing the distance of the source, we can then empirically determine T_eD^2 and $DN_e^2T_e^{-1.45}$, where D is the source diameter. When the source is a point source at 8085 MHz for the NRAO interferometer, this means the angular diameter is less than $0''.5$, which puts an upper limit on D. This, with knowledge of T_eD^2 and $DN_e^2T_e^{-1.45}$, puts lower limits on T_e and N_e. In addition, the fact that radio waves cannot pass through too dense a plasma puts an upper limit on N_e, hence an upper limit on T_e and lower limits on D and the angular diameter θ. For β Persei, with a distance of 25 pc, this means:

$$\theta < 0\overset{''}{.}5 \qquad\qquad N_e < 9 \times 10^{10} \text{ cm}^{-3}$$

$$T_e > 2.5 \times 10^5 \text{ K} \qquad T_e < 2 \times 10^9 \text{ K}$$

$$D < 13 \text{ AU} \qquad\qquad D > 0.1 \text{ AU}$$

$$N_e > 6 \times 10^6 \text{ cm}^{-3} \qquad \theta > 0\overset{''}{.}004$$

The lower limit on T_e is immediately interesting because at temperatures only slightly above that limit, β Persei would be a soft X-ray source. There are two arguments indicating that when flaring most strongly, β Persei is closer to the upper limit on T_e. First, the only way the optical emission lines could be as weak as observed would be if the gas is much hotter than a normal 10^4–10^5 K plasma. Secondly, we can estimate the order of magnitude of all the parameters of the thermal source when at its strongest. As is well known for observations of the solar chromosphere and corona, radio observations of a plasma in the gravitational potential well of a star tend to see radiation from different density levels at different frequencies. It is a well known rule of thumb that such radio emission, at a frequency v, occurs dominantly from regions just out side the level where v equals the plasma frequency. We can therefore argue that, because the β Persei source is optically thick at 2695 MHz when flaring, the radio emission probably comes from regions where $N_e \sim 10^{10}$ cm^{-3} is a good order of magnitude estimate. This then implies that $T_e \sim 10^8$ K, $D \sim 0.3$ AU, and $\theta \sim 0\overset{''}{.}01$. It is unavoidable that β Persei will be a strong X-ray source under these circumstances. The next most interesting of these numbers is the estimated diameter, which is exactly a factor of ten greater than the known sizes of the two stars in the close pair in the β Persei system. It would then make the observed variable radio source of the size scale of the Roche lobe of one of the stars.

Why has β Persei, at least, not been detected[*] as an X-ray source as of the time of this symposium? The most likely answer is in its transience. There is no doubt that the current flaring activity in β Persei is considerably greater than occurred in October–November 1971 (Wade and Hjellming, 1972; Hjellming et al., 1972a), further, the work of Bolton (1972) showing the association of this flaring with optical features never before reported suggests that such activity is rare. All of this suggests that, if they exist, the Algol-type X-ray sources will be typified by transience. Indeed, transient X-ray sources (Cen X-2, Cen X-4, Cet X-2) are known; but one cannot tell yet whether they are related to the predicted Algol-type sources.

---

[*] Several scans across the field containing β Persei with the UHURU satellite during 1971 placed an upper limit of 10^{-4} the strength of Sco X-1 on the X-ray flux (Tananbaum, private communication).

5. Cyg X-3

The observations (Hjellming *et al.*, 1972b) of Cyg X-3 obtained simultaneously at 2695 and 8085 MHz with the NRAO interferometer on May 1, 2, and 3, 1972, are shown in Figure 2. Also shown is the spectral index variation for these days. In

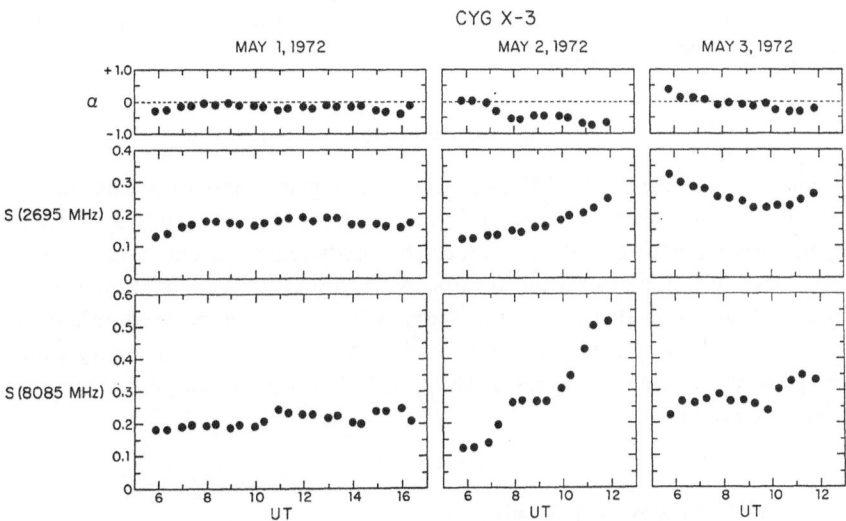

Fig. 2. The spectral index and flux densities at 2695 and 8085 MHz are plotted as functions of UT for Cyg X-3 on May 1, 2, and 3, 1972. The error limits on the flux densities are the size of the filled circles. The flux densities are measured in flux units.

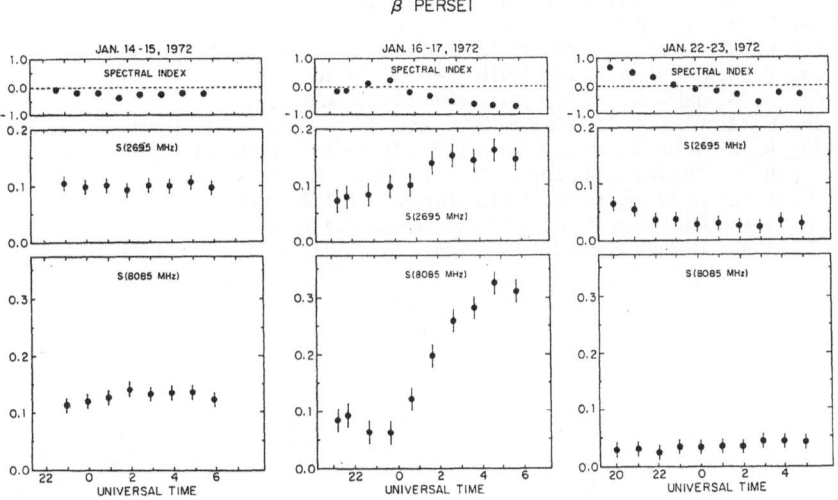

Fig. 3. The spectral index and flux densities at 2695 and 8085 MHz are plotted as functions of UT for β Persei (Algol) on Jan. 14/15, 16/17, and 22/23, 1972.

Figure 3, we show the same data for β Persei on three days which appear, in every important detail, the same as Cyg X-3 events, particularly in the spectral index variation and time scale. The Cyg X-3 flare in May 2, 1972, shows the characteristic evolution of a β Persei flare from an optically thin thermal spectral index to an optically thick thermal spectral index. Furthermore, the non-thermal behavior of Cyg X-3 on May 3, 1972, has an exact counterpart in the only non-thermal event observed for β Persei. It is difficult to avoid the conclusion that the radio emission mechanisms for β Persei and Cyg X-3 are similar.

6. Summary

The radio data indicate that β Persei (Algol) is a proto-type of a class of variable thermal-like radio sources of which β Lyrae, Antares B, and Cyg X-3 are members. It will be important to establish whether the predicted transient X-ray emission of β Persei exists when radio flaring occurs. All radio counterparts show evidence of both thermal and non-thermal components; with the latter being dominant in the Sco X-1 type and the former dominant in Cyg X-3 and the radio binary stars. The on-off and stability characteristics of the Cyg X-1 radio counterpart at the moment are presently unique.

References

Ables, J. G.: 1969, *Astrophys. J.* **155**, L27.
Andrew, B. A. and Purton, C. R.: 1968, *Nature* **218**, 855.
Bolton, C. T.: 1972, *IAU Circ. No. 2388*, 25 February, 1972.
Braes, L. L. E. and Miley, G. K.: 1971a, *Astron. Astrophys.* **14**, 160.
Braes, L. L. E. and Miley, G. K.: 1971b, *Nature* **232**, 246.
Braes, L. L. E. and Miley, G. K.: 1972a, *Nature Phys. Sci.* **235**, 147.
Braes, L. L. E. and Miley, G. K.: 1972b, *Nature*, in press.
Hjellming, Z. M.: 1972, *Nature,* in press.
Hjellming, R. M. and Wade, C. M.: 1971a, *Astrophys. J.* **164**, L1.
Hjellming, R. M. and Wade, C. M.: 1971b, *Astrophys. J.* **168**, L21.
Hjellming, R. M. and Wade, C. M.: 1971c, *Astrophys. J.* **168**, L115.
Hjellming, R. M., Wade, C. M., Hughes, V. A., and Woodsworth, A.: 1971, *Nature* **234**, 138.
Hjellming, R. M., Wade, C. M., and Webster, E.: 1972a, *Nature* **236**, 43.
Hjellming, R. M., Hermann, M., and Webster, E.: 1972b, *Nature*, in press.
Wade, C. M. and Hjellming, R. M.: 1971a, *Astrophys. J.* **163**, L106.
Wade, C. M. and Hjellming, R. M.: 1971b, *Astrophys. J.* **170**, 523.
Wade, C. M. and Hjellming, R. M.: 1972, *Nature* **235**, 270.

9. X-RADIATION FROM SUPERNOVA REMNANTS

K. A. POUNDS

Dept. of Physics, University of Leicester, England

Abstract. Available X-ray observations of supernova remnants are reviewed. The number of SNR seen above 2 keV remains small after inclusion of the UHURU results and for only the Crab Nebula is the data adequate to clearly indicate the radiation mechanism. The increasing importance of low energy X-ray studies (below 1 keV) of older and relatively nearby remnants is noted. Brief discussion is given of the relation of the X-ray data to current ideas of the evolution of SNR.

This meeting marks the 10th anniversary of the discovery of the first cosmic X-ray source, Scorpius X-1. In the following few years it was considered that most of the newly discovered and obviously powerful sources were associated with supernovae and, indeed, the first identification of a cosmic X-ray source was with the Crab Nebula in 1964 (Bowyer *et al.*, 1964). This expectation has not been realised and of the 40 or so Galactic sources known by the end of 1970 only four others had been reliably linked with supernova remnants, these being Cas A, SN 1572 (Tycho's nova), Vel X and Pup A. A comparison of a further forty sources listed in the second UHURU catalogue (Giacconi *et al.*, 1972) with the positions of 120 non-thermal Galactic radio sources (Milne, 1970 and Downes, 1971) (generally thought to be SNR) has revealed only three additional associations, none of which are considered convincing. Conversely, a number of X-ray sources previously associated with nearby SNR such as Nor X-1, Nor X-2 and a source near Doradus in the LMC, have been ruled out by improved UHURU locations.

The available X-ray results on all the above (certain and possible) SNR/X-ray source associations are briefly reviewed below, including recent soft X-ray observations which indicate that SNR sources may play an increasingly important role in the newly emerging regime of X-ray astronomy below 1 keV photon energy.

1. Confirmed Identifications

A. CRAB NEBULA

This is by far the most observed and best understood SNR X-ray source. Figure 7 of Peterson (this volume, p. 51), shows a representative selection of spectral data with good statistical precision, which cover the wide energy range from 1.0 keV, below which interstellar absorption is important, to 500 keV. As the figure shows, the data are fit quite well over this entire energy range by the power law photon spectrum*:

$$dN/dE = 10.5 \, (E/1 \text{ keV})^{-2.25} \text{ photons/cm}^2 \text{ s keV} \quad (1\text{–}500 \text{ keV}).$$

---

* Individual experiments in limited energy ranges have yielded slopes which are slightly flatter, i.e., 2.0 to 2.15 (see Gorenstein *et al.*, 1970a; Ducros *et al.*, 1970; Jacobson, 1968).

Bradt and Giacconi (eds.), X- and Gamma-Ray Astronomy, 105–117. All Rights Reserved.

Integrating from 1–500 keV, this yields the flux

$$5.4 \times 10^{-8} \text{ erg/cm}^2 \text{ s} \quad (1\text{–}500 \text{ keV})$$

and a luminosity, assuming a distance of 1500 pc:

$$L_x = 1.4 \times 10^{37} \text{ erg sV}^{-1} \quad (1\text{–}500 \text{ keV}).$$

A recent detection of an overall X-ray polarisation of $15 \pm 5\%$ by the Columbia University Group (Novick *et al.*, 1972) has provided strong evidence for the common (synchrotron) nature of the X-ray, visible continuum and radio fluxes, as suggested by extrapolation of the Crab Nebula electro-magnetic spectrum (Figure 1). Discovery

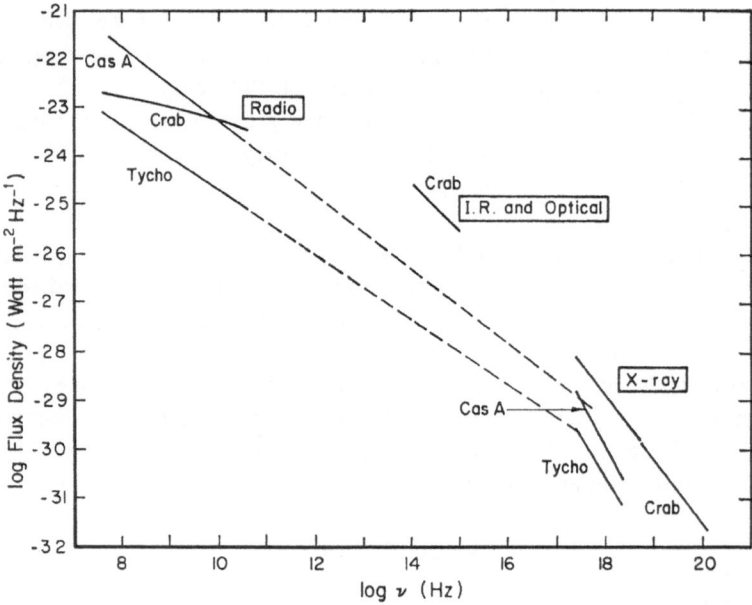

Fig. 1.   Electromagnetic spectra of the Crab Nebula, Cas A and Tycho's Nova over a broad frequency range (from Gorenstein *et al.*, 1970a).

of the shortest period pulsar, NP 0532 – observed also at X-ray energies (see Figure 1 of Fazio, this volume, p. 303) – near the centre of the Nebula has suggested a probable means by which the short-lived high energy electrons ($\sim 10^{13}$ eV to produce keV synchrotron X-rays in a milligauss field) may be continually regenerated. Thus, the $\sim 2'$ extent of the diffuse X-ray component at 1–10 keV (Bowyer *et al.*, 1964 and Oda *et al.*, 1967) may be simply a measure of the life span of these electrons and the source may be considerably smaller at the high photon energies. New observations, with higher spatial resolution (such as may be obtained with planned grazing telescope experiments, or during the series of lunar occultations in 1974–5) should check this and could, for example, yield an 'image' of the magnetic field distribution in the Nebula.

The attenuation of the Crab Nebula spectrum below 1 keV (Fritz *et al.*, 1971) is particularly interesting since it is unlikely to be intrinsic absorption in so extended a source. Thus, detailed observations should provide a measure of the column densities of those elements – particularly helium, oxygen and neon – which contribute most to the photoelectric opacity of the interstellar gas at 0.5–1 keV for a source at the Crab's distance (1.5–2.5 kpc).

B. CAS A AND SN 1572 (TYCHO'S NOVA)

The identification of comparatively weak X-ray sources with these two SNR's was confirmed on an AS&E rocket experiment in 1969 (Gorenstein *et al.*, 1970b). More recent UHURU observations show them to have 2–6 keV fluxes of 5% and 1% of Crab, respectively. The published spectra, from the earlier rocket data, are reproduced

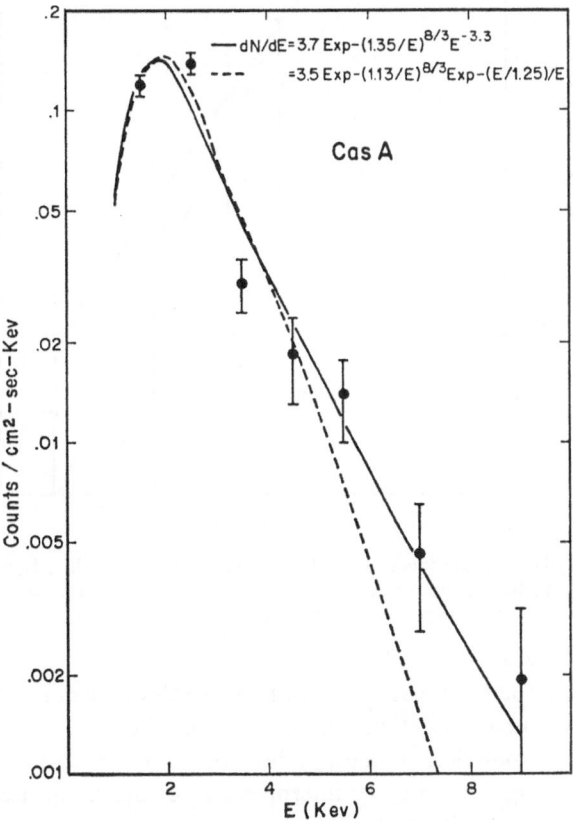

Fig. 2. The measured count rate spectrum of Cas A (circles, with ± 1σ error bars) compared with the best-fit power law (solid line) and exponential (dashed line) input spectra (from Gorenstein *et al.*, 1970a).

in Figures 2 and 3 and are seen to be inadequate to favour power-law or exponential spectral fits. The UHURU results should be of considerably better statistical quality but are probably of insufficient bandwidth to clearly indicate spectral type. Furthermore, as early discussions (Sartori and Morrison, 1967) of the Crab Nebula X-ray source illustrated, even a well-defined power-law can be imitated by exponential (or thermal) spectra from a multi-temperature plasma. Though such an interpretation

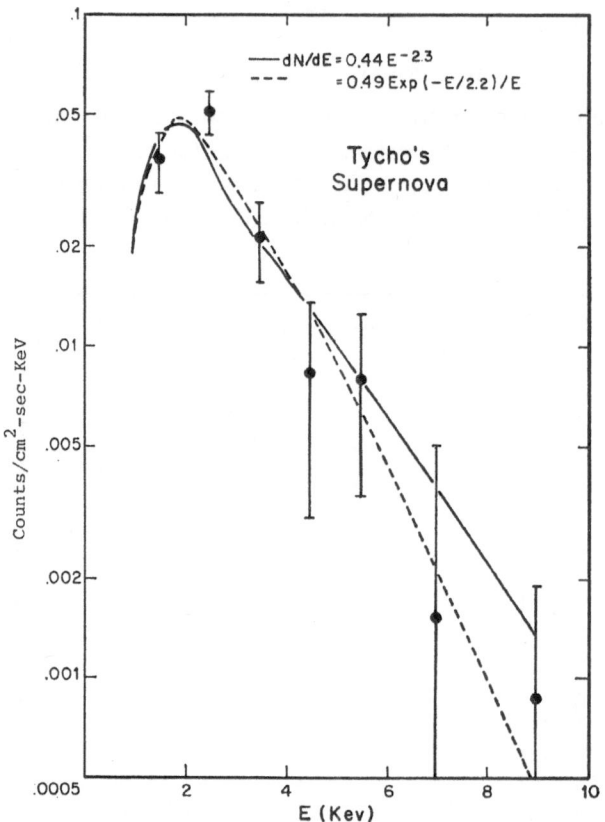

Fig. 3.   The measured count rate spectrum of Tycho's Nova (circles, with $\pm 1\sigma$ error bars) compared
with the best-fit power law (solid line) and exponential (dashed line)
input spectra (from Gorenstein *et al.*, 1970a).

is now unlikely for the Crab source, both bremsstrahlung from a 10^7–10^8 K plasma and electron synchrotron radiation are permitted by the X-ray spectral data for Cas A and Tycho. Extrapolation of the synchrotron radio spectra of these objects to X-ray frequencies (Figure 1) may be interpreted as evidence for a common synchrotron mechanism (Gorenstein *et al.*, 1970b); however, polarized optical radiation has not been reported and this interpretation of the X-ray emission remains unconfirmed. No pulsed component has been observed in the X-ray emission from either source, while useful polarization studies must await the increased sensitivity of satellite observations. The first satellite-borne polarimeters are likely to be those of the Leicester and Columbia Groups, to be flow on UK-5 and OSO-I respectively in late 1973 or early 1974. The alternate possibility of thermal X-ray production could be

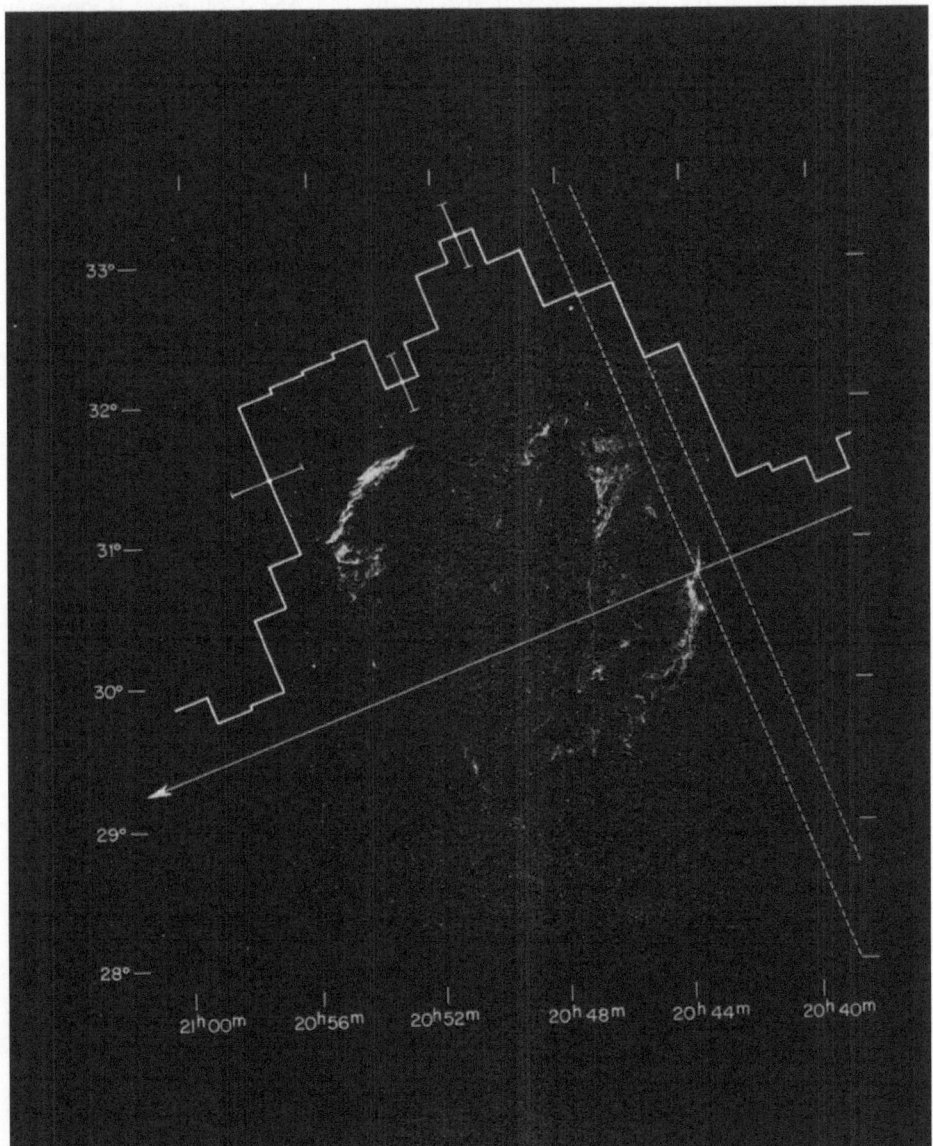

Fig. 4. The one-dimensional X-ray profile of Cygnus loop superimposed on a 48-in. Schmidt red-light photograph of the nebula. The field-of-view of the ASE-Columbia instrument used to obtain these X-ray data is represented by the dashed lines (from Gorenstein *et al.*, 1971).

checked by the same instrument operating as a Bragg crystal spectrometer, to search for emission lines of highly ionised Mg, Si, S, Ca and Fe.

C. CYGNUS LOOP

First identified (Grader *et al.*, 1970) by the LRL Group in 1968 as an intense and extended source at 0.2–1.0 keV, recent surveys by the ASE-Columbia (Gorenstein *et al.*, 1971) and Leiden-Nagoya Groups (Bleeker *et al.*, 1972) have provided further detail on the spatial extent and spectrum of this relatively old SNR X-ray source. Figure 4 reproduces data from the ASE-Columbia survey with a one-dimensional reflecting telescope, having 0.5° spatial resolution. A general extent of ~3° and consistency with a shell-structure emission is indicated, though much better resolution

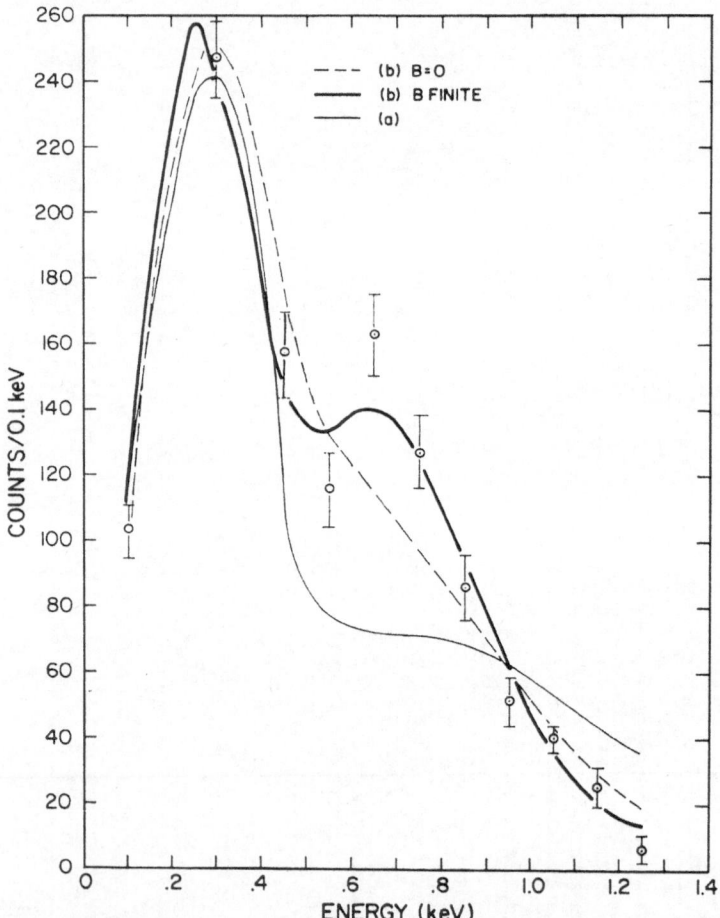

Fig. 5. The measured count rate spectrum of Cygnus loop (circles, with 1σ error bars) compared with the best-fit power law (curve a and exponential (curves b) input spectra. For the solid curve b, a monoenergetic (line) feature is included at 0.65 keV in the input spectrum
(from Gorenstein *et al.*, 1971).

data – in two dimensions – is clearly required, and should be forthcoming quite soon for such an intense source. The relatively crude, proportional counter spectrum obtained in the above experiment has been fitted (Figure 5) with a simple exponential of effective temperature $4 \times 10^6 \, \text{K}$, with a mono-chromatic feature at 0.65 keV. The latter is suggested to be due to O VIII Ly α, which certainly may be expected to be strong in a thermal spectrum of such a characteristic temperature. Though this particular spectral data may be criticised on grounds of inadequate resolution, experiment calibration difficulties and an over-simplified interpretation (several other strong lines would complicate the spectrum and be unresolved, for example), recent observations from a Cal. Tech. rocket experiment, using discrete oxygen and teflon filters to give more certain wavelength discrimination, are consistent with the presence of strong oxygen lines. Direct confirmation of this may come next year with the rocket flight of a Bragg crystal spectrometer by the UCL Group.

D. VELA X AND PUPPIS A

Two rocket flights in May 1970 by the LRL Group, surveyed the Vela region with thin window, large area proportional counters. These had conventional 'egg-crate' collimators, with a $1.3° \times 20°$ field (FWHM) and electrostatic shields for low energy

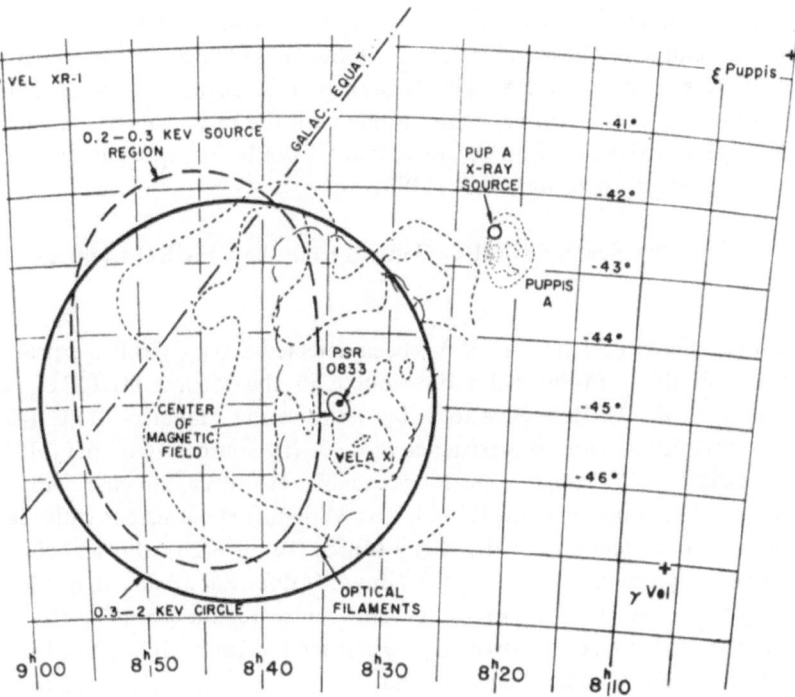

Fig. 6. A map of the Vela-Puppis region showing the radio brightness contours (dashed lines), the pulsar 0833-45 and the optical filaments (wispy lines), together with the extended X-ray source found by the LRL observations in 1970 (Seward *et al.*, 1971). (Figure from Bunner, 1972).

particle rejection. An intense and extended source was found at 0.3–2 keV (Palmieri *et al.*, 1971; Seward *et al.*, 1971), coinciding with the Vela X supernova remnant (Figure 6), believed to be the only SNR younger than 5×10^4 yr as close as 500 pc and possibly the outstanding object in the southern sky. Again, the count rate profile across this source is a flat-topped trapezoid, consistent with the observation of a shell source or alternatively, as the experimenters point out, with 3 or more unresolved point sources. New observations in the near future, employing a mirror system such as that used on Cyg Loop by the ASE-Columbia Groups, or with the UCL-Leicester X-ray package on OAO-C, may be expected to clarify this, providing X-ray maps with 10–20′ resolution. The relation of the (probably) extended source to the Vel X radio object and to the pulsar PSR 0833-45 will be of particular interest in these future observations. From the available LRL data, it is clear that the Vel X spectrum is very soft, but a clear choice between a thermal spectrum ($T \sim 3 \times 10^6$ K) and a steep power law spectrum is not yet possible. As with the Cyg Loop source, the steep spectrum is compatible with the age of the remnant (2×10^4 yr) in terms of either the slowing down of the supernova blast-wave as it pushes against the interstellar gas, thereby yielding a lower temperature for the heated gas, or by the electron spectrum steepening with age and giving a correspondingly steep synchrotron X-ray spectrum.

The Vela observations also detected a soft X-ray source coincident with Pup A, a supernova remnant believed to be at a distance of 1400 pc and somewhat younger than either Cyg Loop or Vel X. The X-ray observations give only an upper limit of 0.5° to the size of this source, again in line with the radio source diameter of 30′. The LRL spectral data on Pup A are best fitted with an exponential distribution, the effective temperature being $\sim 4.5 \times 10^6$ K.

2. Possible New SNR X-Ray Sources from the UHURU Catalogue

A. IC 443

A weak X-ray source (2U 0601 + 21) has been detected at 0.4% Crab and provisionally identified with this non-thermal radio source in the second UHURU catalogue (Giacconi *et al.*, 1972). Reference to Table I shows the similarity of IC 443 to Pup A at radio wavelengths and the existence of an X-ray source of comparable intensity to that associated with Pup A appears reasonable. However, the best X-ray position of 2U 0601 + 21 is some 3° from IC 443, and while this is consistent with the present large uncertainty of location of the X-ray source (being $\sim 2\sigma$), this association must be tentative until a refined location is obtained. It does not appear that a low energy X-ray survey has yet been made of the anti-centre region (IC 443 is at $l^{II} = 188°$) and, if similar to Pup A, this source may be expected to be much brighter below 1 keV than in the UHURU (2–6 keV) energy band.

B. MSH 15–52

The UHURU catalogue includes a weak source (2U 1509–58, 0.7% Crab) within 0.1°

TABLE I

Radio and optical properties of SNR discussed in the paper

	l^{II}	b^{II}	F_{GHz} (f.u.)	D arcmin	d (kpc)	D (pc)	F_{2-6keV} (relative)
Crab Nebula	184.6	− 5.8	1000	4	1.5–2.5	2–3	950
Tycho	120.1	+ 1.4	58	6	2–5	4–9	11
Cas A	111.7	− 2.1	3000	4	3	4	53
Cygnus Loop	74.0	− 8.6	180	180	0.8	43	−
Vela X	263.4	− 3.0	1800	200	0.5	30	10
Puppis A	260.4	− 3.4	145	55	∼ 1.4	24	7.5
IC 443	189.1	+ 2.9	160	40	1.5–2.0	18–24	4
MSH 15−52A	320.4	− 1.0	40	8	∼ 8	19	7
SNR 1811−17	13.4	+ 0.1	∼13	7	∼ 9	19	294
Lupus Loop	330.0	+15.0	340	270	0.8	65	−
HB 21	89.1	+ 4.7	175	120	1.2	42	−
MSH 14−63	315.4	− 2.3	33	40	∼ 2.5	30	−
References	(1)	(1)	(1)	(1)	(1,2,3)		(4)

References (1) Milne (1970)
(2) Downes (1971)
(3) Woltjer (1972)
(4) Giacconi et al. (1972).

of the double non-thermal radio source MSH 15–52. The radio source is again similar to Pup A, though it may be 2 or 3 times more distant (Table I). The X-ray flux at 2–6 keV is similar to Pup A, but examination of the LRL low energy survey of this region (Hill et al., 1972) reveals no signal greater than ∼5% of Pup A at 0.35–2.8 keV. This is generally consistent with the greater interstellar absorption for a source at 3 or 4 kpc and in the Galactic Plane.

C. SNR 1811–17

This is the third SNR which is provisionally listed as a new X-ray identification in the 2U catalogue. The corresponding X-ray source (2U 1811–17, previously Sgr X-2 and GX 13 + 1) is strong, about 0.3 Crab, apparently non-variable, and is located only 0.1° from the non-thermal radio source. If confirmed, this must be an extremely powerful supernova X-ray emitter, as compared with Kepler's supernova, believed to be at a similar distance (8–10 kpc) and having twice the radio flux at 1400 MHz, but remaining undetected at X-ray energies, shows.

3. Possible Soft X-ray Emission from Extended, Nearby Supernova Remnants

A. LUPUS LOOP

The Galactic Plane survey by the LRL Group in May 1970 (Hill et al., 1972) observed a broad increase in counting rate in the direction of the Lupus Loop, a non-thermal radio source having much in common with the Cygnus Loop and also believed to be a relatively old and nearby SNR. Because of a high background nearby, this detection requires confirmation.

B. HB 21

A University of Wisconsin rocket experiment flown in December 1969 detected a further low energy source in Cygnus (Coleman *et al.*, 1971), designated Cyg X-6. The rather large error box ($1.7° \times 6°$ at 1σ) includes a part of the SNR HB 21, a remnant similar in its radio appearance to Cyg Loop. No obvious extension of the X-ray source is seen, which probably sets an upper limit of $1°$ or so to the main emission region. Because of the large positional uncertainty and the crowded area of sky involved this SNR association must be considered very uncertain at present.

4. Possible X-Ray Emission from Possible SNR

A. NORTH POLAR SPUR

A recent University of Wisconsin rocket experiment has revealed (Bunner *et al.*, 1972) a broad ridge of excess radiation in the <0.28 keV and $0.5–1.0$ keV energy bands in general direction of the North Polar Spur (Figure 7), an extended region of non-thermal radio emission which has been suggested (Hanbury Brown *et al.*, 1960 and Elliot, 1970) to be an old and relatively nearby SNR (but see Mathewson, 1968). If confirmed, X-radiation from Galactic radio loops, such as this and the Cetus Arc, could contribute significantly to the soft X-ray background radiation.

B. ETA CARINAE

An intense low energy X-ray source was detected by the 1970 LRL Galactic survey rocket flight at $l^{II} \sim 288°$, $b^{II} \sim 4.0°$, within $0.5°$ of the bright infra-red star η Car which, one school of thought suggests, is a 'slow supernova'. The crude X-ray spectrum is steep and may be interpreted as thermal, with a characteristic temperature of $\sim 3 \times 10^6$ K, or non-thermal with a spectral index similar to that in the infra-red region, suggesting the X-rays could arise by inverse Compton scattering of these low energy photons. (For a distance of 1.6 kpc, $L_{IR} \sim 8 \times 10^{39}$ erg s^{-1} (Westphal and Neugebauer, 1969) and $L_x \sim 2 \times 10^{36}$ erg s^{-1})

C. MSH 14–63

A weak non-thermal radio source, located at $l^{II} = 315.4°$ and $b^{II} = -2.3°$ in the Milne list of probable SNR, lies close to the X-ray source Cen X-1. First seen in 1965 NRL survey (Byram *et al.*, 1966) this faint X-ray source has been observed latterly by the Leicester (Cooke and Pounds, 1971) and LRL (Seward *et al.*, 1971) Groups, though not by UHURU. Since each positive sighting was with plastic window detectors, these apparently conflicting results might be reconciled if the X-ray spectrum is steep, as indeed the 40' extension of the radio source would suggest.

5. Discussion

The currently available X-ray data on Galactic SNR show a gradual 'softening' of

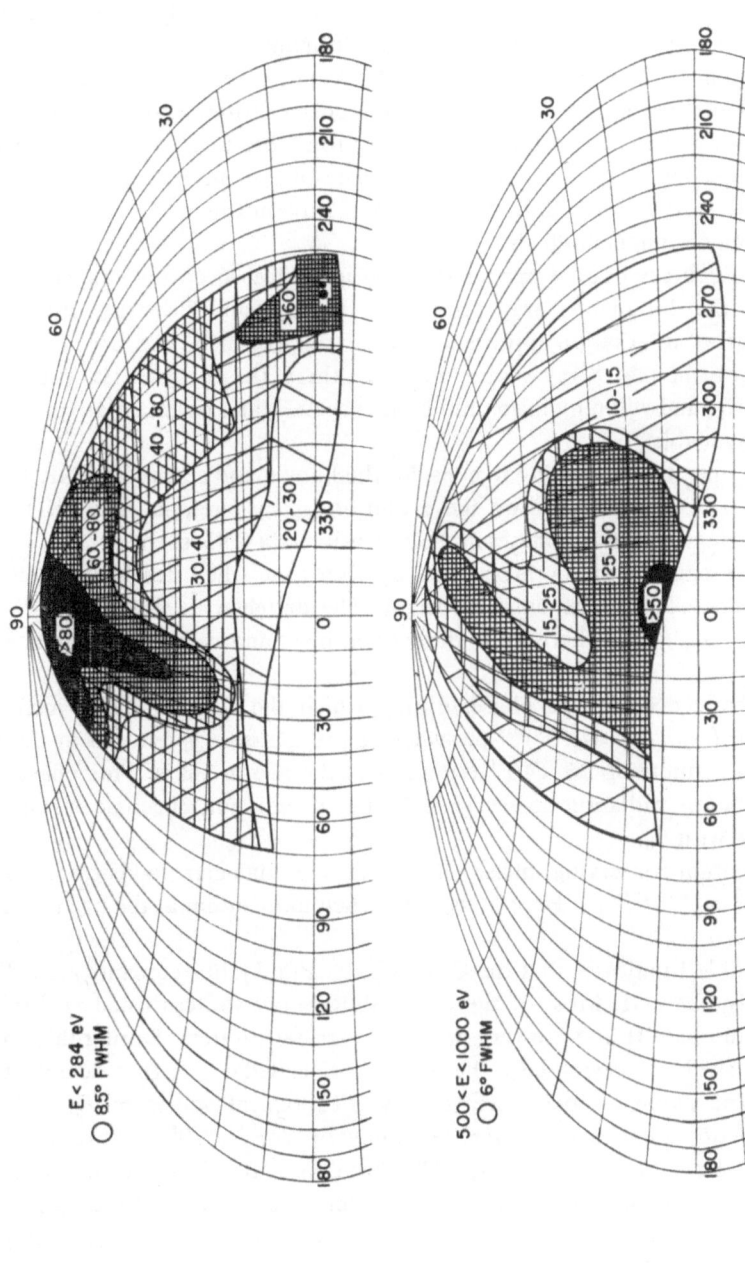

Fig. 7. X-ray isophotes at $E < 284$ eV and $500 < E < 1000$ eV showing a ridge of high intensity near the North Polar Spur (Bunner et al., 1972).

the X-ray emission with increasing size (and age) of the remnant. Thus, the three youngest SNR's listed (Crab, Cas A and Tycho) are bright well into the kilovolt range. Of these, the Crab source is outstandingly bright and hard and this unique property is almost certainly attributable to the existence of the fast pulsar NP 0532, continually injecting 10^{12}–10^{13} eV electrons into the Nebula to maintain the X-ray synchrotron emission. The considerably lower X-ray to radio luminosities of Cas A and Tycho may be due to the absence of a similar central pulsar. Alternatively, it may take several centuries for a pulsar to build up the electron store to its maximum energy (Setti and Woltjer, 1972). In this way the absence of still more luminous objects (there should be ~ 10 younger SNR than Cas A if supernovae occur at a mean rate of 1 per 40 yr) may be understood also.

For only two sources does it appear that the available spectrum is adequate to firmly indicate the dominant X-ray production mechanism. These are the Crab, for which synchrotron radiation appears to be clearly established, and Cygnus Loop, almost certainly a thermal X-ray source. For the others, the X-ray emission process remains uncertain, and will remain so until better spectral (and polarization) measurements become available.

At the same time, it is interesting to consider the known and suspected SNR X-ray sources in terms of current ideas on the evolution of such remnants. Cox (1970), for example, distinguishes three phases in the expansion of a SNR shell:

   I The free expansion phase, proceeding up to a time when deceleration effects of the interstellar gas become important. The diameter at this time is $D = 4$ $(M/M_\odot)^{1/3} \, n_H^{-1/3}$ pc, where M is the ejected mass and n_H the local interstellar density.

   II An adiabatic deceleration phase, during which the shell diameter increases as $D = 26 \, (E_0/n_H)^{1/5} \, (t/10^4)^{2/5}$ pc, where E_0 is the energy released in the SN explosion in units of 7.5×10^{50} erg and t is the age in years. This phase will continue until radiation losses (mainly by X-radiation) become significant compared with E_0.

   III A momentum-conserving phase, in which the diameter increases as $D \sim (M_0 v_0)^{1/4} \, n_H^{-1/4} \, t^{1/4}$ pc. Here $M_0 v_0$ is the shell momentum at the beginning of phase III.

Reference to Table I indicates that the Crab, Cas A and Tycho remnants are still in phase I or early phase II. Most of the other SNR X-ray sources listed in Table I should be within phase II and are very probably thermal emitters. During this phase the expanding shell will shock-heat the ambient interstellar gas to a temperature depending on the squared shock-wave velocity. For typical values of E_0 and n_H, the resulting temperature is given approximately by $T = 3 \times 10^6 \, t^{-6/5}$ K, for t in units of 10^4 yr (Ilovaisky and Bowyer, 1972), yielding soft X-ray spectra strong in the 0.2–2 keV energy band. During this phase the temperature of the emitting plasma will decrease, but the total 'emission measure' $(N_e^2 V)$ will increase considerably as the shell grows larger. Thus, the X-ray luminosity at a given photon energy will at first rise, due to the increasing emission measure and then rapidly fall as the gas continues to cool. Further spectral and spatial observations of these older, extended X-ray sources are needed to establish the validity of this interpretation in each case. As pointed out by Woltjer (1972), correct interpretation of the X-ray production and the stage of evolution reached by a particular remnant will be vital to evaluating its energetics and unravelling its history.

References

Bleeker, J. A. M., Deerenberg, A. J. M., Hayakawa, S., Tanaka, Y., and Yamshita, K.: 1972, submitted to *Astrophys. J. Letters*.

Bowyer, S., Byram, E. T., Chubb, T. A., and Friedman, H.: 1964, *Science* 146, 912.

Bunner, A. N.: 1972, in *The Gum Nebula and Related Problems* (S. Maran and J. Brandt, eds., National Technical Information Service, U.S. Department of Commerce reprint No. N72-11750-774).

Bunner, A. N., Coleman, P. L., Kraushaar, W. L., and McCammon, D.: 1972, *Astrophys. J.* 172, L67.

Byram, E. T., Chubb, T. A., and Friedman H.: 1966, *Science* 152, 66.

Cooke, B. A. and Pounds, K. A.: 1971, *Nature* 229, 144.

Cox, D. P.: 1970, Ph.D. Thesis, Univ. of California, San Diego.

Downes, D.: 1971, *Astrophys. J.* 76, 305.

Ducros, G., Ducros, R., Roccia, R., and Tarrius, R.: 1970, *Astron. Astrophys.* 7, 162.

Elliot, K. H.: 1970, *Nature* 226, 1236.

Fritz, G., Chubb, T. A., Henry, R. C., Friedman, H., and Meekins, G.: 1971, *Astrophys. J.* 164, L61.

Giacconi, R., Gursky, H., Kellogg, E. M., Murray, S., Scheier, E., and Tananbaum, H.: 1972, submitted to *Astrophys. J. Letters*.

Gorenstein, P., Kellogg, E. M. and Gursky, H.: 1970a, *Astrophys. J.* 160, 199.

Gorenstein, P., Kellogg, E., Giacconi, R., and Gursky H.: 1970b, *Astrophys. J.* 160, 947.

Gorenstein, P., Giacconi, R, R., Gursky, H., Harris, B., Novick, R., and Vanden Bout, P.: 1971, *Science* 172, 369.

Grader, R. J., Hill, R. W., and Stoering, J. P.: 1970, *Astrophys. J.* 161, L45.

Hanbury Brown, R., Davies, E. D., and Hazard, C.: 1960, *Observatory* 80, 191.

Hill, R. W., Burginyon, G., Grader, R. J., Palmieri, T. M., Seward F. D., and Stoering, J. P.: 1972, *Astrophys. J.* 171, 519.

Ilovaisky, S. A. and Bowyer, S.: 1972, to be published in *Astron. Astrophys*.

Jacobson, A. S.: 1968, Ph.D. Thesis, Univ. of California, San Diego.

Mathewson, D. S.: 1968, *Astrophys. J.* 153, L47.

Milne, D. K.: 1970, *Austral. J. Phys.* 23, 425.

Novick, R., Berthelsdorf, R., Linke, R., Weisskopf, M. C., and Wolff, R. S.: 1972, *Astrophys. J.* 174, L1.

Oda, M., Bradt, H. V., Garmire, G., Giacconi, R., Gorenstein, P., Gursky, H., Spada, G., Sreekantan, B. V., and Waters, J.: 1967, *Astrophys. J.* 148, L5.

Palmieri, T. M., Burginyon, G., Grader, R. J., Hill, R. W., Seward, F. D., and Stoering, J. P.: 1971, *Astrophys. J.* 164, 61.

Peterson, L. E.: 1972, this volume, p. 51.

Sartori, L. and Morrison, P.: 1967, *Astrophys. J.* 150, 385.

Setti, G. and Woltjer, L.: 1972, to be published.

Seward, F. D., Burginyon, G., Grader, R. J., Hill, R. W., Palmieri T. M., and Stoering J. F.: 1971, *Astrophys. J.* 169, 515.

Westphal, J. and Neugebauer, G.: 1969, *Astrophys. J.* 156, L45.

Woltjer, L.: 1972, Columbia Astrophysics Lab. Cont; No. 59.

10. GALACTIC X-RAY POLARIMETRY AND
HIGH-RESOLUTION X-RAY SPECTROSCOPY

R. NOVICK

Columbia Astrophysics Laboratory, Columbia University, New York, N.Y., U.S.A.

Abstract. Stellar X-ray spectroscopy and polarimetry are discussed in terms of the source parameters that can be determined through such studies and in terms of the constraints that these studies will place on theoretical models. The spectroscopic and polarimetric results that have been obtained to date are reviewed. These include the recent discovery of X-ray polarization in the Crab Nebula and the recent evidence for X-ray coronal line emission in the Cygnus Loop. Finally, the properties and predicted performance of a number of satellite-borne spectrometers and polarimeters are presented.

1. Introduction

While very great progress has been made in X-ray astronomy with very simple observational tools, it is clear that the further development of the subject requires high-resolution spectroscopic and polarimetric observations. The photometric methods that have been used have served to reveal the binary nature of a number of compact X-ray sources. Modulation collimator and lunar occultation data have provided very precise positional data. In a number of cases it has been possible to correlate X-ray sources with particular radio and optical objects. Such correlations have enormously expanded our knowledge of these sources and have yielded some important optical spectroscopic observations on a few of them. Since the X-ray, optical, and radio emission regions may not be identical in the various sources, it is clear that direct X-ray spectroscopic and polarimetric observations must be made if we are to understand the physical conditions in the X-ray emitting regions. Also, since some objects seem to have no observable optical counterpart, we must rely entirely on X-ray observations to understand these objects. Thus it is clear that spectroscopic and polarimetric observations are essential for determining the physical conditions in the X-ray sources and for distinguishing between various theoretical models.

In this review we indicate the nature of the information that we hope to obtain about X-ray sources through spectroscopic and polarimetric observations. We discuss instrumental problems, we indicate the present satellite flight schedule for various spectrometers and polarimeters, and finally we review the results that have been obtained to date.

2. Spectroscopy

Spectral lines will be observable in all thermal X-ray sources. Knowledge of the width of these lines, their intensities, and possible shifts will be essential for the determination of the temperature, density, abundances, mass motions, and gravitational fields within the emitting region. The spectroscopic techniques that are available include the

Bradt and Giacconi (eds.), X- and Gamma-Ray Astronomy, 118–131. All Rights Reserved.
Copyright © 1973 by the IAU.

use of proportional counters, solid-state detectors, filters, gratings, and Bragg crystal spectrometers.

Proportional counters are available with large areas and with low background counting rates. Their energy resolution is limited by the relatively high effective ionization energy of the counting gas. Typically 30 eV are required to produce an ion pair in the gas, and this implies a resolution width of 170 eV at 1 keV photon energy.

The most comprehensive development on proportional counters has been accomplished by Boldt and his collaborators of the NASA Goddard Space Flight Center. This group has pioneered in the development of multiwire, multidepth proportional counters in which they have achieved a resolution of 840 eV with the 6-keV ^{55}Fe X-ray line (Bleach *et al.*, 1972). At longer wavelengths, a resolution of 190 eV has been reported for the 282-eV carbon-K line and 170 eV at the 185-eV boron-K line with simple proportional counters operating with pure methane gas (Yentis *et al.*, 1971).

Solid-state detectors require only a few electron volts per ion pair and therefore exhibit much higher resolution than proportional counters. Goulding and Stone (1970) have reviewed the status of the development of solid-state X-ray detectors. At low photon energies, the resolution is limited by preamplifier noise, and the best reported resolution at long wavelengths is 105 eV (Landis *et al.*, 1970). Unfortunately, it seems to be difficult to produce large-area, high-resolution solid-state detectors, and in addition they must be refrigerated to obtain the best resolution. In view of their small size it appears that solid-state detectors can best be used in conjunction with focusing optics. Very dramatic improvements could be made in X-ray astronomy if large-area, high-resolution, position-sensitive solid-state detectors could be developed. Among other applications, such a detector could be used at the focus of a true X-ray telescope to serve as an image-forming element with reasonable spectral resolution. One advantage of these detectors over proportional counters for this application is the fact that the detectors are relatively thin; they therefore could be used with very high-resolution telescopes which have a limited depth of focus. This application would require spatial resolution on the order of 0.02 mm.

In May, 1971, Singer (1972), of the Los Alamos Scientific Laboratory, flew three lithium-drifted silicon detectors in a rocket designed to study the spectrum of Sco X-1. Each detector had an area of 1.1 cm^2, they were cooled with liquid nitrogen, and the resolution widths, determined by electronic noise, were about 250 eV. One detector provided background information. While the final results have not been published, Singer finds that the energy spectrum is flat from about 700 eV to 4 keV with a constant intensity (within 20%) of 25 ± 5 keV keV^{-1} cm^{-2} s^{-1}. At higher energy, the spectrum falls according to the well-known thermal bremsstrahlung formula with a temperature in the range of about 3 keV. Singer finds no evidence for any lines, but the area of his detector is so small that even with the improved resolution, he would not have been able to detect the 7-keV feature suggested by proportional counter experiments (see below). Singer has stated that his data suggest the presence of an absorption feature at the neon absorption edge, but the statistical significance of this feature is not clear.

Womack and Overbeck (1972), of the Massachusetts Institute of Technology, have also flown a solid-state detector in a balloon, but their results have not yet been published. The X-ray astronomy group of the NASA Goddard Space Flight Center is scheduled to fly a cooled nonimaging solid-state detector at the focal plane of the high-efficiency telescope in the HEAO-C spacecraft.

The theoretical resolution $(\lambda/\Delta\lambda)$ of gratings is equal to the total number of lines in the grating. In practice, the resolution is determined by the width of the entrance and exit slits and by aberrations. Various groups are working on true astigmatic focusing holographic gratings of sufficient size to be useful for nonsolar X-ray astronomy (Pieuchard et al., 1972). Diffraction gratings were successfully used to study the solar corona in the far ultraviolet during the March, 1970, solar eclipse (Speer et al., 1970). The American Science & Engineering (AS & E) X-ray astronomy group is scheduled to fly a normal incidence transmission objective grating on the HEAO-C spacecraft. It is expected that the resolution $(\lambda/\Delta\lambda)$ of this grating will be about 50.

True Bragg crystal spectrometers offer a very great promise in stellar X-ray astronomy. In view of the low fluxes from most sources, it is necessary to employ crystal areas of several thousand square centimeters or more. The X-ray astronomy groups at the University of Leicester, at the Los Alamos Scientific Laboratory, and at the Columbia Astrophysics Laboratory have flown large-area spectrometers in sounding rockets to search for the Fe xxv, the S xvi, the Ne ix, and the O viii lines in Sco X-1.

In the first Leicester flight on March 18, 1970, a lithium fluoride spectrometer with an effective crystal area of 220 cm^2 was used to search for the $(1s^2\,^1S-1s2p^1P)$ line of Fe xxv (6.70 keV) in the X-ray spectrum of Sco X-1 (Griffiths et al., 1971). No line was detected, and an upper limit (at the 3-σ level of confidence) of 1.2×10^{-9} ergs cm^{-2} s^{-1} was established on narrow-line emission from Sco X-1 between 1.83 Å and 1.91 Å. The overall spectrum between 4 and 14 keV was observed with a proportional counter, and it was found that the electron temperature was 6×10^7 K. By combining this result with the above limit on the line flux, it was found that the 3-σ upper limit on the equivalent width of any narrow line between 6.51 and 6.77 keV was 50 eV. The alignment of the crystals of this spectrometer was improved so that the width of the rocking curve was reduced from 10' to 5'. This improved spectrometer was flown on March 11, 1971, in another search for the Fe xxv lines in Sco X-1. Again, no lines were observed, and the 3-σ limit on narrow-line emission from Fe xxv was reduced to an equivalent width of 25 eV when the source temperature was 5.5×10^7 K (Griffiths, 1972).

The Los Alamos Scientific Laboratory flew mica and KAP crystal spectrometers in a rocket to search for the Ne ix and O viii lines in Sco X-1. Each spectrometer employed two crystal panels each with an area of about 290 cm^2. The continuum from Sco X-1 was observed with each spectrometer, but no lines were observed. The authors have not given upper limits on either the line strengths or the equivalent widths of these lines (Argo et al., 1972).

On April 24, 1970, the Columbia Astrophysics Laboratory flew a 2000-cm^2 graphite crystal spectrometer in an Aerobee-170 rocket to search for the S xvi line in Sco X-1 (Kestenbaum et al., 1971). Again, the continuum was detected with a temperature of 8×10^7 K, but no line was observed. The 3-σ upper limits on line strength and equivalent width were 0.08 photons cm^{-2} s^{-1} and 6.7 eV, respectively. In a more recent Aerobee-170 rocket flight on March 22, 1972, a lithium-fluoride spectrometer with 3400-cm^2 crystal area was flown to search for the Fe xxv lines in Sco X-1. No lines were observed, and it was shown that at the 3-σ level the equivalent width of the Fe xxv lines was 3 eV or less (Stockman et al., 1972).

It is generally agreed that the absence of sharp lines in Sco X-1 results from severe broadening in this dense object. The first estimate of these effects was made by Angel (1969a) who showed that at a temperature of 5×10^7 K and a density of 10^{16} electrons cm^{-3}, suggested by the infrared observations, the line strengths will be decreased by a factor of about 16 by electron scattering. These calculations have been considerably refined and extended by Loh and Garmire (1971) and by Felten et al., (1972). These authors include the effects of photoionization by the intense X-ray flux in Sco X-1 and the effect of trapping on the resonance lines. They show that at a temperature of 7×10^7 K and an optical depth of 5 for electron scattering, the equivalent width of the core of the S xvi Lα line of 2.62 keV is 0.32 eV. In the case of the Fe xxv resonance line at a temperature of 5×10^7 K and an optical depth for scattering of 5, the predicted equivalent width of the line core is 1.9 eV. Under the same conditions, the sum of the equivalent widths of the Fe xxv forbidden and intercombination lines is predicted to be 34 eV. Thus, the observations indicate that either the temperature or the density is higher than the values assumed above. For example, according to Felten et al. (1972), if the temperature were 7×10^7 K and the optical depth were 10, then the predicted equivalent width of the Fe xxv resonance and intercombination lines would be about 3 eV, consistent with the observations. Alternatively, the temperature and density may have lower values suggested above, but there may be appreciable mass motion and turbulence in the source. Such motion would, of course, broaden the line core remaining after scattering. Mass motion is expected if the radiation arises in an accretion disk surrounding a dense stellar object. In any case it is clear that any further observations of Sco X-1 require long-term satellite observations. As an example of the power of satellite spectrometers, we note that the Bragg crystal spectrometer on HEAO-B (see below) will have sufficient sensitivity to detect the Fe xxv and Fe xxvi lines in Sco X-1 in one day at the 3-σ level if the equivalent width of the core is 0.2 eV or more. In addition, this spectrometer will have sufficient sensitivity to reveal the O vii coronal lines in Alpha Centauri at the 3-σ level in five days if they have the same strength as those observed on the Sun. The detection of the lines in Sco X-1 and other compact sources is essential for determining the density, temperature, and velocity distribution in the X-ray emitting region.

In addition to the Bragg crystal and solid-state detector studies of Sco X-1 discussed above, numerous observations have been made of this object with proportional counters. Four groups report weak, broad features in the neighborhood of the 7-keV iron lines. The U.S. Naval Research Laboratory reports a 3-σ upper limit on iron line emission from Sco X-1 with an equivalent width of about 50 eV at a temperature of 5.5×10^7 K (Meekins *et al.*, 1969). The NASA Goddard Space Flight Center reported a 3.25-σ iron line feature with an equivalent width of about 50 eV and a temperature of 6×10^7 K (Holt *et al.*, 1969, 1970). The University of Leicester reported a 5-σ upper limit on the iron line with an equivalent width of 100 eV at a temperature of 5.5×10^7 K (Griffiths, 1972). The Lockheed group reported an observation of iron line emission with an equivalent width of 240 ± 140 eV at a temperature of 5.5×10^7 K (Acton *et al.*, 1970). In the case of the Lockheed experiment, it should be noted that the detector used had a very thick window that caused considerable curvature in the observed spectrum at the position of the iron line. Ignoring the Lockheed result, we find that the equivalent width of the iron lines in Sco X-1 as observed with proportional counters is 100 eV or less. Since the resolution width of a proportional counter of 7 keV is about 1 keV, these results cannot be directly compared with the estimates of Felten *et al.* (1972). As noted above, these authors estimated the equivalent width of the cores of the Fe xxv and S xvi lines as well as the equivalent width of the total line. Since the total line energy is spread over several keV (Angel, 1969a), it will be necessary to make detailed predictions of the line shape before making a comparison with theory. As yet no detailed line-shape or continuum-shape predictions have been made for compact X-ray sources.

The HEAO-B satellite will contain two stellar Bragg crystal X-ray spectrometers, one to cover the 1 to 8 keV region and the other to cover the 0.4 to 2 keV region. This will be the first satellite-borne spectrometer to observe the region below 1 keV. The higher-energy spectrometer will contain flat crystals for high resolution such as germanium with subarc-minute rocking curves and crystals for high integrated reflectivity such as graphite to observe broad lines and the detailed shape of continua. Mounted on the back of the high-energy crystal panel will be another crystal such as PET. The panel can be rotated to expose either the PET or the germanium and graphite crystals. The lower-energy spectrometer will use flat crystals of RAP and a thin window counter as detector. It is hoped to collect the radiation reflected by the crystal in paraboloidal mirrors and to focus the radiation into very small detectors in order to reduce the background rate. It is also hoped to put an array of grazing-incidence gratings back to back to the RAP to be able to observe in the energy range 0.1 to 0.5 keV. The area of each crystal panel will be 2300 cm^2. The orientation of both the crystal panels and the detectors can be controlled in orbit in order to perform a Bragg scan. Sources can be observed for several days if necessary.

In addition to HEAO-B, the OSO-I, ANS, UK-5, and the HEAO-C spacecraft will also contain stellar Bragg crystal X-ray spectrometers. The OSO-I instrument will contain a large-area mosaic crystal spectrometer that will be useful for determining the exact shape of the continua from a number of sources, and in addition, it will provide a complete spectral survey of all of the presently known sources. The OSO-I spectrometer consists of two 970-cm^2 graphite crystal panels that view the X-ray sky through the edge of the satellite wheel compartment. The angle between the panels is

40° and located between them is a two-sided proportional counter. The Bragg scan from 9° to 84° (11.8 to 1.86 keV) is accomplished by the rotation of the satellite wheel compartment.

The ANS instrument consists of two 65-cm² EDDT or PET crystal panels oriented at about 45° to the viewing direction. One crystal is adjusted to detect the Si XIII lines and the other the Si XIV line. The spacecraft can be oriented so that a source can be studied for about one-half day. The UK-5 spectrometer will employ graphite and lithium hydride crystal panels. Each panel has an area of about 200 cm². The Bragg angle is adjustable in flight from 25° to 65° and can be set at 45° for polarimetry (see below). The spectral range of the instrument extends from 2 to 8 keV. The viewing direction of the spectrometer is parallel to the spacecraft spin axis so that sources may be observed for extended periods of time. The HEAO-C spectrometers employ very high-resolution, curved crystals at the focus of the high-resolution (1″) telescope. These instruments will permit detailed studies of the line shapes and shifts.

Angel and Weisskopf (1970) have proposed and tested a novel Bragg objective crystal spectrometer that appears to have a great promise. Briefly, the instrument consists of a large-area crystal mounted in front of an X-ray telescope. The angle between the telescope axis and the crystal plane and the angle between the crystal plane and the line of sight to the X-ray star are chosen to satisfy the Bragg condition for a particular X-ray line or multiplet. A spectral line appears in the focal plane of the telescope as an arc of a large diameter circle. The resolution of the instrument is determined by the diffraction width of the individual domains in the crystal panel, and the spectral range is determined by the mosaic spread in the crystal and the field of view of the telescope. Since the entire aperture of the telescope can be illuminated by the diffracted beam, the instrument has very high sensitivity. A schematic diagram of this spectrometer is shown in Figure 1, and in Figure 2 we show the rhodium $L\alpha_1$ and $L\alpha_2$ lines as observed in the laboratory with a small grazing incidence telescope and an objective mosaic graphite crystal. The 3.5-eV linewidth observed here is determined by the natural width of the rhodium lines and is in excellent agreement with the 2.75-eV width obtained with a very high-quality calcite Bragg crystal spectrometer. The energy and Bragg-angle separation of the $L\alpha_1$ and $L\alpha_2$ lines are 4.69 eV and 5.5′, respectively. The mosaic spread of the crystal used in this experiment was 24′. The fact that the $L\alpha_1$ and $L\alpha_2$ lines could be clearly resolved shows that the resolution of the objective crystal spectrometer is determined by the diffraction width of the individual domains rather than the mosaic spread of the crystal.

The only object besides Sco X-1 which has been subject to detailed spectroscopic study is the Cygnus Loop. It is believed that this is a supernova remnant and that the emitting region is a low-density, high-temperature plasma. If this is correct, then coronal lines should be observable in both the X-ray and visible bands. In a collaborative experiment between AS&E and Columbia, the Cygnus Loop was observed with a one-dimensional focusing collector and thin window proportional counter (Gorenstein *et al.*, 1971). The data rule out a power-law spectrum and strongly suggest a thermal bremsstrahlung spectrum with $T = 4.3 \times 10^6$ K and a line at 0.65 keV in the

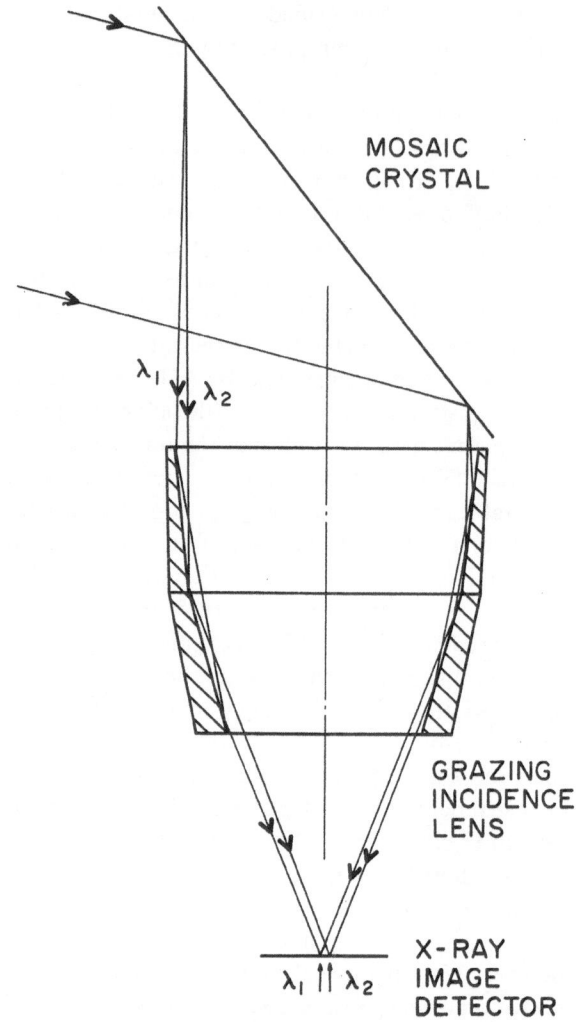

Fig. 1.   Schematic diagram of the Bragg objective crystal spectrometer.

neighborhood of the expected coronal O VII and O VIII lines. Tucker (1971) reinter-
preted these data and suggested that the temperature might be 2×10^6 K. The interpretation of the line depends on our knowledge of the reflection efficiency of the
focusing collector since this instrument employed a chromium surface which has a
strong absorption feature close to the oxygen lines. The reflection efficiency of the
mirror was measured at 10 Å and 44 Å and found to be in good agreement with the
results of Ershov *et al.* (1967). Thus there was no reason to believe that the assumed
reflection efficiency in the neighborhood of 0.65 keV is incorrect. Recently Bleeker
et al. (1972) reported on a very careful proportional counter study of the Cygnus Loop.
They confirmed the thermal spectrum with $T = 2.7 \times 10^6$ K but found that they were

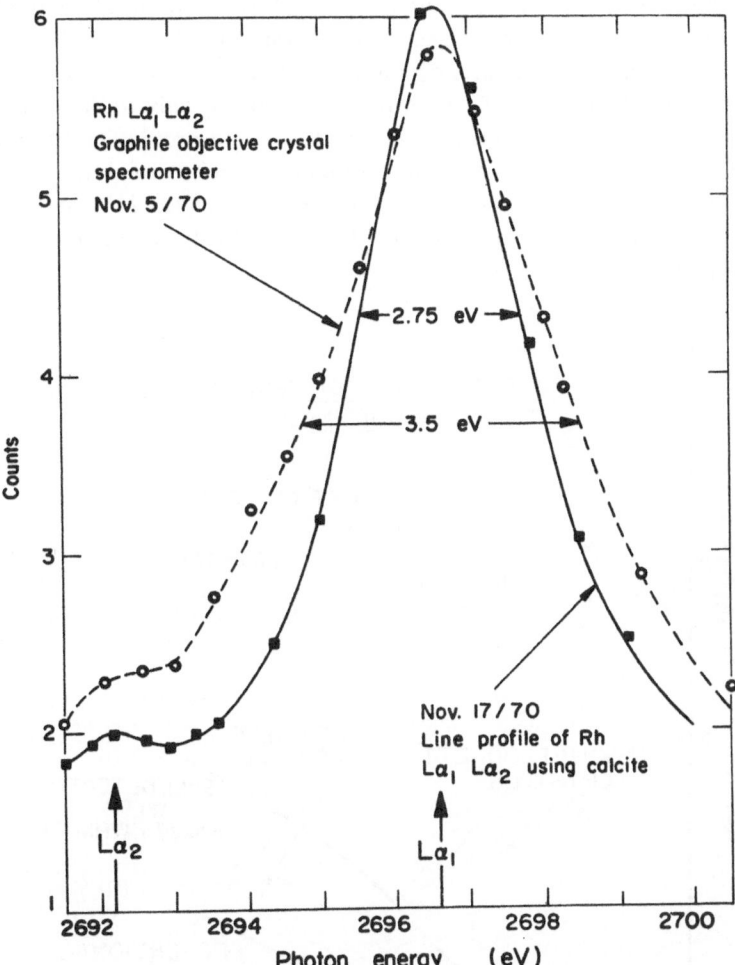

Fig. 2.   Rhodium Lα₁ and Lα₂ lines observed with a small mosaic graphite objective crystal spectrometer and with a conventional high-resolution calcite crystal spectrometer.

not required to assume the existence of X-ray lines to fit their data. More recently, Stevens *et al.* (1972) made a study of this object with proportional counters and balanced absorption filters. One of the filters consisted of a gas cell that was filled with oxygen during the rocket flight. The other was a thin Teflon sheet that provided the fluorine absorption edge. This experiment has provided very strong evidence for the existence of the oxygen lines. Thus, it appears that the Cygnus Loop is the first non-solar object to exhibit X-ray lines.

Kurtz *et al.* (1972) have made extensive searches for the optical coronal Fe XIV line at 5303 Å in the Cygnus Loop, but so far they have failed to detect the lines. Using a filter with 3-Å bandwidth, they have shown that the λ5303 flux from the Cygnus

Loop is less than 5×10^{-9} ergs cm^{-2} sterad^{-1} s^{-1}. This is an order of magnitude less than the predicted total flux for a plasma temperature of 2.7×10^{6} K. This negative result is taken to indicate either broadening of the line profile by mass motion or a higher temperature in the Loop. Clearly, it is essential not only to detect the X-ray and optical lines but also to determine their intensity and profile. This information will provide a direct measure of the temperature and velocity distribution within the Loop and will certainly help to clarify the difference between the 110 km s^{-1} velocity observed in the filaments and the 430 km s^{-1} shock velocity required by the X-ray observations (Woltjer, 1972).

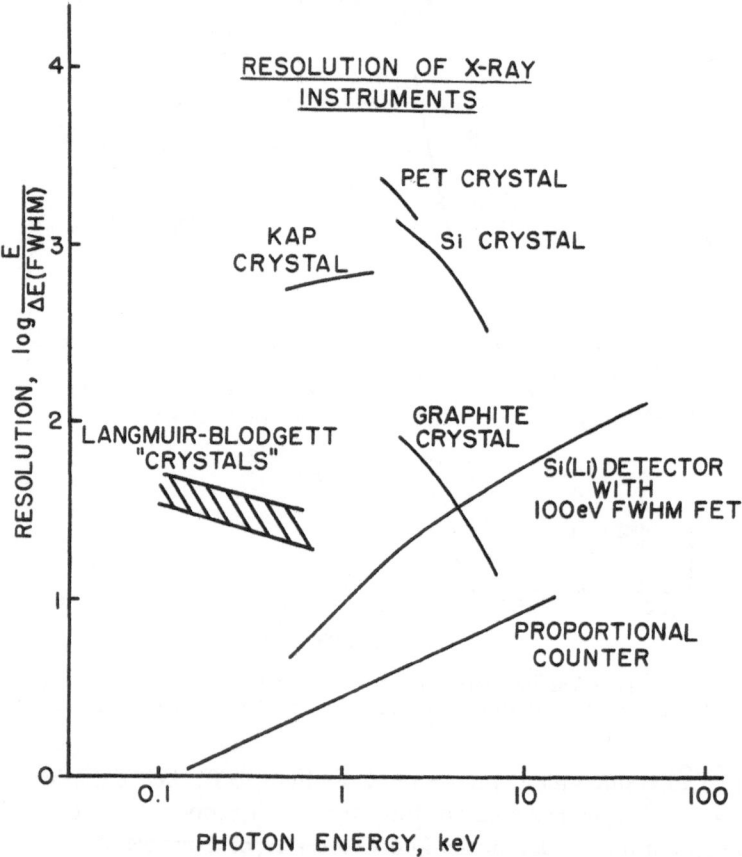

Fig. 3.   Resolution of X-ray instruments.

In this section we have indicated the present status of X-ray spectroscopy, particularly in regard to Sco X-1 and the Cygnus Loop. Also, we have indicated the instrumental developments, and we have briefly indicated the forthcoming satellite X-ray spectroscopy experiments. In Figure 3, we briefly summarize the resolution of various types of spectrometers.

3. Polarimetry

X-ray polarization is expected in any source that is either nonthermal or aspherical. In the case of synchrotron emission, it is well known that linear polarization is observed, that it is independent of energy, and that the polarization pattern within an extended source provides a map of the magnetic field. In the case of a dense aspherical thermal source, polarization of a few percent is expected if there is strong scattering within the source (Angel, 1969b). Recently Pringle and Rees (1972) suggested a model for the pulsating binary X-ray source Cen X-3 in which the emission results from the accretion of matter onto the poles of a rotating magnetic neutron star. If the model is correct, it is expected that the source will exhibit both linear and circular polarization and that the polarization will depend upon the X-ray pulse phase. In this review we will primarily discuss linear polarization, but we note in passing that X-ray circular polarization has been detected by observing the Compton scattering of photons by the polarized d shell electrons in magnetized iron (Goldhaber et al., 1958).

It is, of course, well known that the Crab pulsar NP 0532 exhibits phase-dependent linear polarization in both the radio and optical bands, and it is expected that X-ray polarization will also be observed in this object. In the exploding galaxy M 87, it is known that polarization up to 20% exists in the visible band in some of the 'knots' in the jet. This fact is taken as evidence of synchrotron emission (Hiltner, 1959). If the X-ray emission occurs by the same process, then we might expect similar polarization. Since the average polarization for the entire galaxy is much lower than the above amount, we can only expect to observe polarization with a nonimaging polarimeter if the X-ray emission is primarily confined to one or two 'knots'. One of the favorite models for the compact X-ray source Sco X-1 ascribes the X-ray emission to a high-temperature accretion disk that surrounds a white dwarf member of a contact binary star system. As yet no evidence has been found for binary motion in this source. If the accretion model is correct, then we might expect a few percent X-ray polarization (Angel, 1969b) due to scattering within the disk. Thus, polarimetry may be useful for establishing the true nature of this and other compact X-ray sources.

Finally, it is noted that, in the impact of monodirectional electrons onto matter, linear polarization is observed and that the sign and magnitude of the polarization depend on the photon energy. At low photon energies, the polarization (electric vector) is perpendicular to the direction of electron flow. At photon energies near the upper end of the bremsstrahlung spectrum, the polarization is parallel to the electron flow. This process is known as linear bremsstrahlung and is believed to account for the X-ray polarization observed by Tindo et al. (1971) in solar flares. It is conceivable that a similar process occurs in the flares observed in Sco X-1 and the other highly variable X-ray sources. In this case it is critical to observe the energy dependence of the polarization with a broad-band polarimeter. It is amusing to speculate that the high-energy nonthermal spectral 'tail' observed in Sco X-1 at energies above 40 keV may be produced by the synchrotron process and may exhibit strong polarization. It is clear from the above discussion that polarimetry is essential for the determination of both the structure and X-ray emission mechanisms in a number of X-ray sources.

The techniques that have been used for X-ray polarimetry depend on the angular asymmetry of electron scattering. If the electrons are essentially free (i.e., their binding is small compared to the photon energy), then scattering is observed over a broad energy band, and it has a characteristic dipole form with the axis of the dipole co-aligned with the polarization (electric vector) of the incident polarized photon. If the electrons are bound in a regular crystal lattice, than the scattering (Bragg reflection in this case) is restricted to a narrow energy band given by the mosaic spread of the crystal. The scattering observed with a Bragg angle of 45° is zero if the polarization vector is in the plane of incidence, and it is maximum if it is perpendicular to this plane. In principle, other techniques can be used for polarimetry, but so far none of them have proved as effective as the scattering method. The technique most often considered involves detecting the direction of the primary photoelectron track. Unfortunately, at the energy at which most X-ray sources are most intense (1–20 keV), it is very difficult to detect the direction of an electron track. At energies of 50 keV and higher, this method is probably superior to the scattering method. At much higher energies it will probably be best to determine polarization from the orientation of the plane of the electron pairs formed by a converter in a spark chamber.

All of the X-ray polarization studies that have been made to date have used either Thomson scattering in lithium or beryllium or Bragg reflection at 45° on mosaic graphite crystals. Mosaic graphite is used for polarimetry since it has the highest integrated reflectivity at an energy (2.6 keV) near the maximum in the spectrum of most X-ray sources. The Thomson-scattering polarimeters provide data over a broad band, set at the lower end (about 5 keV in lithium) by photoelectric absorption in the scattering material and at the upper end by the rapid decrease in flux from the X-ray sources. Since the efficiency of all polarimeters is quite low, it is essential to go to great lengths to reduce the non-X-ray background. In the case of Bragg crystal polarimeters, this may be accomplished in part by mounting the crystals on an arc of a parabolic surface so that the reflected X-rays are focused onto a small detector. This results in a substantial reduction in the size of the detector and yields a commensurate reduction in the non-X-ray background event rate. This type of polarimeter has been discussed in detail elsewhere (Weisskopf *et al.*, 1972). The Thomson-scattering polarimeter has been discussed by Angel *et al.* (1969), by Wolff *et al.* (1970), and by Tindo *et al.* (1970). The Thomson-scattering and graphite polarimeters have been flown in several rocket flights by the X-ray astronomy group at Columbia University. This work has culminated in the detection of X-ray polarization from the Crab Nebula as described below. The Thomson-scattering polarimeters constructed by the group at the Lebedev Institute in Moscow have been flown on Intercosmos 1 and 4. These have been used to study solar flares and, as indicated above, this group has detected strong solar flare X-ray polarization. This work has recently been summarized by Tindo *et al.* (1971). Bragg-crystal polarimeters are scheduled to be flown on the UK-5 and OSO-I satellites. The UK-5 instrument also serves as a spectrometer and therefore employs a flat crystal.

The OSO-I instrument uses a parabolic array of mosaic graphite crystals. Two complete polarimeters will be installed on the spacecraft, and their axes will be roughly orthogonal so as to provide both of the Stokes linear polarization parameters simultaneously. The OSO-I instrument is mounted in the bottom of the wheel section of the OSO vehicle with the polarimeter axis parallel to the spin axis of the wheel.

Sources will be selected by pointing the spin axis toward the source; under normal flight conditions, it will be possible to observe a source for several days. The wheel rotation, of course, provides the desired rotation of the polarimeter to help eliminate spurious systematic effects. The predicted performance of the instrument in both first order, 2.6 keV, and second order, 5.2 keV, for several interesting sources is shown in Table I.

TABLE I

Percent minimum detectable polarization at 3-σ, 99 % confidence level

Source	Observing time (days)	2.6 keV only (%)	5.2 keV only (%)	Combined result (%)
Crab Nebula (Tau X-1)	1	1.8	3.8	1.7
Sco X-1	1	0.59	1.1	0.52
Cyg X-1	1	4.4	9.0	4.0
M 87 (Vir X-1)	1	7.6	18	7.3
Nor X-2	1	2.4	3.9	2.0
Sco X-2	1	2.7	3.8	2.2
Crab Primary Pulse	6	7.0	18	7.0
Cen X-3 (each of four 0.25–s bins within pulse)	6	17	6.4	6.1

It has been proposed to fly a focal plane Bragg crystal polarimeter on the HEAO-C spacecraft. The instrument will be located at the focus of the high-efficiency lens, and it can be used for polarization mapping of extended sources.

The nonsolar X-ray polarization observations that have been made so far have been restricted to Sco X-I and the Crab Nebula. It has been shown that polarization of Sco X-1 is less than 20% (Angel et al., 1969). This result is entirely consistent with the view that Sco X-1 is a thermal X-ray source. The determination of whether or not Sco X-1 has a small polarization that might arise from scattering within an accretion disk must await the flights of UK-5 and OSO-I. As indicated above, two rocket experiments that have been performed on the Crab Nebula by Wolff et al. (1970) and Novick et al. (1972) have demonstrated that the X-ray emission from this object is polarized. The magnitude and position angle of the polarization are (15.4±5.2)% and (156 ±10)°, respectively (Novick et al., 1972). This is to be compared to the polarization in the optical of 14% at a position angle of 154° when averaged over the X-ray emitting region, which, for the present purpose, is assumed to be 1′ in radius. The equality of the observed X-ray and optical polarizations and the fact that the spectrum can be represented by a power law confirm the view that the synchrotron process is responsible for the X-ray emission. No X-ray polarization would have been observed if the thermal model of Sartori and Morrison (1967) obtained. Thus, the Crab Nebula is the first X-ray object for which we have confirmed emission mechanism. This observation poses new problems for pulsar theories since it implies that the synchrotron electron spectrum extends to 6×10^4 GeV if the Nebular magnetic field is taken as 10^{-4} G. Since the radiative lifetime for such electrons is a fraction of a year, they must either be continuously generated by the pulsar, or they must be continuously accelerated by magnetic dipole radiation from the pulsar (Gunn and Ostriker, 1971; Rees, 1971) or by some other presently unknown process. Continuous injection of 6×10^4 GeV electrons appears to be impossible since in the high magnetic field (10^{12} G) of the pulsar the electrons will lose their energy by either synchrotron or curvature radiation before they escape from the immediate neighborhood of the pulsar (Ruderman, 1972). Continuous acceleration by magnetic dipole radiation seems to be ruled out by the fact that Landstreet and Angel (1970) failed to find the circular polarization in the optical band that had been predicted by Rees (1971) for this process. Thus, it appears that we must look elsewhere for either an injection or acceleration mechanism.

From the above discussion, it is clear that both spectroscopy and polarimetry are destined to play an important role in X-ray astronomy. However, most of the critically important results must wait until the various satellite instruments are in orbit.

Acknowledgements

The author wishes to thank Dr B. Woodgate for many helpful discussions about X-ray spectroscopy and for having prepared Figure 3. The author is indebted to Dr P. Landecker for having evaluated the OSO-I polarimeter sensitivity given in Table I and to Dr M. C. Weisskopf for having made many helpful comments on this manuscript. Finally, he wishes to acknowledge the enthusiastic work of Mr H. Helava on the objective crystal spectrometer. This work was supported by the National Aeronautics and Space Administration under grants NGR-33-008-102 and NGR-33-008-125. This is Columbia Astrophysics Laboratory Contribution No. 67.

References

Acton, L. W., Catura, R. C., Culhane, J. L., and Fisher, P. C.: 1970, *Astrophys. J. Letters* **161**, L175.
Angel, J. R. P.: 1969a, *Nature* **224**, 160.
Angel, J. R. P.: 1969b, *Astrophys. J.* **158**, 219.
Angel, J. R. P., Novick, R., Vanden Bout, P., and Wolff, R.: 1969, *Phys. Rev. Letters* **22**, 861.
Angel, J. R. P. and Weisskopf, M. C.: 1970, *Astron. J.* **75**, 231.
Argo, H., Evans, D., Bergey, J., and Liefeld, R.: 1972, private communication.
Bleach, R. D., Boldt, E. A., Holt, S. S., Schwartz, D. A., and Serlemitsos, P. J.: 1972, *Astrophys. J.* **171**, 51.
Bleeker, J. A. M., Deerenberg, A. J. M., Yamashita, K., Hayakawa, S., and Tanaka, Y.: 1972, 'Soft X-ray Spectra of the Cygnus Loop and Cyg X-2 in the Energy Range 0.16-6.7 keV', submitted for publication.
Ershov, O. A., Brytov, I. A., and Lukirskii, A. P.: 1967, *Opt. Spectrosc. (U.S.S.R.)* **22**, 66.
Felten, J. E., Rees, M. J., and Adams, T. F.: 1972, 'Transfer Effects on X-Ray Lines in Optically Thick Celestial Sources', submitted for publication.
Goldhaber, M., Grodzins, L., and Sunyar, A. W.: 1958, *Phys. Rev.* **109**, 1015.

Gorenstein, P., Harris, B., Gursky, H., Giacconi, R., Novick, R., and Vanden Bout, P.: 1971, *Science* **172**, 369.

Goulding, F. S. and Stone, Y.: 1970, *Science* **170**, 280.

Griffiths, R. E.: 1972, 'A Further High-Resolution Search for Fe xxv Line Emission from Scorpius X-1', *Astron. Astrophys.* **21**, 97.

Griffiths, R. E., Cooke, B. A., and Pounds, K. A.: 1971, *Nature Phys. Sci.* **229**, 175; corrigendum *ibid.*, **231**, 136.

Gunn, J. E. and Ostriker, J. P.: 1971, *Astrophys. J.* **165**, 523.

Hiltner, W. A.: 1959, *Astrophys. J.* **130**, 340.

Holt, S. S., Boldt, E. A., and Serlemitsos, P. J.: 1969, *Astrophys. J. Letters* **158**, L155.

Holt, S. S., Boldt, E. A., and Serlemitsos, P. J.: 1970, in L. Gratton (ed.), 'Non-Solar X- and Gamma-Ray Astronomy', *IAU Symp.* **37**, 138.

Kestenbaum, H., Angel, J. R. P., and Novick, R.: 1971, *Astrophys. J. Letters* **164**, L87.

Kurtz, D. W., Vanden Bout, P. A., and Angel, J. R. P.: 1972, 'Search for Coronal Line Emission from the Cygnus Loop', submitted for publication.

Landis, D. A., Goulding, F. S., and Jakpevic, P. M.: 1970, *Nucl. Instrum. Meth. (Netherlands)* **87**, 211.

Landstreet, J. D. and Angel, J. R. P.: 1971, *Nature* **230**, 103.

Loh, E. D. and Garmire, G. P.: 1971, *Astrophys. J.* **166**, 301; erratum, *ibid.*, **169**, 447.

Meekins, J. F., Henry, R. C., Fritz, G., Friedman, H., and Byram, E. T.: 1969, *Astrophys. J.* **157**, 197; *Astrophys. J. Letters* **156**, L33.

Novick, R., Weisskopf, M. C., Berthelsdorf, R., Linke, R., and Wolff, R. S.: 1972, *Astrophys. J. Letters* **174**, L1.

Pieuchard, G., Flamand, J., and Laude, J. P.: 1972, 'Recent Developments in the Field of Holo-Gratings', April 21, Instititute of Physics, Spectroscopy Group and Optical Group, Meeting on Diffraction Gratings, Imperial College, London, England, unpublished.

Pringle, J. E. and Rees, M. J.: 1972, *Astron. Astrophys.*, to be published.

Rees, M. J.: 1971, *Nature Phys. Sci.* **230**, 55.

Ruderman, M.: 1972, private communication.

Sartori, L. and Morrison, P.: 1967, *Astrophys. J.* **150**, 385.

Singer, S.: 1972, private communication.

Speer, R. J., Garton, W. R. S., Goldberg, L., Parkinson, W. H., Reeves, E. M., Morgan, J. F., Nicholls, R. W., Jones, T. J. L., Paxton, H. J. B., Shenton, D. B., and Wilson, R.: 1970, *Nature* **226**, 249.

Stevens, J., Garmire, G., and Riegler, G.: 1972, private communication.

Stockman, H. S., Jr., Angel, J. R. P., Novick, R., and Woodgate, B. E.: 1972, to be submitted for publication.

Tindo, I. P., Shuryghin, A. I., Ivanov, V. D., Vasiljev, V. A., Godrnov, D. A., and Komjak, N. I.: 1970, *Kratkie Soobshchenia Po Fisike (U.S.S.R.)*, No. 7, p. 15.

Tindo, I. P., Valnicek, B., Livshits, M. A., and Ivanov, V. D.: 1971, November 15–19, *International Meeting on Solar Activity, Izmiran, U.S.S.R.*

Tucker, W.: 1971, *Science* **172**, 372.

Weisskopf, M. C., Berthelsdorf, R., Epstein, G., Linke, R., Mitchell, D., Novick, R., and Wolff, R. S.: 1972, 'A Graphite Crystal Polarimeter for Stellar X-Ray Astronomy', *Rev. Sci. Instr.*, **43**, 967.

Wolff, R. S., Angel, J. R. P., Novick, R., and Vanden Bout, P.: 1970, *Astrophys. J. Letters* **160**, L21.

Woltjer, L.: 1972, 'Supernova Remnants', *Ann. Rev. Astron. Astrophys.*, to be published.

Womack, E. A., Jr. and Overbeck, J. W.: 1972, private communication.

Yentis, D. J., Angel, J. R. P., Mitchell, D., Novick, R., and Vanden Bout, P.: 1971, in F. Labuhn and R. Lüst, (eds.), 'New Techniques in Space Astronomy', *IAU Symp.* **41**, 145.

PART II

THEORETICAL MODELS FOR COMPACT SOURCES

11. MODELS FOR COMPACT X-RAY SOURCES

E. E. SALPETER

Cornell University, Ithaca, N.Y., U.S.A.

Abstract. A statistical analysis of the UHURU catalogue of X-ray sources leads to the following conclusions. If the weak sources are omitted, there is a strong concentration to low galactic latitude but the absence of a strong background and the presence of some strong sources at low galactic longitude indicate an appreciable number of sources of luminosity $L \sim 10^4 L_\odot$ in the vicinity of the 'nuclear bulge'. This region generally suggests 'stellar population II' and therefore stars of small mass. However, there is some suggestion of a second class of sources, distributed in the galactic plane like 'stellar population I' and suggesting large stellar masses.

There are three possible types of simple X-ray spectra, (a) optically thin bremsstrahlung, (b) black-body spectrum, and (c) power-law spectra. In this talk only theoretical models for type (a) are reviewed, including accretion, rotation and vibration for a white dwarf star and cocoons around a neutron star.

1. Introduction

This is merely the first of three review talks on theoretical models for compact X-ray sources. I will start (Section 2) with a topic which is not strictly theoretical but provides an important input for such models: Over 100 individual X-ray sources (mainly from the UHURU catalogue of sources) are now known and some information about luminosities, lifetimes and stellar populations can already be obtained from statistical data.

A multitude of theoretical models will be discussed and there are even different ways to classify them. Let us start with one type of classification that refers to the observational data rather than models for them, namely the type of X-ray spectral distribution. There are at least three simple kinds of spectra which, in principle, should be readily distinguishable: (a) Optically thin (Bremsstrahlung or 'free-free') thermal emission. The most characteristic feature of such a distribution is that it is 'flat-topped', but at high frequencies it falls off exponentially and the 'Boltzmann factor' determines a temperature T_x. (b) Black-body radiation, characterized by temperature T_x. (c) An inverse power-law spectrum with some index n, characteristic of synchrotron radiation.

The type of X-ray spectrum refers not so much to the primary energy source, but to conditions in the gaseous region where this energy is converted into X-ray emission, and depends on the size of this region. The most exciting speculations concern the recently discovered periodic X-ray sources and the easiest kinds of theoretical models to build for such sources involve small emitting regions, not much larger than the size of a neutron star. Such models will be discussed in the later reviews and usually will predict complicated spectra, but somewhat resembling types (2) and (3) above. Unfortunately, rather little is known directly about the spectra of the most interesting sources. However, for at least one source, Sco-X1, the spectrum is known to be of type (1) – optically thin Bremsstrahlung from regions of size comparable with (or slightly larger than) radii of white dwarf stars. I will

Bradt and Giacconi (eds.), X- and Gamma-Ray Astronomy, 135–142. All Rights Reserved.
Copyright © 1973 by the IAU.

discuss (Section 3) models only for sources of this relatively large size (10^4 to 10^5 km), even though it may turn out that none of the periodic sources are of this size.

2. Statistical Data

At the moment we have no accurate distance-indicators for *individual* X-ray sources, but we shall nevertheless be able to draw some statistical conclusions about the overall spatial distribution of galactic sources. Such conclusions, although preliminary, are useful in at least two ways: (1) If one can establish a rough distance-scale one also has a rough luminosity function and some information on general energy requirements. (2) If one can identify the spatial distribution of galactic X-ray sources with that of one of the stellar populations, one has some indirect indications about the masses of such sources: It seems reasonable to assume that X-ray emission occurs during a relatively short-lived phase in the evolution of a single or binary star. Stars of population II in an active evolutionary phase have relatively small masses, $M \sim M_\odot$. If some class of X-ray sources is distributed with population II it is then likely that the star (or at least one of the two stars in the case of a binary) required to initiate the X-ray phenomenon need not be very massive.

The UHURU catalogue (Giacconi *et al.*, 1972) of X-ray sources lists intensity I (in the 2–6 keV energy range in counts $s^{-1}/840$ cm^2 i.e. units of approximately 1.7×10^{-11} erg $s^{-1}/$cm^2) and angular position for about 125 sources. This list contains some extragalactic sources, but contamination due to them is unimportant for intense, low-galactic latitude sources: We restrict ourselves to sources with $I > 5$ and eliminate a few positively identified extragalactic sources (in the Virgo, Coma and Perseus clusters, in NGC 5128 and 5 sources in the Magellanic Clouds). About 75 sources remain and their distribution in galactic latitude b and longitude l is indicated in Table I. Half the total solid angle away from the galactic plane ($|b| > 30°$) has only about 3 sources and extragalactic sources are indeed unimportant in this list.

TABLE I

The stronger sources in the UHURU catalogue, with numbers in bins of galactic latitude b and longitude l[a]

$b(°)$ l(or $360°-l$)	0 to 1.25	1.25 to 2.5	2.5 to 5	5 to 10	10 to 20	20 to 30	>30	
0–45°	11	7	6	10	4	2	0	40
45–90°	6	3	1	2	2	2	2	18
90–180°	4	2	5	1	3	1	1	17
								75

[a] Note that these numbers do not take into account the variable exposures to the various celestial regions. However, the uniformity of the survey was sufficient to permit the conclusions drawn in this paper.

The sources are highly concentrated to low galactic latitudes (median $|b| \sim 4°$) and one might at first expect the sources to be distributed like extreme population I stars. This extreme assumption would have the following consequences which we shall see are unlikely: The median galactic height of extreme population I (see Figure 1) is $|z| \sim 80$ pc; if we average the data in Table I over galactic longitude l,

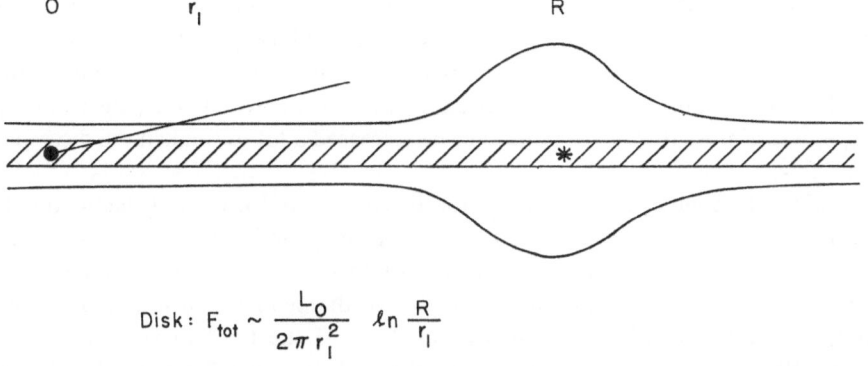

$$\text{Disk}: F_{\text{tot}} \sim \frac{L_0}{2\pi r_1^2} \, \ln \frac{R}{r_1}$$

Fig. 1. A schematic view of the galactic plane (shaded region) occupied by stellar population and the outline of the 'intermediate population II' distribution including the 'nuclear bulge'.

the latitude distribution then gives us information on distances to the brighter sources. Assuming the same absolute luminosity L_X for all sources, the latitude distribution of the 18 brightest sources coupled with the assumption of $|z| \sim 80$ pc would then require a source density of about one source per cylinder of radius 250 pc and the intensity distribution (of the ~ 18 sources with $I > 200$) would require a mean X-ray luminosity of $L_X \sim 100\, L_\odot$ ($\sim 4 \times 10^{35}$ erg s^{-1}). For resolved sources located in the thin galactic plane, the number $N(I)$ with intensity exceeding I should be of form

$$N(I) = b/I \tag{1}$$

with $b \approx 4,000$. A source with $I \sim 200$ is at a distance of $r_1 \sim 1.1$ kpc and the weakest source admitted in our Table I with $I \sim 5$ is at ~ 7 kpc (compared with a distance of ~ 10 kpc to the center of our Galaxy). We have heard from Tananbaum that (for $I \lesssim 200$ where the statistics is good enough so we need not worry about fluctuations) $N(I)$ does not follow Equation (1) at all but approximates $N(I) \propto I^{-0.4}$. The failure to resolve a large number of sources with $I < 200$ would then represent a discrepancy; further, the unresolved weaker sources would contribute to a strong background (highly concentrated to very low latitudes $|b|$): With the radial distance to the galactic center assumed to be $R_G \sim 10$ kpc, this background would have to be $B \sim 3000$ counts per radian (s 840 cm^2), roughly given by

$$B \sim 2b \ln (R_G/r_1). \tag{2}$$

The observed background in the galactic plane reported by Cooke *et al.* (1969) and Gursky (1972) is much smaller than that and Clark mentioned an even smaller value at this Symposium.

Arguments of this kind (Ryter, 1970; Setti and Woltjer, 1970) were already suggestive before the UHURU data and now seem quite convincing. Although we have assumed above a single absolute luminosity, our arguments would not be weakened appreciably if we allowed a more general luminosity function as long as we assumed a median height of $|z| \sim 80$ pc and have to fit the observed latitude dependence. If one assumed a larger median height $|z|$ (but still assumed a disk without any 'bulge'), the calculated distance r_1 would increase in proportion to $|z|$ but (since our value of R_G/r_1 was almost 10) $|z|$ would have to be increased a lot to weaken the discrepancy appreciably.

In reality, the sources in Table I are not distributed uniformly in galactic longitude *l*, but have a concentration within $30°$ or $45°$ of the galactic center (especially for the stronger sources). The angular distribution of this concentration is similar to that of the 'nuclear bulge' (the innermost few kpc in diameter and a few hundred pc in height, see Figure 1) which contains about 80% of the total mass of our Galaxy (Inanen, 1966). Much of the total X-ray output of the whole Galaxy is thus likely to come from about 20 to 50 intrinsically bright sources with a mean luminosity of $L_X \sim 2 \times 10^4 L_\odot \sim 10^{38}$ erg s^{-1}, with a spatial distribution through the Galaxy similar to the overall distribution of ordinary stars. The five sources in the Magellanic Clouds (total mass $\sim 10\%$ of our Galaxy) corroborate this picture (Leong *et al.*, 1971).

The indirect arguments above then suggest that these bright X-ray sources in the 'nuclear bulge' do not require particularly massive stars as progenitors. This argument cannot be made rigorous at the moment because a concentration of massive stars in the 'nuclear bulge' of our *own* galaxy cannot be ruled out with certainty. However, data from other galaxies (and the distribution of neutral hydrogen gas in our own) make such a concentration unlikely.

Although there is probably no physical connection, the bright X-ray sources have a similar luminosity and spatial distribution to planetary nebulae but are less numerous by a factor of about 100. X-ray sources with known, rapid intensity-variations represent an interesting but numerically small subclass of sources. Although one cannot be quantitative at the moment because of the small numbers involved, the angular distribution of the fluctuating sources is strikingly different: No concentration near the 'nuclear bulge' is apparent and the data is consistent with an 'extreme population I' distribution (e.g. with a concentration in spiral arms). It is then quite likely that the rapidly fluctuating sources are associated with massive, young stars.

The distribution of received intensity *I* for all the galactic X-ray sources seems to suggest a break in the distribution. Interstellar absorption of soft X-rays (Seward *et al.*, 1972) gives distance estimates for some of the galactic sources and hence gives absolute luminosities: This data gives firmer evidence for a second group of weaker sources with $L_X \sim (10^{35}$ to $10^{37})$ erg s^{-1} in addition to the strong sources with $L_X \sim 10^{38}$ erg s^{-1} already mentioned. Unfortunately we do not have enough data yet to tell whether the second group is an entirely distinct one with a separate spatial distribution (as is probably the case for the fluctuating sources) or merely the low-luminosity tail of one wide distribution. Incidentally, if there is a separate class of source with 'extreme population I' distribution then it should contribute appreciably to the source counts at the faint end. If the integrated intrinsic luminosity function of the strong sources in the 'nuclear bulge' is of form $N(L_X) \propto L_X^{-\alpha}$, then a comparison with the observed overall $N(I) \propto I^{-0.4}$ shows that α must be less than 0.4.

Some theoretical comments and questions seem in order: The group of bright sources associated with the nuclear bulge probably have a star of mass $M \sim M_\odot$ at the center of each source and their luminosities L_X cluster close to but probably do not exceed $\sim 10^5 L_\odot$. For a given stellar mass M and opacity coefficient κ, there is a critical luminosity

$$L_{\rm crit} \sim 4\pi c G M / \kappa \tag{3}$$

such that radiation pressure would overcome gravity if L_X exceeded $L_{\rm crit}$. With Thomson scattering opacity and $M \sim M_\odot$, $L_{\rm crit}$ is close to the observed upper range of $\sim 10^5 L_\odot$ and this coincidence might be of physical significance. Much of the theoretical discussion will be about compact X-ray sources with $R \lesssim 10^5$ km. This is certainly the case for the rapidly fluctuating sources, but for only a fraction of the brightest sources (near the nuclear bulge) have rapid fluctuations been confirmed so far. Are the remaining bright sources in fact also compact or are these a completely different class, possibly associated with supernovae remnants? For the genuinely compact sources, on the other hand, we do not know their mass distribution nor whether they represent a homogeneous class.

3. Models for Optically Thin Emitters

A large number of rival models for various kinds of galactic, compact X-ray sources have already been proposed and more are likely to follow. When more observational data is available on the X-ray (and optical) spectra, on time variations of intensity and spectrum and on interaction with any companion, hopefully the choice will narrow. At the moment the number of models is large because each model is complex and at least a three-dimensional classification-scheme is needed even to characterize a model qualitatively:

(1) A compact, stable type of star at the center of the emitting region is usually invoked, to provide a gravitational field to contain the emitting gas and/or to provide an energy source. Since the emitting regions have sizes $\ll 10^6$ km, ordinary stars are ruled out, but white dwarfs, neutron stars, black holes (and some further variants) are all possibilities. (2) Some form of primary energy source is required, which could be gravitational energy released by accretion, rotational kinetic energy of the central star, or nuclear energy production or vibrational energy (acting as an intermediary). (3) The primary energy source somehow has to energize the material which eventually emits the X-rays. Here there is even more variety in the 'transmitting agent' (relativistic

particles, low-frequency electromagnetic radiation, compression, direct heating, etc.) and in the type, location and size of the final X-ray emitter (small, optically thick stellar atmosphere or large, dilute optically thin gas cloud or a relativistic plasma emitting non-thermally, etc.).

The emitted X-ray spectrum depends most directly on the choice in (3), but indirectly also on (1) and (2). For a few sources, most notably for Sco-X1, we have fairly good observational evidence for an optically thin thermal Bremsstrahlung spectrum with temperature of order $T_X \sim 10^8$ K from a region typically of size $R \sim (10^4$ to $10^5)$ km. I will briefly review only models relevant for emitting regions of this size. We do not know whether any of the sources with intricate periodic intensity variations (and suggestions of a binary system) are of this type. If one only has to fit luminosity L_X, temperature T_X and size R (and irregular intensity fluctuations over seconds or longer), one finds that most of the models have sufficient parameters to enable a fit. The overall energy requirements for the whole Galaxy are also not too severe: The number of observed sources could be accounted for if, for instance, an appreciable fraction of stars at some stage of their evolution emitted $L_X \sim 10^4$ $L_\odot$ for a few hundred years; the required amount of energy could be supplied by burning $\sim 10^{-4}$ $M_\odot$ of hydrogen or gravitationally by adding a mass of $\sim 10^{-2}$ $M_\odot$ to a white dwarf. I cannot give a *critical* review – each of n^3 differerent models works just fine in the absence of more observational data – but will at least list those models that have already been discussed.

We are considering emitting regions of size comparable with (or slightly larger than) that of typical white dwarfs ($\sim 10^4$ km). The simplest kind of models are then likely to be those which invoke a white dwarf as the central star. Of the various primary energy sources, gravitational energy released by accretion is probably the easiest to visualize (but not the easiest to calculate with, nor the most economical of mass). Accretion has been most popular with champions of neutron stars (Shklovsky, 1967; Shwarzman, 1970; Sofia, 1970), in which case the main emitting region would be much smaller. However, mass-loss from a larger companion-star in a close binary system which gets accreted onto a white dwarf star does give emitting regions of the size we are considering (Cameron and Mook, 1967; Prendergast and Burbidge, 1968). If the accreted atoms moved under free fall conditions and then gave up their energy in just a few collisions, color-temperature of the emitted radiation would be of order $T_X \sim 10^9$ K rather than $\sim 10^8$ K. In reality, the infalling material has to give up angular momentum and flows through a rotating disk of gas, giving up its energy more slowly so that a smaller value of T_X seems reasonable. Rates for angular momentum transfer and accretion rates for a given binary system cannot be calculated in an honest, a priori manner at the moment, but the rate required to give the observed luminosity L_X is of course known. As mentioned, L_X is close to L_{crit} so that radiation pressure may even act as a servomechanism to control the accretion rate to keep L_X/L_{crit} small but not too small.

If sufficient material ($> 10^{-4}$ $M_\odot$) still rich in hydrogen should be accreted onto the surface of a white dwarf, the hydrogen may ignite and the *H–He* conversion could release appreciably more energy than was released in gravitation. The vibration period of a dense white dwarf can be as short as a few seconds (Gribbin, 1971). A model has been suggested (Blumenthal *et al.*, 1972) where kinetic energy of large amplitude vibrations drives shockwaves out from the atmosphere of a white dwarf and heats outer layers to X-ray temperatures. More detailed calculations (Katz, 1972) show that sufficient X-ray emission is obtained from shock-heating only if the vibrations are violent enough to disrupt much of the atmosphere. Nevertheless, hydrogen-burning in intermittent flashes might power violent vibrations and accretion might replenish lost material. Magnetically controlled flares (somewhat analogous to solar flares) could also give non-thermal (inverse power-law) X-ray emission from a white dwarf atmosphere (Blumenthal and Tucker, 1972).

Rapidly rotating white dwarfs with a strong magnetic field emit 'pulsar radiation' which could (via the intermediary of relativistic particles) heat material to X-ray temperatures near the 'speed-of-light circle'. For ordinary, dense white dwarfs the emitting region is then rather large, $R \sim 10^6$ km (Apparao, 1971). However, rapidly rotating condensed stars of mass somewhat greater than the Chandrasekhar limit are also theoretically possible and would have a rotation period of only a few seconds (Ostriker and Bodenheimer, 1968). Pulsar radiation from such an object (Van Horn and Lamb, 1972) could give somewhat smaller emitting regions; on this model periodic intensity variations would reflect the rotation period which *shortens* as the star loses energy (in contrast with ordinary pulsars).

A number of models invoke 'pulsar radiation' derived from the rotational kinetic energy of a neutron star (as in 'ordinary pulsars') as an intermediary to power some emission mechanism. In most models the emitting region is either well inside the 'speed-of-light circle' (Coppi and Treves, 1971) or near it (Apparao, 1972), so that the emission radius R is typically less than 10^4 km. In one model (Davidson *et al.*, 1971) the emission region is a gaseous coccoon held out at distances R much larger than the speed-of-light radius by the Poynting-Robertson effect, so that R is large even for a rapidly rotating neutron star. In this model (and probably also for a rapidly rotating white dwarf) one has a gaseous 'doughnut' in the equatorial plane which favors the ejection of particles or of electromagnetic radiation (as proposed recently by Rees) along the two polar directions to form a double radio source (as reported for Sco X-I). Finally, one model (Jackson, 1972) uses relativistic electrons accelerated by a rotating neutron star to make nonthermal X-rays by the inverse Compton-effect on optical photons coming from a companion B-star.

Although it seems hard to select out 'the best' model at the moment, hopefully the situation will change with further observational data. Spectral details and their theoretical analysis (Felten and Rees, 1972) will help, both for the continuous and the line spectrum. For instance, the present upper limit on the iron-line intensity for Sco X-I already rules out models with large emitting regions ($> 10^5$ km), the positive identification of this line would rule out very small ($< 10^4$ km) emitters. More information on time-variations of intensity and spectrum, especially when there are periodic variations, will also help.

Acknowledgements

I am grateful to J. Katz, J. Ostriker, F. Pacini and M. Rees for interesting discussions. This work was supported in part by National Science Foundation Grant GP-26068.

References

Apparao, K. M.: 1971, *Nature Phys. Sci.* **232**, 153.
Apparao, K. M.: 1972, contributed paper, this symposium.
Blumenthal, G., Cavaliere, A., Rose, W., and Tucker, W.: 1972, *Astrophys. J.* **173**, 213.
Blumenthal, G. R. and Tucker, W. H.: 1972, *Nature* **235**, 97.
Cameron, A. G. and Mook, M.: 1967, *Nature* **215**, 464.
Cooke, B., Griffiths, R., and Pounds, K.: 1969, *Nature* **224**, 134.
Coppi, B. and Treves, A.: 1971, *Astrophys. J.* **167**, L9.
Davidson, K., Pacini, F., and Salpeter, E.: 1971, *Astrophys. J.* **168**, 45.
Felten, J. E. and Rees, M. J.: 1972, *Astron. Astrophys.* **17**, 226.
Giacconi, R., Murray, S., Gursky, H., Kellogg, E., Schreier, E., and Tananbaum, H.: 1972, *Astrophys. J.*, in press.
Gribbin, J.: 1971, *Nature Phys. Sci.* **233**, 19.
Gursky, H.: 1972, private communication.
Inanen, K. A.: 1966, *Astrophys. J.* **143**, 153.
Jackson, J. C.: 1972, unpublished work.
Katz, J.: 1972, unpublished work.
Leong, C., Kellogg, E., Gursky, H., Tananbaum, H., and Giacconi, R.: 1971, *Astrophys. J. Letters* **170**, L67.
Ostriker, J. P. and Bodenheimer, P.: 1968, *Astrophys. J.* **151**, 1089.
Prendergast, K. H. and Burbidge, G. R.: 1968, *Astrophys. J. Letters* **151**, L83.
Ryter, C.: 1970, *Astron. Astrophys.* **9**, 288.
Setti, G. and Woltjer, L.: 1970, *Astrophys. Space Sci.* **9**, 185.
Seward, F., Burginyon, G., Grader, R., and Hill, T.: 1972, *Bull. Astron. Astrophys. Soc.* **4**, 221.
Shklovsky, I.: 1967, *Astrophys. J. Letters* **148**, L1.
Shwarzman, V. F.: 1970, *Astron Zh.* **47**, 824.
Sofia, S.: 1970, *Astrophys. Letters* **5**, 45.
Van Horn, H. M. and Lamb, D. Q.: 1972, to be published.

12. MODELS FOR COMPACT PULSING X-RAY SOURCES

JEREMIAH P. OSTRIKER and KRIS DAVIDSON

Princeton University Observatory, Princeton, N.J., U.S.A.

Abstract. Cen X-3 is probably a neutron star, releasing the infall energy of accreted matter. Sufficient material for accretion will be provided by a conventional stellar wind from its more massive companion star. That star is not likely to rotate synchronously; therefore a 'Roche lobe' analysis of the eclipses is not valid. A 'tidal lobe' analysis allows the neutron star to have a mass of the order of one solar mass. Overflow of the 'Roche lobe' is neither necessary as a source of mass nor probable in view of the observed stellar line widths of the two identified X-ray companions.

The mass flow onto the condensed star is very small in all cases. It is limited, for an object of m solar masses by the Eddington Limiting Luminosity to $\dot{M}_{ac} < 10^{-7.4} \, m(M_\odot \, \text{yr}^{-1})$, which limit applies even if the accreting object contacts or traverses its companion star.

The observed 4.84 s rotation period of the Cen X-3 neutron star is very simply explained as the critical value where a centrifugal barrier regulates the rate of infall to the surface. The X-ray spectrum is understood as blackbody radiation coming from a well-defined area near each magnetic pole of the neutron star.

1. General Remarks: Accretion as a Source of Energy

We restrict attention to those X-ray sources which show significant variations on time scales of one second or less and which do not indicate pulsed radio emission. Since rotation, pulsation, and orbital periods are usually limited to $P \geqslant (G\varrho)^{-1/2}$, we require objects with mean densities greater than $10^7 \, \text{g cm}^{-3}$ – which, by the usual arguments, restricts consideration to degenerate ('white') dwarfs, neutron stars, and black holes.

We focus primarily on the class of X-ray variables whose prototypes are the sources Cen X-3 and Her X-1; these emit fairly regular X-ray pulses with periods of 4.842 and 1.238 s, show regular phase variations, and have apparent eclipses with periods of 2.087 and 1.700 d (Schreier *et al.*, 1972; Tananbaum *et al.*, 1972). According to convincing interpretations of the observations by the above authors, both objects are in nearly circular orbits around stars (which remain unidentified at this time); the mass functions in the two cases are 15 $M_\odot$ and 0.85 $M_\odot$. Estimates for the distances and hence the X-ray luminosities of the variable sources are highly uncertain, but statistical considerations (Salpeter, 1972) and absorption of the lower-energy X-rays seem to indicate typical distances of 1–5 kpc and luminosities of 10^{36}–$10^{38} \, \text{erg s}^{-1}$.

If these guesses are even approximately correct, the periodic sources cannot be degenerate dwarfs: for although very short period regular variations due to either pulsation or rotation are possible in massive, rapidly rotating *cold* dwarfs (Ostriker and Tassoul, 1968; Gribben, 1971), it is easy to show that the high surface temperature required for an X-ray source would produce a distended atmosphere which would lengthen the minimum rotation and pulsation periods to values in excess of one second. Salpeter (1972) has given other arguments against the pulsating white dwarf hypotheses.

The periodic sources may be neutron stars or black holes with modulation due to rotation or to orbital motions; but orbital motions would lead to period changes

Bradt and Giacconi (eds.), X- and Gamma-Ray Astronomy, 143–154. All Rights Reserved.

(Ostriker, 1968) in excess of those allowed by observations, and non-axisymmetric black holes can have only a transient existence (if any). Thus we are led to rotating neutron stars by almost the same chain of arguments which indicated these hypothetical objects as the bases of the radio 'pulsars'.

In a brief assay of the statistical problem, Gott et al., (1970) estimated that perhaps $\frac{1}{15}$ of all pulsars should be found in binary systems. This was probably a severe underestimate, since they neglected the likelihood that mass transfer in a close binary system causes the first supernova explosion to occur in the *less* massive star. Consequently, more than half of the mass is typically retained in the system, and then, as pointed out by van den Heuvel and Heise (1972), the newly created neutron star will usually remain bound to its more massive companion. Yet none of the 60-odd known radio pulsars seems to be a member of a binary system. Evidently neutron stars in binary systems must do something else – and they may become the periodic X-ray systems.

The X-ray luminosities in these systems cannot, however, derive from rotation, as is thought to be the case in the Crab Nebula. If rotation were the energy source, then, since $L_x \sim 10^{37}$ erg s^{-1} for both the Crab Nebula and an X-ray pulsar, we must have

$$(I\Omega \dot{\Omega})_X \approx (I\Omega \dot{\Omega})_{\text{Crab pulsar}} .$$

Therefore, the characteristic time for period increase is

$$\begin{aligned}
\tau_X &\equiv (\Omega/\dot{\Omega})_X \approx (\Omega/\dot{\Omega})_{\text{Crab}} (\Omega_X/\Omega_{\text{Crab}})^2 \\
&\approx (3 \times 10^{-5}) \tau_{\text{Crab}} \\
&\approx 0.06 \text{ yr} ,
\end{aligned}$$

which is orders of magnitude shorter than the observationally permitted rate of period change. In other words, the observed energy output is far more than that available from a relatively slowly rotating neutron star. Internal stored thermal energy is likewise probably insufficient, though not by so large a margin. After considering and discarding other potential sources of energy, we are left with accretion of matter unto the neutron star's surface as the only plausible energy source (for an excellent review of the accretion phenomenon see Zel'dovich and Novikov, 1971). The total energy released per unit time is the rate of proton accretion times $dE(B)/dB$, where $E(B)$ is the neutron binding energy as a function of baryon number.* A one-solar mass neutron star requires an accretion rate of about $10^{-9} M_\odot$ yr$^{-1} = 10^{16.8}$ g s^{-1} to release a total of 10^{37} erg s^{-1}.

2. The Source of the Mass Flow

The required mass flux, while small, is more than can plausibly be expected to accrete from the interstellar medium (Ostriker et al., 1970). However, long before the dis-

---

* Many authors have incorrectly used the surface potential rather than $dE(B)/dB$, neglecting the additional energy liberated as the neutron star adjusts to its increased mass, as well as the correction for the 'redshift' of the escaping energy in the observer's frame. However, the general-readjustment energy is likely to appear in a different form from the initial-infall energy.

covery of the objects considered here, Zel'dovich and Novikov had suggested (1964) that condensed stars would be most easily found as X-ray sources accreting matter from binary companions. Material can become available for accretion by at least two processes: first, many types of *single* stars lose mass at significant rates; in particular, early-type stars are observed to be ejecting mass at rates typically exceeding 10^{-7} $M_\odot$ yr^{-1}, with velocities of the order of 10^3 km s^{-1}. This phenomenon is easily discerned in violet-shifted emission lines of C IV and S IV which appear in the rocket UV spectra by Morton (1967, 1969) and Carruthers (1968), and may occasionally be seen in ground-based spectra (Beals, 1951). Such a star, in a binary system, would provide a source of mass for accretion by its companion. Alternatively, in a very close binary, tidal action may assist in the removal of material from the surface of a star which would not otherwise be losing matter. In an extreme case, if the stellar spin rate is equal to the orbital revolution rate, a star might thus 'overflow its Roche lobe'. We note, however, that there is in fact no reason to suppose that the star must rotate synchronously.

In this connection, consider the upper limit which has recently been suggested by several authors (cf. Wilson, 1972; Leach and Ruffini, 1972; van den Heuvel and Heise, 1972) for the mass of the accreting object in Cen X-2. The chain of argument has three links:

(1) The duration of the X-ray eclipse determines a function of the ratio $a/R_* =$ (binary separation)/(radius of the large eclipsing star) and of the orbital inclination i. (The inclination i is $90°$ if the observer is in the plane of the orbit.)

(2) The sinusoidal Doppler variation of the 4.84 s period, as the X-ray source orbits around the large star, determines the 'mass function', $(M.F.) = M_*^3 \sin^3 i/(m_x + M_*)^2$, where m_x is the mass of the X-ray source and M_* is the mass of the larger companion.

(3) The requirement that m_x must not remove mass from M_* too quickly by tidal forces, *under the assumption that M_* rotates synchronously with the orbital period* – i.e., the larger star must not be larger than its Roche lobe – determines the maximum acceptable value of the mass ratio m_x/M_* as a function of a/R_*. This, with the mass function, implies a maximum possible m_x for a given inclination i.

The lower curve in Figure 1 shows θ_0, the angle in the orbital plane subtended by the larger Roche lobe as seen from m_x, as a function of M_*/m_x. The upper curve shows θ_0 for a tidal lobe, in the case where the larger star is *not rotating* and the Roche potentials have little significance (see Jeans, 1919 and 1928). For a given inclination i, θ_0 may be estimated from the apparent eclipse angle θ, and then the minimum value of M_*/m_x may be found from Figure 1. It is clear that larger values of m_x are permitted in the non-rotating case; and as a rule, early-type stars in close binary systems do *not* rotate synchronously (van den Heuvel, 1970), probably because tidal forces take longer to operate than the stellar lifetimes. We also note that the two known or probable companions of X-ray sources which have been spectroscopically studied, HDE 226868 (Bolton, 1971; Webster and Murdin, 1972) and HD 77581 (Hiltner et al., 1972), do not show lines wide enough to indicate synchronous rotation. Therefore, slow rotation seems to be appropriate.

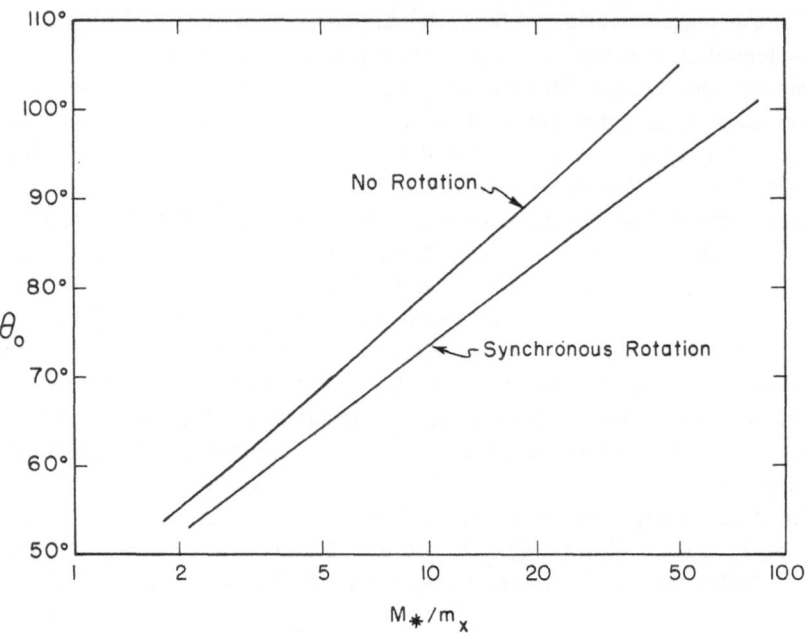

Fig. 1.   Angle in orbital plane eclipsed by larger tidal lobe, as seen from smaller companion,
for various mass ratios.

According to Schreier *et al.* (1972), Cen X-3 is totally eclipsed for 0.488 d during
the orbital period of 2.087 d, indicating $\theta \approx 84°$. Following Wilson (1972), one might
increase this to about 96° by including the transition periods at the beginning and end
of each eclipse, but these are evidently caused by an extended atmosphere, which may
be moving dynamically outside the critical tidal lobe. Therefore, adopting the smaller
value and the observed mass function, M.F. = 15 $M_\odot$ (Schreier *et al.*, 1972), one finds

TABLE I

Maximum masses in Cen X-3

i	θ_0	R^*/r	d_{crit}/r	$M^*/M_\odot$	$m_x/M_\odot$
90°	84°.5	0.67	0.21	17.2	1.24
80°	86°.4	0.68	0.20	17.8	1.13
70°	91°.9	0.72	0.17	19.8	0.90
60°	100°.3	0.77	0.14	24.3	0.65
50°	110°.9	0.82	0.10	34.2	0.43

i = inclination of orbit;
θ_0 = angle eclipsed, in plane of orbit;
r = separation of the two components;
R^* = average radius of larger component;
d_{crit} = distance from X-ray source to the cusp of the larger star's tidal lobe;
m_x = maximum mass of X-ray source;
M^* = maximum mass of larger component.

the maximum acceptable masses listed in Table I, for various inclinations. Unless the inclination is small (in which case M_* would be surprisingly large), m_x may have a very conventional neutron-star mass. In the case of Her X-1, where the mass function is only 0.85 $M_\odot$ and θ is 60° or less (Tananbaum *et al.*, 1972), plausible values (e.g., $M_* \sim 2\ M_\odot$, $m_x \sim 1\ M_\odot$) are possible if i is not much less than 90°.

While not excluding the possibility that tidal forces may drive some mass outflow, here we shall investigate models in which the material to be accreted appears in the form of a stellar wind.

3. Accretion Rate and Luminosity

Consider a neutron star of mass m_x, travelling at velocity v_{rel} through a gas of density ϱ. if v_{rel} is much larger than the sound speed, the gas flows in the manner described by Bondi and Hoyle (1944). If some gas initially flows within a cylinder of radius a_{accr} $\approx 2Gm_x/v_{rel}^2$, then it loses enough energy in passing through the tail shock to fall back towards the star (see Figure 2). The complicated flow near the rotating magnetic neutron star will be discussed in the next section. The accretion rate is then

$$\dot{M}_{accr} = \pi\zeta a_{accr}^2 v_{rel}\varrho, \tag{1}$$

where ζ, which is less than unity, is a factor correcting for the repulsion of infalling material by radiation pressure. Radiation pressure, acting on electrons, provides

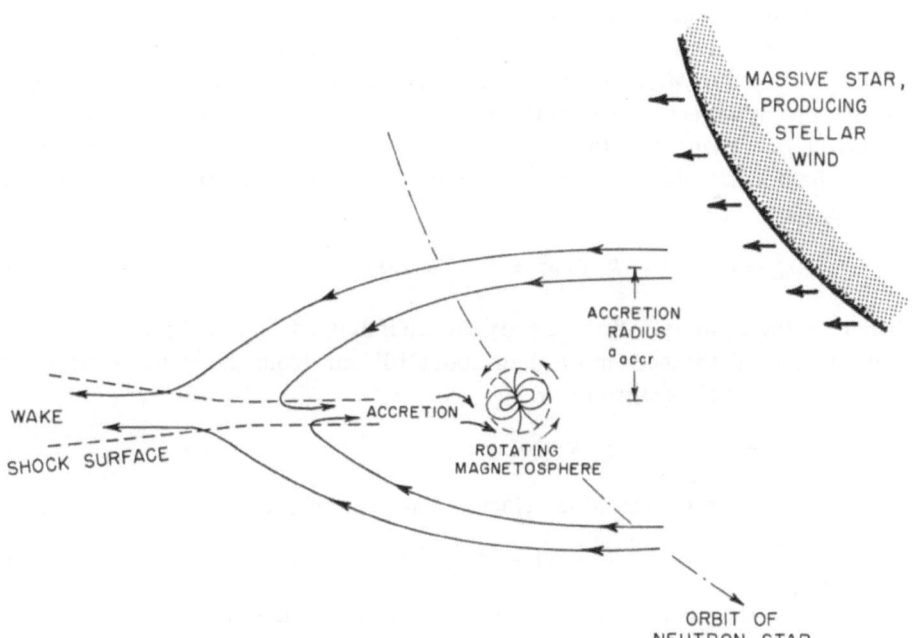

Fig. 2. Streamlines of stellar-wind material, in frame of an accreting neutron star. Relative dimensions are not to scale.

a $1/r^2$ repulsive force which effectively reduces the gravity of the central object:

$$\zeta = [1 - 10^{-4.8} (1 + X) (L_x/L_\odot)/(m_x/M_\odot)]^2 , \tag{2}$$

where X is the fractional abundance of hydrogen in the gas and L_x is the luminosity of the accreting object (Eddington, 1926). Equation (2), of course, is valid only if the radiation is isotropic, the surrounding medium optically thin, and the radiation L_x not so hard that the Thomson cross-section is appreciably reduced ($h\nu < 0.5$ MeV).

The total luminosity resulting from accretion is, as mentioned earlier,

$$L_x = (\dot{M}_{accr}/m_H) [dE(B)/dB] \approx (10^{46} \text{ erg s}^{-1}) (\dot{M}_{accr}/M_\odot \text{yr}^{-1}), \tag{3}$$

where m_H is atomic mass unit. We see from Equations (1), (2), and (3) that radiation pressure acts to limit the luminosity and accretion rate:

$$L_x < L_{crit} = (10^{4.8} L_\odot) (m_x/M_\odot)/(1 + X), \tag{4}$$

$$\dot{M}_{accr} < \dot{M}_{crit} = (10^{18} \text{ g s}^{-1}) (m_x/M_\odot)$$
$$= (10^{-7.4} M_\odot \text{yr}^{-1}) (m_x/M_\odot), \tag{5}$$

which is probably low enough to preclude significant change of the mass m_x during the lifetime of a massive stellar system. Note that the limiting process makes the luminosity depend only very weakly on the details of the accretion mechanism as the limiting luminosity is approached.

We now consider the nature of the stellar wind. According to Lucy and Solomon (1970), the mass flux $\dot{M}_w$ and general characteristics of the flow are established by the time the sonic point is reached. At this point, which occurs not far ($\sim 10^{10.5}$ cm) from the surface of a giant star, the velocity is small ($v_s \approx 20$ km s^{-1}). In the supersonic regime, the flow simulates pressure-free particles moving in an inverse-square repulsive force field. Then,

$$v_w^2 = v_s^2 + (1 - R_s/r) v_\infty^2 \approx (1 - R_*/r) v_\infty^2 , \tag{6}$$

where v_∞ is the asymptotic flow velocity toward infinity. The second form of Equation (6) is suitable at distances more than about 10^{11} cm from the stellar surface. The orbital velocity of the neutron star,

$$v_x^2 = GM_*/r = v_{esc}^2 R_*/2r , \tag{7}$$

must be included to get the wind velocity relative to the neutron star:

$$v_{rel}^2 = v_x^2 + v_w^2 = [1 - (1 - v_{esc}^2/2v_\infty^2) (R_*/r)] v_\infty^2 . \tag{8}$$

The density in the stellar wind is, from the continuity equation,

$$\varrho = \frac{\dot{M}_w}{4\pi r^2 v_w} = \frac{(\dot{M}_w/4\pi v_\infty)}{r^2 (1 - R_*/r)^{1/2}} . \tag{9}$$

Combining the above equations gives

$$\dot{M}_{\text{accr}} = \dot{M}_w \frac{4\,(m_x/M_*)^2\,\varepsilon^2}{(x-1)^{1/2}\,(x-1+\varepsilon)^{3/2}} \left[1 - \frac{L_x}{L_{\text{crit}}}\right]^2, \tag{10}$$

where $\varepsilon = v_{\text{esc}}^2/2v_\infty^2$ and $x = r/R_*$.

We may now construct a few exemplary quantities to show what Cen X-3 might be like. Suppose that $m_x = 1\ M_\odot$, $M_* = 17\ M_\odot$, $R_* = 9 \times 10^{11}$ cm, and $r = 1.3 \times 10^{12}$ cm, consistent with observations and with Table I; then $x = 1.4$ and $v_{\text{esc}} = 500$ km s^{-1}. Suppose that $L_x = 10^{37}$ erg s$^{-1} = 0.1\ \dot{M}_{\text{accr}}\,c^2$ (only part of this may be observable; see Section 4 below); so we require $\dot{M}_{\text{accr}} = 10^{17}$ g s^{-1} while $\zeta \doteq (1 - L_x/L_{\text{crit}})^2 = 0.8$. Then, necessary values of the stellar mass-loss rate M_w, according to Equation (10), are shown in Table II for several values of v_∞. These rates are of the order of 10^{-6} $M_\odot$ yr^{-1} – not at all exorbitant, especially if tidal forces assist the ejection of material from the massive star.

TABLE II

Necessary stellar mass-loss rates
(If the *total* X-ray production in
Cen X-3 is 10^{37} ergs s^{-1})

v_∞ (km s^{-1})	ε	M_w (g s^{-1})	($M_\odot$ yr^{-1})
500	0.5	2.0×10^{19}	3×10^{-7}
1000	0.125	1.4×10^{20}	2×10^{-6}
1500	0.056	5.5×10^{20}	9×10^{-6}

It is also relevant that, since the X-ray source is rather close to the surface of the larger star, the X-ray flux, incident from above, may perturb the stellar atmosphere in a small region around the sub-neutronstar point. Some of the lower-energy X-rays are likely to be absorbed in or above the photosphere, thereby heating the region and provoking additional mass-ejection.

4. Behavior of Gas near the Accreting Object

The accreted gas falls toward the neutron star within a narrow, conical tail shock (see Figure 2) until it reaches a rather small radius. Then, if the neutron star has a dipole-like magnetic field, the infalling material must encounter a region where its pressure is exceeded by magnetic pressure: the magnetosphere (see Figure 3). Within this region, matter is constrained to follow the co-rotating field lines. Gas is likely to collect around the boundary of the magnetosphere, whose size is readily estimable.

Denote the radius of the magnetosphere by R_m; while R_0 is the radius of the neutron star, $R_G = Gm_x/c^2$ is half the Schwarzschild radius of the neutron star, and $R_c = c/\Omega$ is the radius of the speed-of-light cylinder where the angular rotation speed is Ω. If B_0 is the surface magnetic field, then the field at the magnetopause is approximately

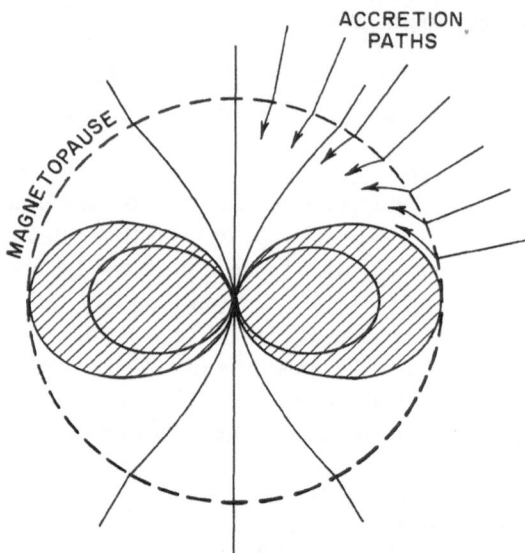

Fig. 3. A simplified dipole magnetosphere around an accreting neutron star.

$B_m = B_0 R_0^3 / R_m^3$ and the magnetic pressure is $B_m^2/8\pi$. We approximate the gas pressure at R_m by a rampressure $\varrho v_{\text{infall}}^2$, where $v_{\text{infall}}^2 \approx R_G c^2 / R_m$. Then (ram-pressure) $\approx$ $\approx \dot{M}_{\text{accr}} v_{\text{infall}} / 4\pi R_m^2 = (\dot{M}_{\text{accr}} c / 4\pi) R_G^{1/2} R_m^{-5/2}$. Equating magnetic and gas pressures, we find

$$R_m \approx \left[\frac{B_0^2 R_0^6}{2 \dot{M}_{\text{accr}} c R_G^{1/2}} \right]^{2/7} = (3 \times 10^8 \text{ cm}) \left(\frac{B_0}{10^{12} \text{ G}} \right)^{4/7} \left(\frac{R_0}{10 \text{ km}} \right)^{12/7} \times$$
$$\times \left(\frac{\dot{M}_{\text{accr}}}{10^{17} \text{ g s}^{-1}} \right)^{-2/7} \left(\frac{m_x}{M_\odot} \right)^{-1/7} \quad (11)$$

If Cen X-3 has a surface field of the order of 10^{12} G (like the pulsars), and $R_0 \approx 14$ km, we expect $R_m \approx 5 \times 10^8$ cm $\ll R_c$.

We now preface the spin-angular momentum problem by noting a suggestive coincidence. The radius of a synchronous orbit, where the gravitational orbital velocity would be in co-rotation with the spinning neutron star, is

$$R_{\text{sync}} = R_G^{1/3} R_c^{2/3}$$
$$= (1.5 \times 10^8 \text{ cm}) (m_x/M_\odot)^{1/3} (P/1 \text{ s})^{2/3}, \quad (12)$$

where P is the period of rotation. For Cen X-3 $(P = 4.84 \text{ s})$, $R_{\text{sync}} \approx 4.3 \times 10^8$ cm, which is very close to R_m. This is suggestive because it means that *if Cen X-3 were to rotate faster, accretion to the surface would become impossible* due to the centrifugal barrier in the co-rotating magnetosphere.

Now consider the transfer of angular momentum near the accreting object. Infalling matter, as it reaches the magnetosphere, is accelerated until is co-rotates. This process removes angular momentum from the spinning neutron star. If the matter then falls to the surface, the angular momentum is replaced and no spin-down results; but if the rotation is so fast that complete accretion is prevented by centrifugal forces, then this form of drag serves to slow the rotation. The torque may be roughly $B_m^2 R_m^2 = = B_0^2 R_0^6 R_m^{-3}$, which is larger than the vacuum pulsar-radiation torque by a factor of the order of $(R_c/R_m)^3$. This would be sufficient to appreciably alter the rotation of Cen X-3 in only a few hundred years. Evidently, such braking provides a suitable mechanism for slowing the spin until $R_m \sim R_{\text{sync}}$.

Another mechanism prevents the rotation from decreasing beyond that point. The pattern of accretion shown in Figure 2 is unlikely to have perfect symmetry; since the density of the stellar wind decreases outward while its speed increases, we might expect more gas to be accreted on that side of the flow pattern which faces the larger star. Consequently, if all of the accreted material were to reach the surface of the neutron star without somehow losing angular momentum, it would exert a torque whose order of magnitude is $(\dot{M}_{\text{accr}} v_{\text{rel}} a_{\text{accr}}^2 / r)$. This would alter the spin of Cen X-3 in 10^4 or 10^5 yr.

Thus we are led to a picture in which the rotation rate of the neutron star is held near an equilibrium value, such that the centrifugal barrier acts as a valve regulating the flow of angular momentum to the surface. Hot gas may collect around the magnetopause, occasionally escaping to carry away excess angular momentum. The rotation rate is determined chiefly by the surface magnetic field and the accretion rate. In equilibrium the period $P \propto (\dot{M}_{\text{accr}})^{3/7}$ may be expected to decrease as the accretion rate increases.

Actual production of the X-ray luminosity remains to be discussed. The efficiency of energy release mentioned in Section 1 probably involves two distinct stages: (1) infalling material releases its kinetic energy, about $(R_G/R_0) (\dot{M}_{\text{accr}} c^2/2) \approx 0.05 \, \dot{M}_{\text{accr}} c^2$, as it strikes the surface; and (2) a comparable amount of energy is released as the neutron star adjusts internally to its increasing mass. The latter process is like an internal energy source, which probably serves to maintain an effective surface temperature near 8×10^6 K that emits $\sim 10^{37}$ erg s^{-1} in a mildly directional fashion. The immediate process, which occurs in a localized area of the surface, may account for the observed X-ray pulses. Figure 3 shows schematically how the infalling material must be funneled along the field lines toward the magnetic polar regions. It is easy to estimate the radius of each polar funnel at the surface, by tracing the critical field lines shown in Figure 3 (assuming an approximately-dipolar field); the result is nearly

$$r_{\text{pole}} \approx R_0^{3/2} R_m^{-1/2} \approx 0.8 \text{ km}. \tag{13}$$

The situation at the base of the funnel is shown in Figure 4. Since the electron density in the funnel approaches 10^{21} cm^{-3} near the surface of the neutron star, the gas is optically thick to ordinary electron scattering (though the strong magnetic field must cause the opacity and the behavior of the gas to be very complicated). Radiation pressure may cause the infalling matter to release its kinetic energy over a distance comparable to r_{pole}. The radiation will diffuse out through a cylindrical surface with height $\approx$ radius $= r_{\text{pole}}$. Supposing this surface to radiate like a black body, its temperature T_x may be found from

$$2\pi r_{\text{pole}}^2 \sigma T_x^4 \approx L_{x,\text{pole}}. \qquad (14)$$

If $r_{\text{pole}} = 0.8$ km and $L_{x,\text{pole}} = 10^{36.5}$ erg s^{-1}, we find $T_x \approx 3.4 \times 10^7$ K, which is in excellent agreement with the observed spectrum (Giacconi *et al.*, 1971) – although admittedly, we have neglected to correct for the 10% gravitational redshift!

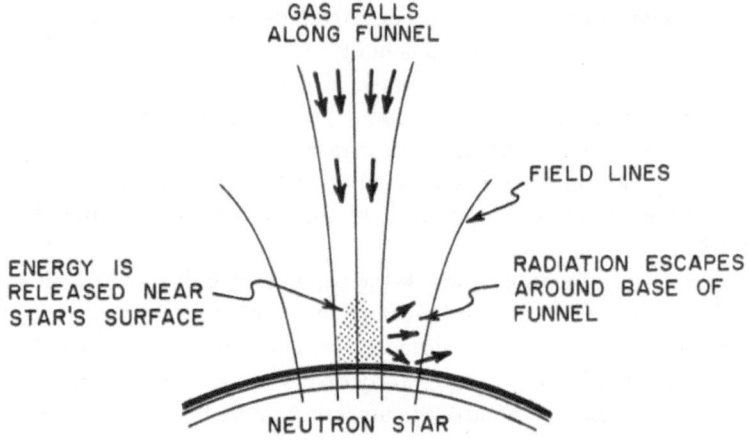

Fig. 4.   Infalling material striking the surface near one magnetic pole of a neutron star.

As for the pulses, Figure 4 shows that the escaping radiation tends to move sideways relative to the funnel – favoring distant observers who are not far from the neutron star's magnetic-equatorial plane. However, since the emitting gas is moving downward at nearly a third the speed of light, there is a tendency for the radiation to move *downward*, to be reflected from the surface of the neutron star and then move upward in the form of a broad, hollow cone. If the neutron star has an extended 'atmosphere' of high-energy particles, refraction effects could narrow the cone. Hence, an observer may see pulses from the above model, and in any case will see a doubly periodic X-ray luminosity as the two magnetic poles rotate through the line of sight.

5. Conclusion

This relatively simple stellar wind driven accretion model appears to satisfactorily account, in a quantitative but not precise way, for most of the observed features of Cen X-3; in particular the luminosity, spectrum, and period of the pulsed X-rays are calculated with reasonable accuracy. The non-pulsed X-ray component described by Tananbaum *et al.* (1972) is most likely radiation which has been scattered off electrons in the stellar wind, since the scattering optical depth outside the orbit of the neutron stars is of the order of 0.1 for the stellar wind described in Section 3.

We have not explained the pulse shape in detail, nor do we propose a reason for the 'on' and 'off' modulation period of about thirty days (Tananbaum *et al.*, 1972), which may have something to do with the outer layers of the massive star and its rate of mass ejection.

The total number of X-ray sources of the type discussed is small even if it is several times larger than discovered so far by the UHURU survey, due to radiation beaming effects, and the consequent total galactic X-ray luminosity very small compared to $10^{10} L_\odot$ – the galactic luminosity. This new class of objects may prove very important, however, as astrophysical laboratories to study high density objects. In particular when the companion stars are discovered they will presumably also be found to be single line spectroscopic binaries with measurable mass functions. Then a determination of the inclination of the system (possible in principle by several means) will allow a direct experimental determination of a neutron star mass. If, further, we were lucky enough to discover a low mass third body orbiting the close binary as a distant visual companion, a direct test of the scaler tensor theory would be possible. This follows since Kepler's laws are not satisfied in the scalar tensor theory, the discrepancy for the neutron star being of order $R_g/R_\odot \sim 0.1$. The reader may invent at leisure other experiments possible in a system involving a clock in a deep potential well in nearly grazing orbit about a massive star.

Acknowledgements

The authors are pleased to acknowledge useful conversations with A. Hiltner, M. Rees, E. E. Salpeter, and many others. J.P.O. gratefully acknowledges the support he has received as an Alfred P. Sloan Fellow, 1970–72, and K.D. the support from National Aeronautics and Space Administration grant NGL-31-001-007.

References

Beals, C. S.: 1951, *Publ. Dom. Astrophys. Obs.* **9**, 1.
Bolton, C.: 1971, *Bull. Am. Astron. Soc.* **3**, No. 4.
Bondi, H. and Hoyle, F.: 1944, *Monthly Notices Roy. Astron. Soc.* **104**, 273.
Carruthers, G. R.: 1968, *Astrophys. J.* **151**, 269.
Eddington, A.: 1926, *The Internal Constitution of the Stars*, (reprinted by Dover, New York, 1959).
Gott, J. R., Gunn, J. E., and Ostriker, J. P.: 1970, *Astrophys. J.* **160**, L91.
Gribben, J.: 1971, *Nature* **233**, 19.
Hiltner, W. A., Werner, J., and Osmer, P.: 1972, *Astrophys. J. Letters*, (in press).
Jeans, J. H.: 1919, *Problems of Cosmogony and Stellar Dynamics*, Cambridge Univ. Press.
Jeans, J. H.: 1928, *Astronomy and Cosmogony*, Cambridge Univ. Press.
Leach, R. and Ruffini, R.: 1972, (preprint).
Lucy, L. B. and Solomon, P. M.: 1970, *Astrophys. J.* **159**, 879.
Morton, D. C.: 1967, *Astrophys. J.* **147**, 1017.
Morton, D. C.: 1969, *Astron. Space Sci.* **3**, 117.
Ostriker, J. P.: 1968, *Nature* **217**, 1227.
Ostriker, J. P., Rees, M., and Silk, J.: 1970, *Astrophys. Letters* **6**, 179.
Ostriker, J. P. and Tassoul, J.-L.: 1968, *Nature* **219**, 577.
Salpeter, E. E.: 1972, this volume, p. 135.
Schreier, E., Levinson, R., Gursky, H., Kellogg, E., Tananbaum, H., and Giacconi, R.: 1972, *Astrophys. J.* **172**, L79.

Tanabaum, H., Gursky, H., Kellogg, E. M., Levinson, R., Schreier, E., and Giacconi, R.: 1972, *Astrophys. J. Letters*, (in press).
van den Heuvel, E. P. J.: 1970, in A. Slettebak (ed.), *Stellar Rotation*, D. Reidel Publ. Co., Dordrecht-Holland, p. 178.
van den Heuvel, E. P. J. and Heise, J.: 1972, *Nature*, (in press).
Webster, B. L. and Murdin, P.: 1972, *Nature* **235**, 37.
Wilson, R. E.: 1972, *Astrophys. J.* **174**, L27.
Zel'dovich, Ya. B. and Novikov, I. D.: 1964, *Sov. Phys. Dokl.* **9**, 246.
Zel'dovich, Ya. B. and Novikov, I. D.: 1971, *Relativistic Astrophysics*, Vol. 1, Univ. of Chicago Press, Chicago, Chap. 13.

13. BLACK HOLES IN BINARY SYSTEMS: OBSERVATIONAL APPEARANCES*

N. I. SHAKURA

Sternberg Astronomical Institute, Moscow, U.S.S.R.

and

R. A. SUNYAEV

Institute of Applied Mathematics, U.S.S.R. Academy of Sciences, Moscow, U.S.S.R.

Abstract. The outward transfer of angular momentum of accreting matter can lead to the formation of a disk around the black hole. The structure and radiation spectrum of the disk depends, in the main, on the rate of matter inflow $\dot{M}$ into the disk at its external boundary. Dependence on the efficiency of mechanisms of angular momentum transport (connected with the magnetic field and turbulence) is weaker. If $\dot{M} = 10^{-9}$–3×10^{-8} M$_\odot$/yr, the disk around the black hole is a powerful source of X-radiation with $h\nu \sim 1$–10 keV and luminosity $L \sim 10^{37}$–10^{38} erg s^{-1}. If the flux of the accreting matter decreases, the effective temperature of radiation and the luminosity will drop. At the same time when $\dot{M} > 10^{-9}$ M$_\odot$ yr^{-1}, the optical luminosity of the disk exceeds the solar one. The main contribution to the optical luminosity of the black hole is due to the re-radiation of that part of the X-ray and ultraviolet energy which is initially produced in the central high temperature regions of the disk and which is then absorbed by the low temperature outer regions. The optical radiation spectrum of such objects must be saturated by the broad emission recombination and resonance lines. Variability is connected with the character of the motion of the black hole and the gas flow in binary systems and possibly with eclipses. For well defined conditions, the hard radiation can evaporate the gas. This can counteract the matter inflow into the disk and lead to autoregulation of the accretion.

If $\dot{M} \gg 3 \times 10^{-8}$ $(M/M_\odot)$ M$_\odot$ yr^{-1}, the luminosity of the disk around the black hole is stabilized at the critical level of $L_{\mathrm{cr}} = 10^{38}$ $(M/M_\odot)$ erg s^{-1}. A small fraction of the accreting matter falls under the gravitational radius whereas the major part of it flows out with high velocities from the central regions of the disk. The outflowing matter is opaque to the disk radiation and completely transforms its spectrum. As a consequence, a black hole in a supercritical regime of accretion can appear as a bright star with a strong outflow of matter.

A black hole (collapsar) does not radiate either electromagnetic or gravitational waves (Zel'dovich and Novikov, 1967). Therefore, a black hole can be observed only through its gravitational influence on a neighboring star or on the ambient gas medium, since the gas must be accreted with great energy release (Salpeter, 1964; Zel'dovich, 1964).

In many papers, it has been proposed to search for collapsars in binary systems. It is often considered that a collapsar should appear as a 'black' object which emits no appreciable radiation. In this paper, it is pointed out that the outflow of matter from the surface of the visible component and its accretion onto the black hole should lead to observable emission. In systems with outflow of matter $dM/dt = \dot{M} \gtrsim 10^{-12}$ M$_\odot$ yr^{-1}, the total luminosity of the disk around the black hole formed by the accreting matter can be comparable and even exceed the luminosity of the visible component. In the typical case, the main part of this luminosity is in the range of $h\nu \sim 100$–10^4 eV. However, as will be shown below, the optical and ultraviolet luminosities (responsible for

---

* This paper was presented by J. E. Pringle, in lieu of an invited talk by Ya. B. Zel'dovich. This paper is similar in form to the introductory part of a paper to be published in full in *Astron. Astrophys.*

Bradt and Giacconi (eds.), X- and Gamma-Ray Astronomy, 155–164. All Rights Reserved.
Copyright © 1973 by the IAU.

the formation of a Strömgren region) are also high. Therefore, it is entirely possible that black holes are among the optical objects, soft X-ray sources and the harder X-ray sources now being intensively investigated. The radiation connected with accretion on black holes in binary systems has, in fact, distinctive features. However, they are not so astonishing as is usually assumed; thus, black holes can be hidden among the known objects. A collapsar in a binary system will be a 'black' object only if the system is remote with a weak stellar wind from the visible component.

Up to 50% of the stars are in binary systems (Martynov, 1971). A sufficiently massive $(M > 2\,M_\odot)$ star, being part of a binary system is able to evolve up to the moment when it loses stability and collapses*. In this case, it is possible that an appreciable number of binary systems will not be destroyed and the stars will remain physically bound. These statements are, of course, controversial. However, if we are reminded that the total number of stars with $M > 2\,M_\odot$ which have existed in the Galaxy is of the order of 10^9 (Zel'dovich and Novikov, 1967), then it becomes reasonable to assume that the number of binary systems including a black hole can be very large, up to 10^6–10^8.

The outflow of matter from the surface of the visible companion – the stellar wind – is one of the main properties of stars. Depending on the type of star, the rate of mass loss ranges from $2 \times 10^{-14}\,M_\odot\,\mathrm{yr}^{-1}$ for the Sun up to $10^{-5}\,M_\odot\,\mathrm{yr}^{-1}$ for the nuclei of the planetary nebulae, Wolf-Rayet stars, MI supergiants and O stars of the main sequence (Pottash, 1970). In binary systems, an additional strong matter outflow connected with the Roche limiting surface is possible. At a definite stage of evolution, for example after leaving the main sequence, the star begins to increase in size so that, after filling the Roche cavity, an intensive outflow of matter occurs mostly through the inner Lagrangian point (Martynov, 1971).

What will be the consequences of a black hole in a binary system with a visible companion which has strong matter outflow? Some fraction of the matter flowing out from the normal star must come into the sphere of influence of the gravitational field of the black hole, accrete onto it and as the final result, fall under its gravitational radius (Figure 1). In a regime of free radial falling (initially the matter was at rest and there was no magnetic field) the cold matter accretes onto the black hole without any energy release and noticeable observational appearance (Zel'dovich and Novikov, 1967). However, in a binary system, the matter flowing out from the normal star and falling onto the black hole has considerable angular momentum relative to the latter, which prevents the free falling of the matter. At some distance from the black hole, centrifugal forces are comparable to the gravitational forces and the matter begins to rotate in circular orbits. The matter is able to approach the gravitational radius only if there exists an effective transport mechanism of angular momentum outward.

The magnetic field, which must exist in the matter inflowing into the disk, and turbulent motions of the matter, give rise to the transfer of the angular momentum outward. The efficiency of the mechanism of angular momentum transport is charac-

---

* It is possible that M_{min} considerably exceeds $2\,M_\odot$ (Zel'dovich and Novikov, 1967).

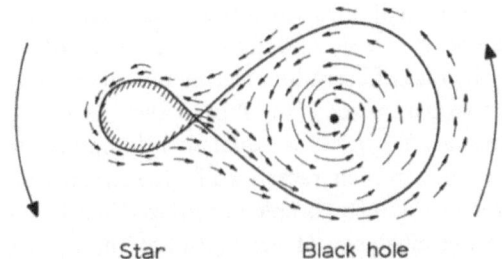

Star   Black hole

Fig. 1a-b. Two regimes of matter captured by a collapsar; (a) the normal companion fills up its Roche cavity, and the gas outflow goes mainly through the inner Lagrangian point; (b) the companion's size is much less than the Roche cavity; the outflow is a stellar wind. The matter loses part of its kinetic energy in the shock wave; there after the gravitational capture of accreting matter becomes possible.

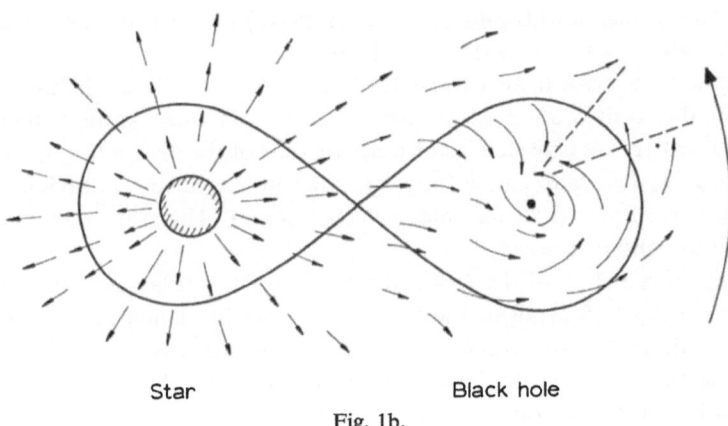

Star   Black hole

Fig. 1b.

terized by the parameter

$$\alpha = \frac{v_t}{v_s} + \frac{H^2}{4\pi\varrho v_s^2}$$

where

$$\frac{\varrho v_s^2}{2} = \frac{3}{2}\frac{\varrho kT}{m_p} + \varepsilon_r$$

is the matter thermal energy density, ε_r is the energy density of radiation, v_s is the sound velocity, and v_t is the turbulent velocity. In the following, we take $\alpha \lesssim 1$ (see Shakura and Sunyaev, 1972).

The picture of accretion with the formation of a disk around a black hole is the most probable (Gorbatsky, 1965; Prendergast and Burbidge, 1968; Lynden-Bell, 1969; Shakura, 1972). The particles in the disk due to the friction between the adjacent layers lose their angular momentum and spiral into the black hole. Gravitational energy is released during this spiraling. Half of this energy increases the kinetic energy of rotation and the second half turns into thermal energy and is radiated from the disk surface. The total energy release as well as the spectrum of the outgoing radiation are determined mostly by rate of accretion, i.e., by the rate of inflowing of matter into the disk*. The distinctive parameter is the flux of the matter $\dot{M}_{cr}$ at which the total energy released by the disk $L = \eta c^2 \dot{M}$ is equal to Eddington's luminosity $L_{cr} = 10^{38}$ $(M/M_\odot)$ erg s^{-1}, which in turn is characterized by the equality of the force associated with the radiation pressure on the completely ionized matter and the gravitational attraction forces (η is the efficiency of gravitational energy release; in the case of Schwarzchild's metric $\eta = 0.06$; in a Kerr black hole η can attain 40%). For a black hole having a mass M, the critical flux is equal to

$$\dot{M}_{cr} = 3 \times 10^{-8} \, (M/M_\odot) \, 0.06/\eta \quad M_\odot \text{ yr}^{-1}.$$

There is no particular reason for considering a rate of accretion exactly equal to $\dot{M}_{cr}$. A subcritical regime of the disk accretion is possible as well as a flux of matter into the disk exceeding many times the critical value.

At essentially subcritical fluxes $\dot{M} \sim 10^{-12}$–10^{-10} $M_\odot$ yr^{-1}, the luminosity of the disk is of the order of $L = 10^{34}$–10^{36} erg s^{-1}. Maximal surface temperatures $T_{eff} \simeq 3 \times 10^5$–$10^6$ K are expected in the inner regions of the disk, where the main fraction of energy is released. This energy, radiated mainly in the ultraviolet and soft X-ray bands, is absorbed in the interstellear medium. Therefore these sources are inaccessible for observations**.

The observed radiation of the disk is formed in the upper layers of its atmosphere. The local spectrum F_ν depends on the distance from the black hole and the distribution of the matter along the Z-coordinate. The form of the local spectrum reduces to one of the four characteristic distributions (Figure 2). The integral spectrum (Figure 3) is determined by the expression $J_\nu = 2\pi \int F_\nu(R) R dR$.

For the disk accretion, one typically expects for $h\nu < kT_{max}$, a weak dependence of the radiation intensity on the frequency, $F_\nu \sim \nu^{-\beta} (-1/3 < \beta < 1)$. As a result, the optical luminosity of the black hole may be appreciable. Our estimates show (Shakura and Sunyaev, 1972) that for black holes with $M = 10 \, M_\odot$, even if $\dot{M} = 10^{-9} \, M_\odot$ yr^{-1}, one may expect an optical luminosity of the order of the solar luminosity.

In fact, the optical luminosity can be much higher, since it is connected with the reradiation of the hard radiation of the hot central regions of the disk by the outer cold layers. The thickness of the disk increases with distance from the black hole (Figure 4). This is why the outer regions of the disk must effectively absorb the X-ray flux from central regions of the disk and reradiate the absorbed energy in the ultra-violet and optical bands. Thus, from 1.0% to 10% of the total luminosity of the disk

---

* The efficiency α of the angular momentum transport mechanism is assumed to be constant along the disk (v_t and H vary in accordance with the change of ϱv_s^2). The observational appearance of the disk, i.e., the spectrum of its radiation and the effective temperature of its surfaces does not strongly depend on the chosen value of α. However at supercritical accretion this dependence becomes dominant.

** However they may sufficiently contribute to the galactic component of soft X-ray background and to the thermal balance of interstellar medium. Their radiation must ionize and heat neutral interstellar hydrogen.

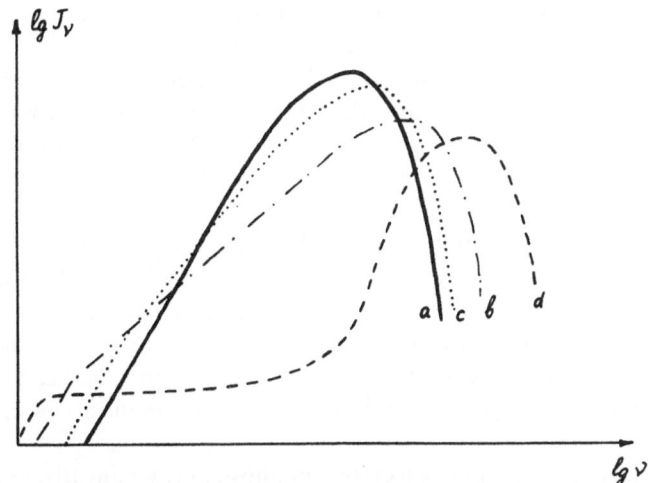

Fig. 2. Characteristic local spectra of radiation in the disk atmosphere: (a) black body spectrum $\varepsilon_r = bT^4$, (b) radiation spectrum of an isothermic homogeneous medium where the main contribution to the opacity is given by scattering $\varepsilon_r = cT^{2.25}$, (c) the same in an isothermic exponential atmosphere, $\varepsilon_r = dT^{2.5}$, (d) the spectrum forming as a result of Comptonization $\varepsilon_r = eT^4$. The intensity is chosen in such a way that the energy density of radiation ε_r is equal in all four cases. The change of effective radiation temperature is easily seen. (Details are given in Shakura and Sunyaev, 1972).

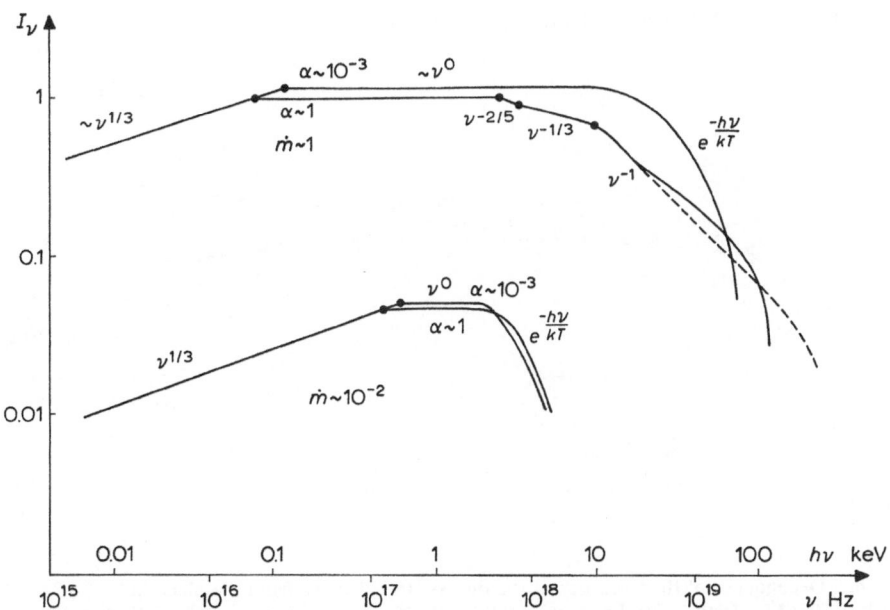

Fig. 3. An integral radiation spectrum of the disk for $M = M_\odot$, $\dot{M} = \dot{M}_{cr}$ and different α. The spectrum of an opaque disk at low α is analogous to the bremsstrahlung radiation spectrum of optically thin plasma, i.e. to the spectra of Sco-X1 type thermal X-ray sources.

can be reradiated (Shakura and Sunyaev, 1972). The hard radiation must be reradiated mainly in the lines of the different elements.

Strong recombinational fluorescence of hydrogen must be observed with no apparent ionization source and possibly also lines of highly ionized heavy elements. All these lines must be broad because the matter in the disk has large rotational velocities ($\gtrsim 100$ km s^{-1}). The density of matter in the disk is high and forbidden lines should be absent.

Considerable ultraviolet luminosity of the disk can lead to the formation of Strömgren region which distinguishes a black hole from normal optical stars with similar optical luminosity. In certain conditions, the hard radiation of the central regions of the disk can heat the matter in the outer regions up to high temperatures and evaporate the disk, decreasing the inflow of matter into the black hole. Such an autoregulation of accretion can essentially influence the luminosity of the disk around the black hole.

When the rate of accretion increases, the luminosity grows linearly, and the effective temperature of radiation rises (Figures 5 and 6). At fluxes $\dot{M} = 10^{-9}$–10^{-8} $M_{\odot}$ yr^{-1}, the black hole is found to be a powerful X-ray star with luminosity $L = 10^{37}$–10^{38} erg s^{-1} and an effective temperature of radiation $T_{\mathrm{eff}} \simeq 10^7$–$10^8$ K. The star radiates also in the optical and ultraviolet spectral bands. In a close binary system, a significant part of the X-radiation of the black hole can hit the surface of the normal star and be reradiated by its atmosphere, which can lead to an unusual optical appearance in such a system (Shklovsky, 1967; Shakura and Sunyaev, 1972). This effect is observed now in Hz Her = Her X 1 system (Lutiy et al., 1972).

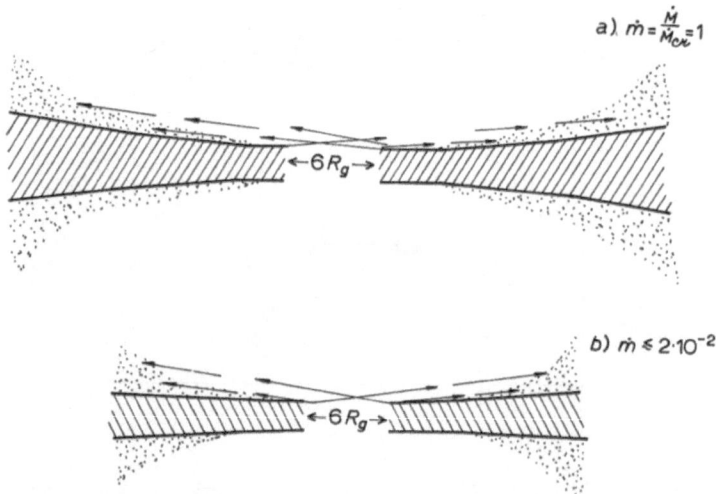

Fig. 4.   Dependence of the thickness of the disk on the distance from the black hole: (a) $\dot{M} = \dot{M}_{\mathrm{cr}}$, (b) $\dot{M} < 10^{-2} \dot{M}_{\mathrm{cr}}$. In the central zone where $R < 3R_g$, Newtonian theory is not applicable. The trajectories of X-ray and ultraviolet quanta leading to evaporation and heating of the matter in the outer regions of the disk are indicated by the arrows. The disc corona formed by hot evaporated matter is denoted by points.

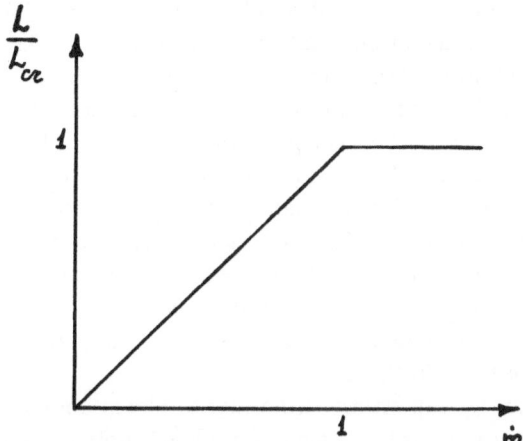

Fig. 5. Dependence of luminosity of the disk around the collapsar on the flux of the matter
entering the external boundary.

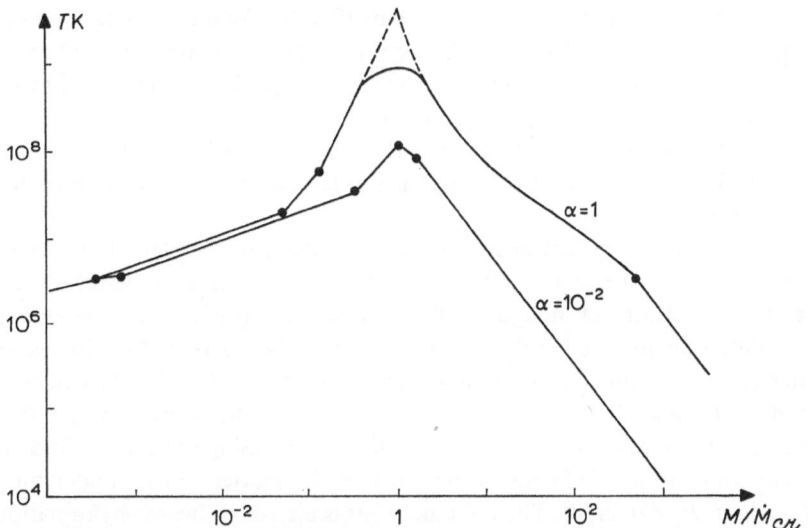

Fig. 6. Maximal effective temperature of the radiation from the disk as a function of the flux of
matter inflowing into it for the different efficiencies α of mechanisms of the
angular momentum transfer.

Aperiodic variability of certain properties such as fluctuations of brightness, principally connected with the variability of infalling matter flux and its non-homogeneity,
should distinguish some collapsars. In remote systems, a collapsar in the perigee of its
orbit gets into the more dense matter flux outflowing from the visible component.
Therefore, periodic variability of luminosity (non-sinusoidal in general) should be
expected. Moreover, eclipses of radiation of the central source by the disk itself are

possible when it is viewed edge-on, if its plane does not coincide with the plane of rotation of the system. Such an orientation of the disk must take place, for example, when the matter flows through the inner Lagrangian point from a non-synchronously rotating star, whose axis of rotation is inclined with respect to the plane of rotation of the system. Taking into account the eclipse of the X-ray source by the adjacent star and by the disk, the number of eclipses amounts to 3 for each period of rotation. If the plane of the disk coincides with the plane of the system, then only the disk can be occulted or darkened, whereas on the other side of the orbit, the thin disk covers only an insignificant fraction of the star surface. However, the most typical characteristic of the black hole in a close binary system should be its X-ray radiation. The detection of compact X-ray stars having a mass $M > 2\ M_\odot$ in binary systems will be the proof of existence of black holes in the Galaxy.

For a neutron star, we can estimate (in order of magnitude) that the energy release of infalling matter per gram is the same as for a black hole ($\eta \sim GM/R_{ns}c^2 \sim 10\text{--}20\%$). However, the accretion on a neutron star in a binary system has its own peculiarities. In the case of a neutron star without a magnetic field, the disk extends to its surface. One half of the entire energy released is radiated by the surface of neutron stars in the form of X-ray quanta having energy less than 10 keV. Compton interaction of radiation and matter in the disk decreases the maximum photon temperature in the disk to $T_{eff} \sim 2 \times 10^7\,\mathrm{K}$, whereas in the neighborhood of a black hole regions having temperatures up to $10^9\,\mathrm{K}$ are possible, in principle.

Accretion on a rotating neutron star with a magnetic field, the direction of which does not coincide with the axis of rotation, can lead to the phenomenon of an X-ray pulsar (Shvartzman, 1971b).

At subcritical inflow of matter into the Roche cavity of the black hole, we may assume that the main part of the inflowing matter is accreted. In supercritical regimes, both in the case of a black hole and of a neutron star* (there is no fundamental difference here), a qualitatively different picture should be realized. An effective outflow of matter under the influence of radiation pressure appears to take place in the region surrounding the black hole. The outflow begins close to the radius at which the forces associated with radiation pressure and gravitation, pressing the matter into the disk plane, are comparable. Only the critical flux of the matter can go under the radius $R = 3R_g$ where $R_g = 2GM/c^2$. The remaining gas is ejected outwards by radiation pressure.

The integral luminosity of such objects is limited by the value of the critical Eddington luminosity. The band of the electromagnetic spectrum where this energy is mostly radiated, depends strongly on the density of the outflowing matter. The density of the matter, in turn, is a function of the ratio $\dot{M}/\dot{M}_{cr}$ and the efficiency α of the transport mechanism of angular momentum which determines the velocity of the outflowing gas. If $\dot{M}$ exceeds $\dot{M}_{cr}$ only slightly and $\alpha \sim 1$, the emission of the disk is reradiated by the outflowing gas practically without change of the spectral properties, i.e., the object

* Critical accretion on a neutron star must lead to its collapse after 3×10^7 yr.

is a source of X-ray radiation as before. If the ratio $\dot{M}/\dot{M}_{cr}$ increases and α decreases, the opacity of the outflowing matter grows, the radiation of the disk is re-emitted as quanta of smaller energy. If $\dot{M} > 10^3 \dot{M}_{cr} (\alpha M_\odot/M)^{2/3}$, a black hole turns into a bright optical star. The less the parameter α the greater the effective radius of the radiating envelope.

Since the angular momentum of the ejected matter must be preserved, a strongly anisotropic outflow of matter can be observed. The hot plasma is ejected with great velocity into a narrow cone of angles near the axis of rotation. The optical depth of the outflowing gas in this cone is not great and at certain orientations of the binary system relative to the observer, the X-ray radiation of the black hole together with the optical should be observed.

The observational appearance of a black hole in a strongly supercritical regime of accretion can be characterized as follows: the luminosity is fixed by the Eddington critical limit $L_{cr} = 10^{38} (M/M_\odot)$ erg s^{-1}; the major portion of the energy is radiated in the ultraviolet and optical regions of the spectrum; in the upper, rarefied layers of the outflowing matter, broad emission lines are formed. There is a strong mass outflow from a hot star with velocities $V \sim \alpha 10^5 (\dot{M}_{cr}/\dot{M})^{1/2}$ km s^{-1}, and the star is surrounded by a colder disk where the accreting matter enters the collapsar. Eclipses of the black hole by the normal component are possible as well as an eclipse of the star by the matter outflowing from the black hole. The latter is opaque with respect to Thomson's scattering at great distances away from the black hole ($R_t \sim 10^{10}$–10^{12} cm). In the radio range, the hot outflowing matter becomes opaque far from the binary system. It can be a source of appreciable thermal radiation with a smooth dependence of intensity on the frequency ($J_\nu \sim \nu^{2/3}$).

In the radio range, for both subcritical and supercritical accretion, non-thermal radiation mechanisms connected with the existence of the magnetic fields (which may achieve $H \sim 10^5$–10^7 G) and beams of the fast outflowing particles can also appear.

Apparently, the 'Quiet' disk, radiating only due to thermal mechanisms, can really exist at low values of the parameter α. If $\alpha \sim 1$, the new important effects (connected with turbulent convectivity; plasma turbulence; reconnection of magnetic field lines through neutral points, leading to solar type flares; the acceleration of particles) and nonthermal radiation can appear. The flares and hot spots on the rotating disk surface must lead to the short term fluctuations of radiation flux in some spectral bands. The variability may have both a stochastic nature (Schvartzman, 1971a) and may be quasiperiodic. The quasiperiod of these fluctuations must be of the order of rotational period $t \sim 2\pi R/v_\varphi \sim 6 \cdot 10^{-4} (R/3R_g)^{3/2} (M/M_\odot)$ s and depends on the distance of the hot spot from the collapsar. Minimal quasiperiod in the case of Kerr metric is 8 times less than in the case of non-rotating black hole with equal mass (Sunyaev, 1972).

Acknowledgements

The authors wish to thank Ya. B. Zel'dovich, A. F. Illarionov, D. Ya. Martynov and

L. R. Yangurasova for consultations and the participants of the 1972 winter school on astrophysics in Arhyz for numerous discussions.

References

Gorbatsky, V. G.: 1965, *Proc. Leningrad Univ. Obs.* **22**, 16.
Lutiy, V. M., Sunyaev, R. A., and Cherepashchuk, A. M.: 1973, *Astron. Zh.* **50**, N1, preprint (in Engl.).
Lynden-Bell, D.: 1969, *Nature* **223**, 690.
Martynov, D. J.: 1971, *Course in General Astrophysics,* Nauka, Moscow.
Pottash, S. R.: 1970, in H. J. Habing (ed.), 'Interstellar Gas Dynamics', *IAU Symp.* **39**, 272.
Prendergast, K. H. and Burbidge, G. R.: 1968, *Astrophys. J.* **151**, L83.
Salpeter, E. E.: 1964, *Astrophys. J.* **140**, 796.
Shakura, N. I.: 1972, *Astron. Zh.* **49**, N5, 921.
Shakura, N. I. and Sunyaev, R. A.: 1972, *Astron. Astrophys.* (in press). Preprint IPM N28 (in Engl.).
Shklovsky, I. S.: 1967, *Astrophys. J.* **148**, L1.
Shvartzman, V. F.: 1971a, *Astron. Zh.* **48**, 479; *Soviet Astron.* **15**, 377.
Shvartzman, V. F.: 1971b, *Astron. Zh.* **48**, 438; *Soviet Astron.* **15**, 343.
Sunyaev, R. A.: 1972, *Astron. Zh.* **49**, N6, preprint (in Engl.).
Zel'dovich, Ya. B.: 1964, *Dokl. Acad. Sci. U.S.S.R.* **155**, 67; *Soviet Phys. Dokl.* **9**, 195.
Zel'dovich, Ya. B. and Novikov, I. T.: 1967, *Relativistkaya Astrofizika,* Izdatel'stvo 'Nauka', Moscow (Engl. transl.: *Relativistic Astrophysics,* Chicago Univ. Press, Chicago, 1971).

14. PULSARS AND X-RAY SOURCES

F. PACINI

Laboratorio di Astrofisica Spaziale, Frascati, Italy

Abstract. We summarize some theoretical aspects of the X-ray emission from genuine pulsars and their relation to the more general problems of X-ray sources.

Most galactic X-ray sources appear to be connected with the final stages of stellar evolution. This is evident for the sources identified with SN remnants; however, even in the case of objects like Sco X-1 or Cen X-3, spectral information and/or the presence of fast periodicities clearly imply that one is dealing with condensed stars such as white dwarfs, neutron stars or, perhaps, black holes.

When discussing the basic nature of these objects the question naturally arises of a possible link between X-ray sources and the pulsar mechanism. The possibility of this connection has been discussed before the discovery of periodic X-ray sources but interest in the subject has been greatly enhanced by the recent observational results.

In the following we comment briefly about some aspects of this relationship.

1. X-Ray Emission from Genuine Pulsars

Only one genuine pulsar, NP 0532, has been detected as an X-ray source. This indicates that the X-ray output of pulsars dies out very rapidly when the star slows down. This observational evidence can be understood if the optical and the X-ray luminosity of pulsars arise through the synchrotron emission by particles moving in the proximity of the speed of light distance. It has been shown (Pacini, 1971) that in this case the synchrotron output evolves roughly like $M^4 \cdot P^{-10}$ (M is the magnetic moment of the star and P the rotation period). This accounts for the lack of optical and X-ray emission from most pulsars.

It cannot be ruled out, however, that some genuine very fast pulsars exist among the known X-ray souces. The lack of a SN remnant around these sources would not pose any particular problem if their age considerably exceeds 10^4 yr. The existence of pulsars born with 'weak magnetic fields' ($\lesssim 10^{11}$ G) would naturally lead to the existence of fast, old pulsars. This could form a population of compact non-thermal X-ray sources similar to NP 0532 and with a typical output $\simeq 10^{36}$–10^{37} erg s^{-1}. The space distribution of these sources would resemble that of normal radio pulsars.

Genuine pulsars with very long periods $\gtrsim 1$ s could only arise around a rotating, magnetic white dwarf (the larger value of M could compensate for the long period!)

Bradt and Giacconi (eds.), X- and Gamma-Ray Astronomy, 165–167. All Rights Reserved.
Copyright © 1973 by the IAU.

2. Pulsars as Motors for X-Ray Sources

Genuine pulsars produce relativistic particles and can provide the energy source for the X-ray emission from SN remnants. This is the well known case of the pulsar NP 0532 in the Crab Nebula. It is not yet clear whether the X-ray emission from some other SN remnants has a non-thermal origin: if so, this would require the existence of a central (as yet undiscovered) pulsar.

Most of theoretical discussion of pulsar electrodynamics is centered upon the unipolar inductor theory (Goldreich and Julian, 1969) or the magnetic dipole radiation theory (Pacini, 1967, 1968; Ostriker and Gunn, 1969). The great success of these theories is due to the fact that they explain the overall energetics of the pulsar and also provide a simple and attractive scheme for investigating pulsar electrodynamics.

In particular both theories account for the production of very fast particles. However, there is a deep disagreement between the observed energy spectrum of the particles injected into the Crab Nebula (a power law between roughly 10^8 eV and 10^{14} eV) and the monoenergetic character of the spectrum predicted by the simple versions of the above mentioned theories. There is little doubt that future investigation of the acceleration should abandon the 'test particle' approach and rather deal with the more complicated problem of the mutual interaction of plasmas and rotating electromagnetic fields. (Both in the near zone and in the far-wave region).

Finally, we recall (Davidson *et al.*, 1971) that sources like Sco X-1 could perhaps be energized by a central pulsar hidden within a cocoon. If this is the case, and if the motor for thermal sources is indeed of a highly non-thermal nature, the heating of cocoon could be accompanied by a conspicuous flux of non-thermal γ-rays produced by bremsstrahlung or by nuclear processes. The detection of γ-rays from thermal X-ray sources would represent an important indication in favour of the existence of a central non-thermal motor.

3. Periodic X-Ray Sources

The recent discovery of the periodic sources Cen X-3 and Hercules has attracted a great interest. Possible explanations are discussed elsewhere in this volume. It appears clear that these sources have little to do with genuine pulsars like NP 0532. Their mechanism seems instead intimately related to the evolution of a close binary system when one of the two stars reaches the end point of stellar evolution and accretes mass from the companion.

Acknowledgements

I am especially grateful to Drs. A. Cavaliere and E. E. Salpeter for a very stimulating collaboration and for interesting discussions. This work has been partly supported by NATO Grant 601.

References

Davidson, K., Pacini, F., and Salpeter, E. E.: 1971, *Astrophys. J.* **168**, 45.
Goldreich, P. and Julian, W.: 1969, *Astrophys. J.* **157**, 869.
Ostriker, J. and Gunn, J.: 1969, *Astrophys. J.* **157**, 1395.
Pacini, F.: 1967, *Nature* **216**, 567.
Pacini, F.: 1968, *Nature* **219**, 145.
Pacini, F.: 1971, *Astrophys. J. Letters* **163**, L17.

PART III

EXTRAGALACTIC SOURCES

15. UHURU RESULTS ON EXTRAGALACTIC X-RAY SOURCES

EDWIN M. KELLOGG

American Science and Engineering, Cambridge, Mass., U.S.A.

Abstract. Data from the UHURU satellite have provided a list of more than forty high latitude sources ($|b| > 20°$). X-rays have been detected from among the nearest normal galaxies, giant radio galaxies, Seyferts, QSOs and clusters of galaxies. The cluster sources appear to be extended by several hundred kiloparsecs as well as being very luminous. These cluster sources have systematic differences in their X-ray spectra from individual galaxies.

About twenty sources are not reliably identified so far. A few of these are located near undistinguished 3C or MSH radio sources. The rest are either located near distant clusters or undistinguished bright galaxies, or are too far south, so that we have not sufficient optical data to allow a thorough search for possible association with clusters or unusual individual galaxies.

The luminosity function for weak, high latitude X-ray sources is determined, and the contribution of sources just below the UHURU threshold of detectability to observed fluctuations in the diffuse X-ray background is evaluated. The total contribution of all observed types of extragalactic sources to the X-ray background is estimated.

1. Introduction

The UHURU satellite, launched in December 1970, has had a great impact on extragalactic X-ray astronomy. Before data from this satellite became available, there were only a few extragalactic X-ray sources known. Sources near M 87 (Byram *et al.*, 1966 and Bradt *et al.*, 1967) and 3C 273 (Bowyer *et al.*, 1970) and NGC 1275 (Fritz *et al.*, 1971) had been observed definitely, and ones near NGC 5128 (Bowyer *et al.*, 1970), the Coma Cluster (Meekins *et al.*, 1971) and the Large Magellanic Cloud (Mark *et al.*, 1969) had been reported using sounding rocket data. The first fairly complete sky survey with UHURU has revealed more than 40 X-ray sources off the galactic plane by 20° or more. Figure 1 shows the locations in galactic coordinates of the sources

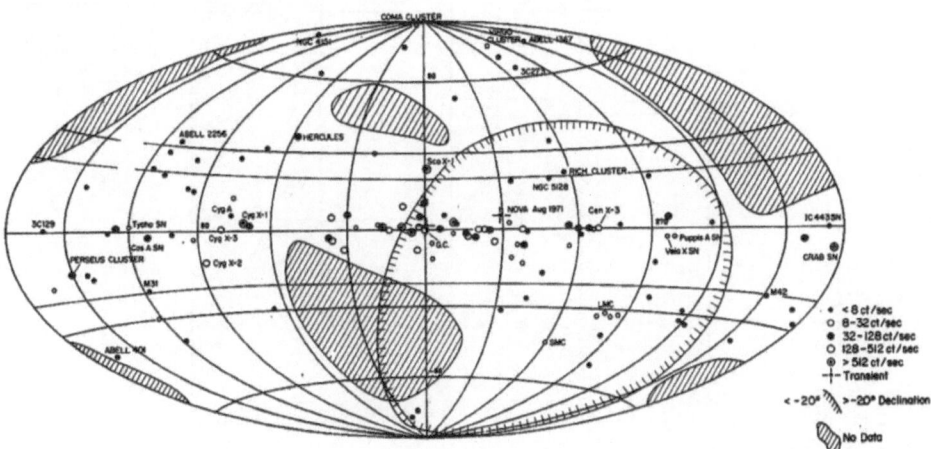

Fig. 1.   The X-ray sources in the UHURU Catalog, plotted in galactic coordinates.

Bradt and Giacconi (eds.), X- and Gamma-Ray Astronomy, 171–183. All Rights Reserved.
Copyright © 1973 by the IAU.

(Giacconi *et al.*, 1972) and it shows some shaded areas we have not yet surveyed. At least fifteen of the high latitude sources are already identified with usual galaxies, or with clusters of galaxies. We believe that most of the remaining unidentified high latitude sources will turn out to be extragalactic.

As was expected, ordinary galaxies are detected as X-ray sources at comparable luminosity to our own galaxy. We also confirm that unusual active galaxies have enhanced X-ray luminosity. The most exciting discovery we have made is that sources in clusters of galaxies are very luminous and are extended, with sizes as large as a few megaparsecs. We also find that the luminosity function of high latitude sources is consistent with a uniform volume distribution as expected for distant extragalactic sources; also that the diffuse background might be the superposition of many distant discrete sources.

2. Identified Sources

Figure 2 shows the identified extragalactic sources by class with their distances, ranging from 50 kpc for the Magellanic Clouds to 600 Mpc for the most distant objects. The top of each shaded area represents the distance of the nearest member of each class. The nearest ordinary galaxies are the Magellanic Clouds, located at the top of the shaded area for the class of ordinary galaxies. The Small Cloud is dominated by emission from a single binary X-ray star (Leong *et al.*, 1971 and Schreier *et al.*, 1972.)

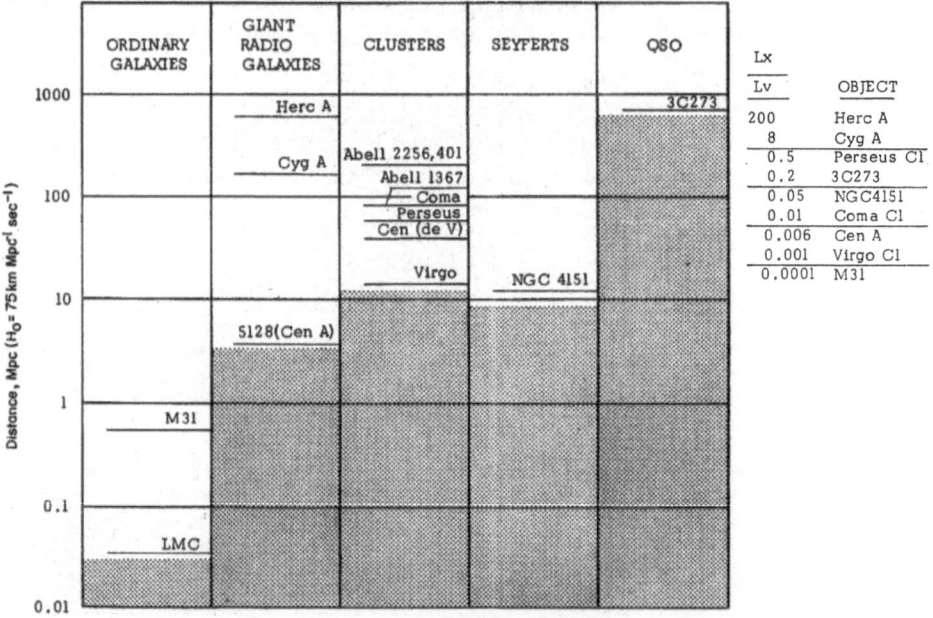

Fig. 2.   Identified extragalactic X-ray sources. The top of the shaded region for each class indicates the distance of the nearest known member of that class. The Seyfert galaxies are the only group where the nearest member, NGC 4051, is not detected, but a more distant member is detected. The table at the right lists the ratio of X-ray to optical luminosity for representative identified sources.

The emission from the Large Cloud is due to four discrete sources (Giacconi *et al.*, 1972). The ratio of X-ray to optical luminosity for ordinary galaxies is 10^{-4} or less, as listed in the table at the right. We have detected the nearest giant radio galaxy, Cen A (Kellogg *et al.*, 1971a and Tucker *et al.*, 1972), located at the top of the shaded area for radio galaxies, and two much more luminous distant radio galaxies, Cyg A (Giacconi *et al.*, 1972) and Herc A. Figure 3 shows our best estimates of the locations for Cyg A and Herc A. The updated Herc A location contains more recent data than that included in the UHURU Catalog (Giacconi *et al.*, 1972). The X-ray to visible

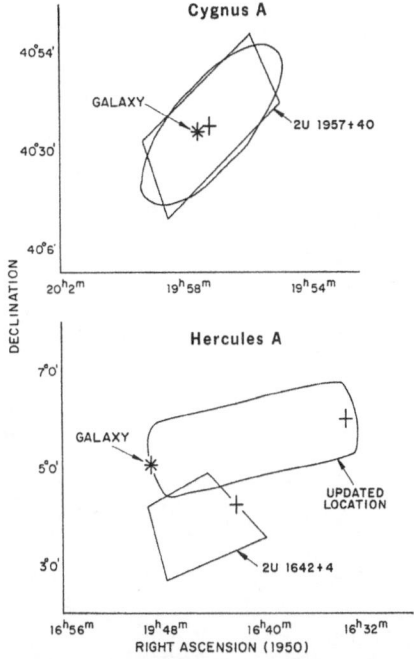

Fig. 3.   Locations of two UHURU sources identified with distant radio galaxies. For Cygnus A, the quadrilateral is the error box listed in the UHURU Catalog. The ellipse is a more exact location contour from which the UHURU error box was obtained. The peak of the location probability distribution is shown by a vertical cross. For Hercules A, new data obtained since the UHURU catalog was generated were added to obtain the updated location.

ratios have a large range from 6×10^{-3} for Cen A to about 200 for Herc A. However, the X-ray to radio luminosity ratio does not vary so widely for these galaxies. We have detected the nearest great cluster in Virgo, and several more distant rich clusters. Luminosity ratios range from 10^{-3} for Virgo to about one in Perseus. We see only one Seyfert galaxy, NGC 4151. The top of the shaded area is at the distance of NGC 4051, the nearest Seyfert, which is not a source. NGC 4051 and NGC 1068 are less than $\frac{1}{4}$ as luminous as 4151. We also see the nearest QSO, 3C 273. If these classifications are physically meaningful, we'd expect each class to have a typical luminosity; then the nearest ones would be detectable first. It appears that only the ordinary galaxies are well behaved in that sense, and possibly QSOs. The other classes show a wide range of luminosities. The enhanced X-ray to visible luminosity ratios for radio galaxies, clusters, the Seyfert and the QSO suggest that the X-rays in these objects are not produced in stars as they are in our own galaxy, but by some other mechanism.

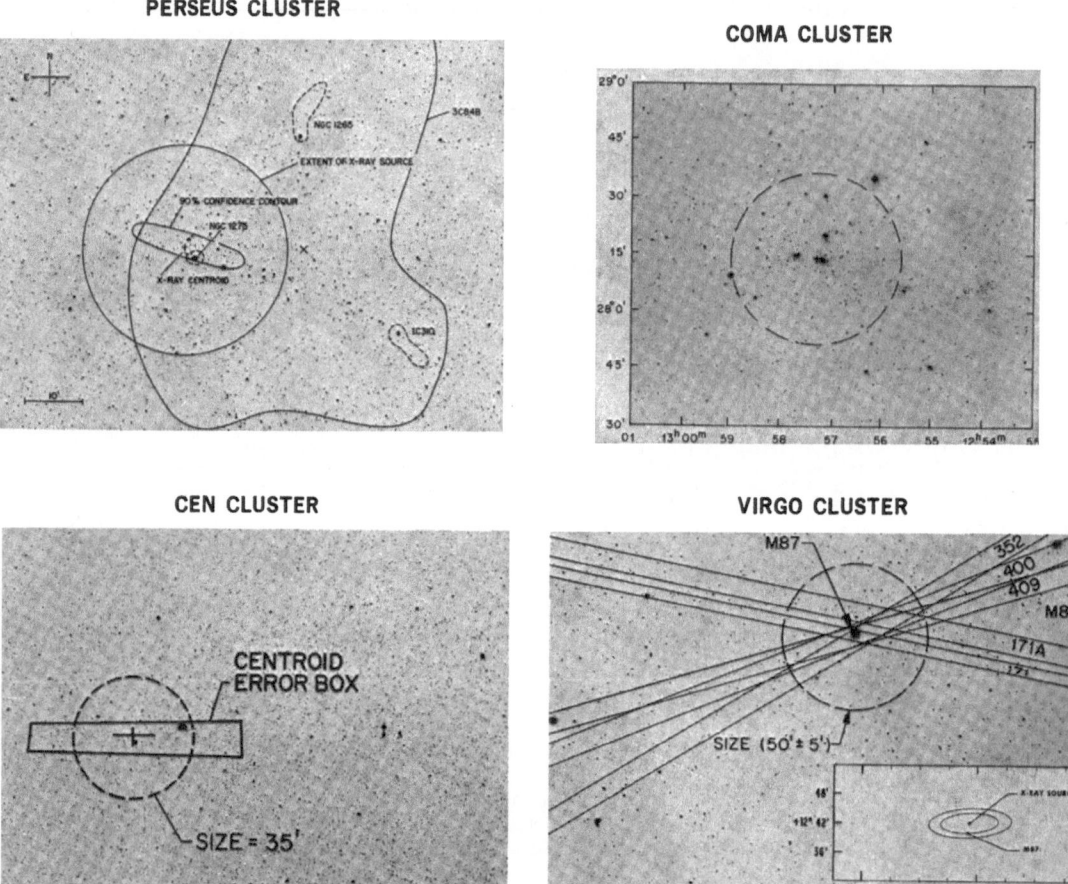

Fig. 4.   X-ray sources in clusters of galaxies. These are the four strongest cluster sources. In each case, we indicate the extent of the source by the simplest figure, a circle, since we have no information on the details of the shape. The Virgo picture also shows the one-directional locations for the centroid of the extended source obtained from five scans, labelled by the orbit number pertaining to each scan. The inset is the two-dimensional error box obtained by combining the five one-directional locations. The inner contour corresponds to 68 % confidence and the outer corresponds to 90 % confidence. The Centaurus source, 2U1247-41 contains NGC 4696 within its centroid error box. The Perseus source contains NGC 1275, the exploding Seyfert galaxy, within the centroid error box. In Coma, both the giant elliptical NGC 4874 and the kinematic center of the cluster lie within the error box. (Gursky et al., 1971; Forman et al., 1972; Kellogg et al., 1972).

The second major finding from UHURU is that sources in clusters are very luminous and extended. We have now identified about 8 sources associated with clusters. In Figure 4, we see four such sources, from the great clusters in Virgo, Cen, Perseus and Coma. The Virgo source is centered on M 87, the active giant elliptical galaxy (Kellogg *et al.*, 1972). The bands are the source centers obtained from 5 different UHURU scans. They intersect at M 87. The inset shows the error box obtained by combining the 5 scans. M 87 is within 3′ of the centroid of the extended X-ray source. Its size is shown by the circle. The true shape of each of these four extended sources is probably not circular, but we cannot tell from our data, so we represent the source shape in the simplest way possible, a circle whose description requires only a single parameter, its diameter.

The Cen cluster is about 3 times further away than Virgo, and the source is weaker. We have just developed a new technique for measuring the size of weaker sources, and we find it to be extended. The giant elliptical galaxy NGC 4696 is located in the error box for the centroid of the extended X-ray source. The size is again shown by a circle, 35′ in diameter. The Perseus cluster source is centered very close to NGC 1275, a violently active Seyfert galaxy (Forman *et al.*, 1972). The source in Coma may be centered on either the giant elliptical galaxy NGC 4874, or on the kinematic center of the cluster which is very close to 4874 (Forman *et al.*, 1972).

We have tried to measure the angular extent of nine sources located in clusters (see Gursky *et al.*, 1972 for the list of cluster sources). Five of these are definitely extended, as summarized in Figure 5; the other four were too weak to measure sizes. They are Abell 401, Abell 1367, 3C 129, and one Zwicky cluster, and ZW 0444.7 + 0828. The last two sources listed are identified with individual galaxies not located in clusters, and are found not to be extended. Abell 2256 is the most distant extended source we have yet detected and the most luminous. These data show that the clusters have a large range of sizes and luminosities. In fact, there appears to be some correlation between the X-ray luminosity of a cluster and its velocity dispersion, as shown

	SIZE		L_x
	angular	kpc	(erg/sec)
ABELL 2256	35′ ± 15′	2800	5×10^{44}
PERSEUS - NGC 1275	35′ ± 3′	740	3×10^{44}
COMA	36′ ± 4′	1050	2×10^{44}
CEN - NGC 4696	37′ ± 8′	500	2×10^{43}
VIRGO - M87	50′ ± 5′	200	7×10^{42}
NGC 4151	≤ 15′	≤ 60	1×10^{42}
NGC 5128	≤ 10′	≤ 20	6×10^{41}

Fig. 5. Sizes and X-ray luminosities of extragalactic X-ray sources. Five extended cluster sources and two compact sources not in clusters are listed. The cluster sources are much more luminous than the compact sources.

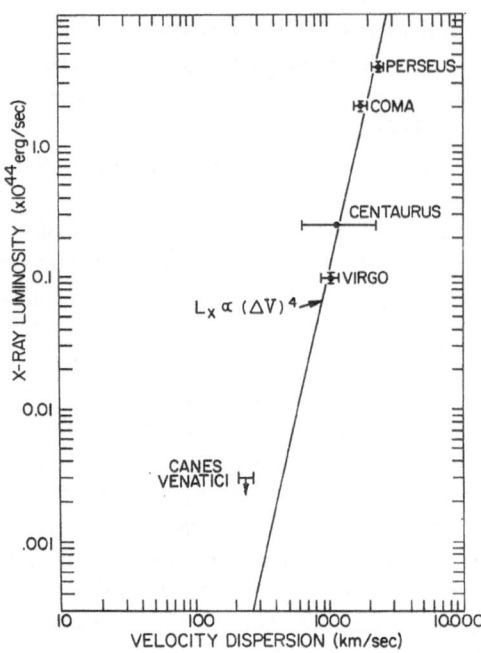

Fig. 6.   Luminosity versus velocity dispersion for X-ray sources in clusters. The curve is arbitrarily normalized. For the three clusters with well determined velocity dispersions, the correlation between ΔV and L_x is undeniable. More detailed analysis may prove that the data fit a curve slightly different from the $(\Delta V)^4$ curve shown, however. (Solinger and Tucker, 1972).

in Figure 6. This effect was noticed by Solinger and Tucker (1972). The observed luminosities are consistent with a fourth power dependence on ΔV. If this type of correlation holds up under future observations, we must interpret that there is a connection between X-ray emission and cluster dynamics involving the presence of some fraction of the virial mass in intergalactic gas, and that thermal bremsstrahlung is the X-ray emission mechanism. This would be the first unambiguous observation of intergalactic gas, whose presence has been suspected in the past, but not confirmed.

Some new UHURU observations of the spectra of extragalactic sources are shown in Figure 7. The ordinates are logarithmic and represent relative counting rates. The abscissae represent photon energy. In the upper right corner, is the spectrum of the Crab Nebula, which we use as a standard. We assumed interstellar absorption corresponding to $(1.6\pm0.16)\times10^{21}$ H atoms/cm^2 as observed in radio, which corresponds to $E_a=0.45\pm0.02$ keV. The spectral index was allowed to vary, with the best fit value being 0.99 ± 0.05. The histogram for the Crab represents both the observed count rate distribution in the UHURU pulse height analyzer and the calculated fit, because the statistical errors are very small. The middle column shows the count rate distributions from cluster sources as histograms with 1σ error bars. The circles are the computed fits. Perseus – NGC 1275 is similar to the Crab which has a power law index of 1.

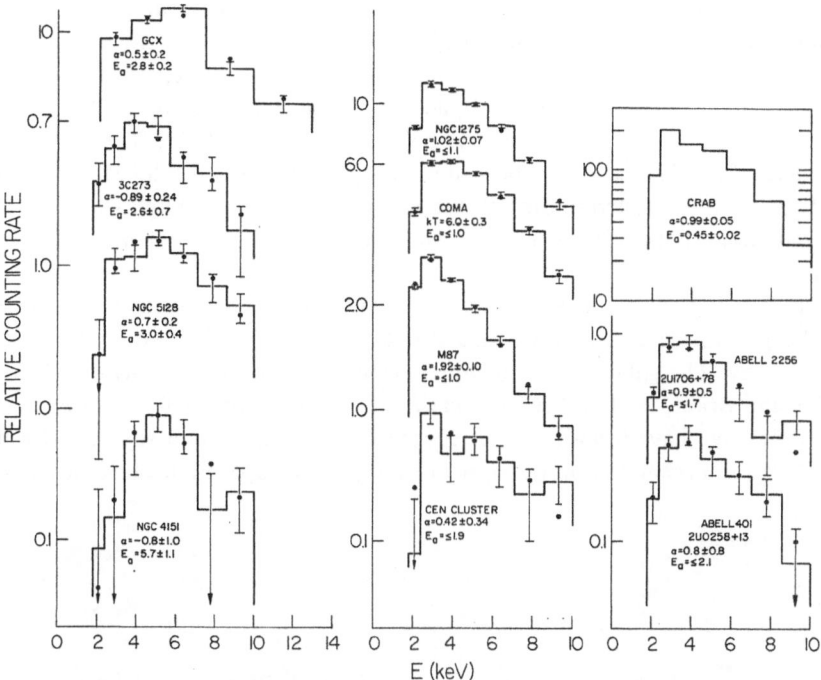

Fig. 7. Spectra of extragalactic sources. Counting rates versus energy are plotted as histograms with one sigma error bars. The closed circles are the predicted count rates assuming model spectra as inputs and computing the count rate after accounting for counter efficiency, pulse height resolution and fluorescence escape effects. The slope is given as α for the energy index of a fit to a power law, or kT for an exponential. E_a is the low energy cutoff in keV, assuming photoelectric absorption. The four spectra in the left column show a drastic turnover at low energies. The spectra in the center column and the two lower ones in the right column are from sources located in clusters, and show no cutoff. The spectrum of the Crab Nebula is shown at the top of the right column for comparison.

M 87 has a steep spectrum with a power law index close to two. The spectrum of the Coma cluster fits better to isothermal bremsstrahlung than a power law; kT is 6.0 ± 0.3 keV. The spectra of the Cen cluster and Abell 2256 and 401 are similar to Coma and NGC 1275 in that they are significantly flatter than that of M 87. We see no definite low energy cutoff in any cluster source.

The left column shows data from individual galaxies and 3C 273. They all have a cutoff. The top spectrum is from the source at the nucleus of our own galaxy, GCX (Kellogg *et al.*, 1971b). The suggestion is that we are seeing X-rays from the nuclei of these other galaxies as well. Since we see no cutoffs in the cluster sources, we are led to believe that the X-ray emission mechanism is different for clusters than for individual galaxies. In particular, it appears that the cluster sources are truly diffuse, and not just a collection of active galaxies.

As an example of how we can use these observed cutoffs, consider 3C 273. We find a cutoff of 2.6 keV. This is probably not due to intergalactic material since such material at the critical density would not have the heavy element abundance required to give that much absorption; also it must be very hot if present, so it would be ionized and absorb still less. The absorption is most likely at the source, and corresponds to 5×10^{22} atoms/cm^2 assuming normal galactic abundance ratios in 3C 273.

If we assume that the X-rays and the point radio source 3C 273B are coming from the nucleus of a galaxy, then the lack of a radio cutoff down to $\cong 80$ MHz means that the cloud of material surrounding the source has a radius > 30 kpc along the line of sight, assuming a temperature of 10^4K. This, in turn, places 3C 273 more than 30 kpc from the Sun.

Figure 8 shows NGC 5128 or Cen A. This is a giant radio galaxy with two lobes several degrees in size. There are also two inner lobes just 7' apart centered on the optical galaxy. The galactic nucleus apparently has been observed as an IR hot spot (Kunkel and Bradt, 1971 and Becklin *et al.*, 1971) and a compact radio source (Wade *et al.*, 1971). We have found that the X-rays are coming from a source $\leqslant 10'$ in size whose location is centered on the optical galaxy. Our upper limit on X-rays from the

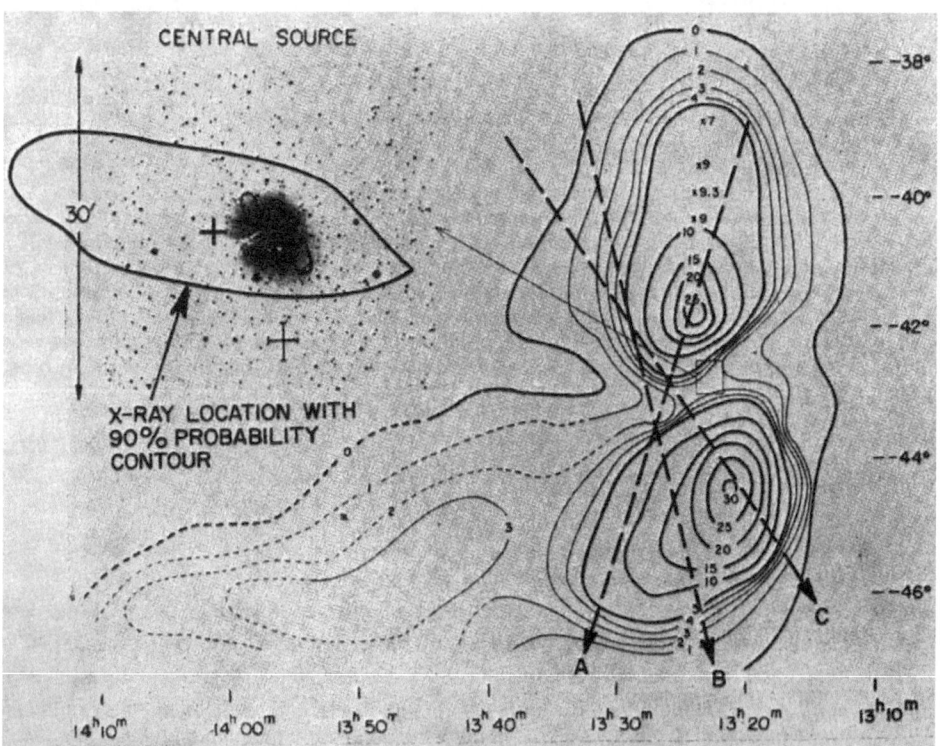

Fig. 8.   Centaurus A. The optical galaxy NGC 5128 is located within the X-ray source error box. The size of the source is less than 10'. This still allows the X-ray emission to be coming from the inner radio lobes shown in the inset, or from the optical galaxy. The nucleus of the optical galaxy is the most likely location of the source, since it is so strongly cut off at low X-ray energies, due to absorption.

outer lobes sets an upper limit on the temperature of the universal blackbody radiation of 5.1 K, assuming a magnetic field strength of 4×10^{-6} G in the lobes which corresponds to equipartition between magnetic and electron energy density. Or, if we fix the blackbody radiation at 2.7 K, we obtain a lower limit of 1×10^{-6} G in the giant lobes of Cen A (Tucker *et al.*, 1972). The X-ray source is probably coming from the nucleus of NGC 5128, since we see it strongly cut off. This must be the center of the high energy activity which has produced the lobes in the past, but until recently has been obscured from us by the great amount of dust in NGC 5128.

The unidentified high latitude sources are listed in Figure 9. The first group are located near 3C or MSH radio sources. The second group are not associated with strong radio sources. The six coincidences between X-ray sources and strong radio

1. NEAR STRONG RADIO SOURCES (≥ 30 F.U.)

2U	RADIO
0410 + 10	3C113 (WEAK)
0515 - 34	MSH 05 - 310
0525 - 38	MSH 05 - 36
0426 - 63	MSH 04 - 64
1253 - 28	MSH 12 - 212
2128 + 81	3C435.1

2. UNCERTAIN I.D.

2U	CANDIDATE
0043 + 32	ZW 0041.1 + 3235 OR OB + 368
0426 - 10	(SOUTH)
0440 + 7	CLUSTER?
0544 - 39	(SOUTH)
0628 - 54	(SOUTH)
1231 + 7	IC 3576?
1420 - 02	NGC 5604
1443 + 43	CLUSTER?
1808 + 50	ABELL 2298?
1843 + 67	CLUSTER?
1849 - 77	(SOUTH)
1954 - 68	(SOUTH)
2134 + 11	CLUSTER?
2346 - 32	NGC 7793?
2358 - 29	(SOUTH)

Fig. 9. Unidentified high latitude X-ray sources ($|b| > 20°$).

sources are somewhat higher than would be expected on a chance basis and may be due to a real physical association. Among the others there are probably many chance coincidences, such as the three sources near undistinguished NGC or IC galaxies.

The latitude distribution of these sources is shown in Figure 10. This has been corrected for nonuniform sky coverage. The horizontal line is the mean for all latitudes above 20° and the dashed lines are $\pm 1\sigma$. The apparent excess in the 20 to 43° zone is not statistically significant, being only 1.4σ. Therefore, to the best of our knowledge, they are distributed uniformly on the sky. Suppose these sources were very close to us, within the disk of our galaxy. Then we would have observed the integrated effect of other sources similar to them but further away as a galactic ridge of X-ray intensity. Such a strong ridge has not been seen. Therefore, we believe these unidentified sources are a new class of distant extragalactic object.

Figure 11 shows the luminosity function of high and low latitude sources (Matilsky *et al.*, 1972). The low latitude sources were discussed by Tananbaum (1972). The high

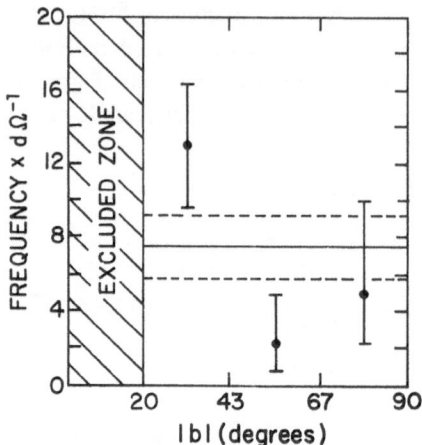

Fig. 10.   Galactic latitude distribution of unidentified X-ray sources. This is a plot of the distribution of the high latitude sources listed in Figure 9, with galactic latitude. Corrections have been made for the solid angle, sky coverage in each zone and sensitivity versus source strength. The solid horizontal line is the mean, and the dashed lines are one sigma errors in the mean.

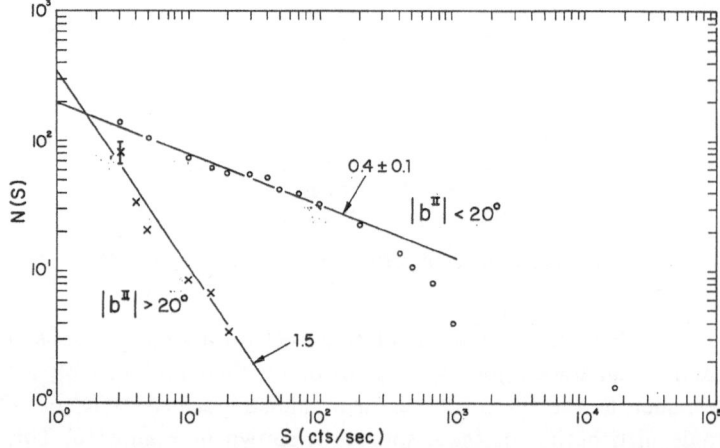

Fig. 11.   Integral distribution of number of sources versus strength, *S*. The sources are those given in the UHURU Catalog; the numbers given have been corrected to complete sky coverage and to 100% source detection efficiency down to the limit of 3 counts s⁻¹. The high latitude sources are consistent with a distribution of constant luminosity sources distributed uniformly throughout space, giving a distribution with a slope of 1.5 as shown by the solid curve drawn through the X's.

latitude sources agree well with a 1.5 power law expected for a uniform volume distribution, with constant space density and luminosity.

Does the luminosity function continue to weaker sources than we have yet observed?
Figure 12 shows the fluctuations in the background counting rate obtained as differences in adjacent 5° regions in the sky. The curve is that expected from counting
statistics. The excesses at 1.6 to 2.4 counts s^{-1} are most likely due to sources below
our present threshold of detectability. The estimated number at this intensity level
fits well on the logN logS curve, which predicts that we will observe about 300 high
latitude sources down to 1 count s^{-1} with UHURU eventually.

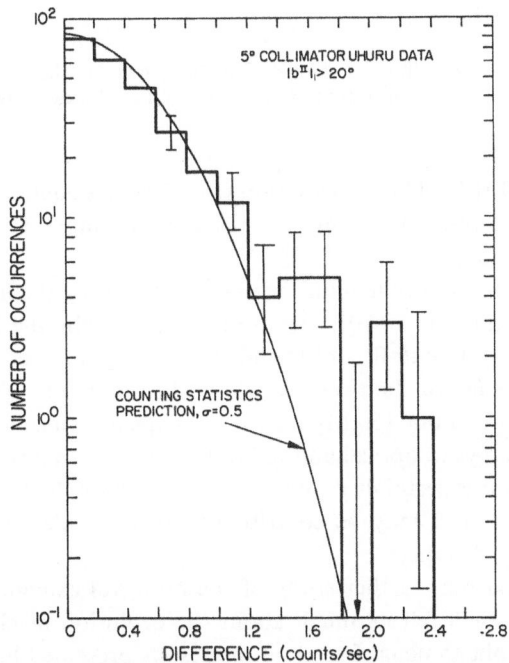

Fig. 12.   Point-to-point Count Rate Differences. Data from each great circle scan at high latitude are
analyzed for fluctuations by taking the difference in absolute magnitude of count rate between adjacent 5° bins. The frequency distribution of these differences is shown here. If the differences were due
only to counting statistics, we would expect the solid gaussian curve. We observe excesses at about
1.6 counts s^{-1} and 2.2 counts s^{-1}. These are undoubtedly due to weak sources just below the
UHURU threshold of detectability.

With this positive indication that still weaker sources than we have yet observed
exist, we can integrate the contributions of the various types of sources to the X-ray
background. Figure 13 shows the estimated contributions from identified classes of
X-ray sources. About equal contributions are obtained from Seyferts and clusters.
We assume that one third of all Seyferts are X-ray sources, based on our results for

TYPE	L_x erg/sec	ρ (Mpc^{-3})	df erg/cm sec ster
NORMAL GALAXIES	2×10^{39}	.03	$.24 \times 10^{-8}$
RADIO GALAXIES	6×10^{41}	3×10^{-5}	$.07 \times 10^{-8}$
SEYFERTS	10^{42}	3×10^{-4}	1.2×10^{-8}
QSOS	3×10^{45}	10^{-8}	$.08 \times 10^{-8}$
CLUSTERS	2×10^{44}	10^{-6}	$.8 \times 10^{-8}$

$$\text{TOTAL} = 2.4 \times 10^{-8}$$

$$\text{OBSERVED} = 3 \times 10^{-8} \text{ ergs/cm}^2 \text{ sec ster}$$

Fig. 13.   Contributions to the diffuse X-ray background (2–6 keV). The two types of source which could contribute significantly to the background are Seyferts and clusters.

NGC 4151, 4051 and 1068. The contribution from Seyferts could be much less than we have calculated if the fraction of Seyferts which are strong X-ray sources like NGC 4151 is much less than $\frac{1}{3}$.

We could approach the problem differently by integrating the log N log S curve out to a faint source limit of 5×10^{-3} counts s^{-1} to explain the background. If we assume that this intensity cutoff corresponds to the edge of the observable universe, we have a distance scale for the twenty or so weak sources already detected at about 3 counts s^{-1}. They are about 150 Mpc away, perhaps not close enough to have been noticed in earlier surveys in optical and radio. It seems that a large fraction of the all-sky X-ray background is certainly contributed by distant rich clusters of galaxies and possibly Seyferts. The rest may be contributed by the unidentified class of sources extrapolated to great distances.

It is to be expected that further study of the powerful extended X-ray sources in clusters and galaxies will tell us much about the evolution of clusters and perhaps about cosmological phenomena. Based on the results presented here, one may speculate that the Virgo and Perseus clusters are similar in that they contain an active galaxy on which an extended X-ray source is centered; the source has a power law spectrum. Also, these two clusters are irregular. In Coma, we do not see such a prominent active galaxy. The source may be centered on the kinematic center of the cluster, and seems to show a spectrum closer to that of isothermal bremsstrahlung at a temperature corresponding to the RMS velocity of galaxies in the cluster. Also, the cluster has a more symmetrical appearance, and is believed by many to be bound. If the X-ray emission from these clusters is due to hot gas, we may be seeing the active galaxies in Virgo and Perseus heating up the gas contemporaneously, whereas in Coma we see the thermalized remnant of a heating process which ended more than 10^9 yr ago.

The rich clusters, being very luminous and of great size, will be observable with finite angular size out to distances where redshift effects dominate. At 3200 Mpc, with

$Z = 1.4$, sources like the Coma cluster will have an angular size of one arc minute and should be observable with large orbiting telescopes of the near future. We may even be able to detect clusters out to $Z = 3$, and learn about evolution of clusters in the early stages of the formation of the universe.

Acknowledgements

I wish to acknowledge the assistance of all the members of the UHURU group at AS & E, in the preparations for this paper, especially Stephen Murray, Terry Matilsky, Herbert Gursky and Riccardo Giacconi; also William Forman of the Harvard College Observatory.

References

Becklin, E., Frogel, J., Kleinmann, D., Neugebauer, G., Ney, E., and Strecker, D.: 1971, *Astrophys. J. Letters* **170**, L15.
Bowyer, C., Lampton, M., Mack, J., and deMendonca, F.: 1970, *Astrophys. J. Letters* **161**, L1.
Bradt, H., Mayer, W., Naranan, S., Rappaport, S., and Spada, G.: 1967, *Astrophys. J. Letters* **161**, L1.
Byram, E. T., Chubb, T. A., and Friedman, H.: 1966, *Science* **152**, 66.
Forman, W., Kellogg, E., Gursky, H., Tananbaum, H., and Giacconi, R.: 1972, to be published.
Fritz, G., Davidsen, A., Meekins, J., and Friedman, H.: 1971, *Astrophys. J. Letters* **164**, L81.
Giacconi, R., Murray, S., Gursky, H., Kellogg, E., Schreier, E., and Tananbaum, H.: 1972, *Astrophys. J.* (Dec., to be published).
Gursky, H., Kellogg, E., Murray, S., Leong, C., Tananbaum, H., and Giacconi, R.: 1971, *Astrophys. J. Letters* **167**, L81.
Gursky, H., Solinger, A., Kellogg, E., Murray, S., Tananbaum, H., Giacconi, R., and Cavaliere, A.: 1972, *Astrophys. J. Letters* **173**, L99.
Kellogg, E., Gursky, H., Leong, C., Schreier, E., Tananbaum, H., and Giacconi, R.: 1971a, *Astrophys. J. Letters* **165**, L49.
Kellogg, E., Gursky, H., Murray, S., Tananbaum, H., and Giacconi, R.: 1971b, *Astrophys. J. Letters* **169**, L99.
Kellogg, E., Gursky, H., Tananbaum, H., Giacconi, R., and Pounds, K.: 1972, *Astrophys. J. Letters* **174**, L65.
Kunkel, W. and Bradt, H.: 1971, *Astrophys. J. Letters* **170**, L7.
Leong, C., Kellogg, E., Gursky, H., Tananbaum, H., and Giacconi, R.: 1971, *Astrophys. J. Letters* **170**, L67.
Mark, H., Price, R., Rodrigues, R., Seward, F., and Swift, C.: 1969, *Astrophys. J. Letters* **155**, L143.
Matilsky, T., Gursky, H., Kellogg, E., Tananbaum, H., Murray, S., and Giacconi, R.: 1972, *Astrophys. J. Letters*, to be published.
Meekins, J., Gilbert, F., Chubb, T., Friedman, H., and Henry, R.: 1971, *Nature* **231**, 107.
Schreier, E., Giacconi, R., Gursky, H., Kellogg, E., and Tananbaum, H.: 1972, *Astrophys. J. Letters*, to be published.
Solinger, A and Tucker, W.: 1972, *Astrophys. J. Letters* **175**, L107.
Tananbaum, H.: 1972, this volume, p. 9.
Tucker, W., Kellogg, E., Gursky, H., Giacconi, R., and Tananbaum, H.: 1972, *Astrophys. J.*, to be published.
Wade, C., Hjellming, R., Kellermann, K., and Wardle, J.: 1971, *Astrophys. J. Letters* **170**, L11.

16. THE PROPERTIES OF EXTRAGALACTIC X-RAY SOURCES FROM VISIBLE LIGHT OBSERVATIONS

WALLACE L. W. SARGENT

Hale Observatories, California Institute of Technology, Carnegie Institution of Washington, U.S.A.

Abstract. We describe the optical properties of the radio galaxy NGC 5128, the Seyfert galaxy NGC 4151 and the QSO 3C 273 all of which appear to be point sources of X-rays. We emphasize how the X-ray observations, particularly the low energy absorption cutoff, may help us to understand the detailed structure and source of energy in these diverse objects.

The clusters of galaxies in Virgo, Perseus, Coma and Centaurus, all associated with extended X-ray sources are described. They have diverse shapes, central concentrations and galactic populations, but all contain a radio galaxy and, in several cases, a low frequency radio halo around it. It is concluded that the X-ray emission is likely to be non-thermal in origin.

1. Introduction

The remarkable observations of extragalactic X-ray sources made recently by UHURU and earlier by rocket-born detectors have added to the rich variety of phenomena manifested by the violent events in galactic nuclei. While the recent X-ray observations have so far led to little new theoretical insight into these phenomena, it does appear that extragalactic X-ray sources can be divided into two categories – individual galaxies with active nuclei (including radio galaxies, Seyfert galaxies and quasars) and clusters of galaxies. These cluster sources appear to be extended and may themselves be due ultimately to active individual galaxies. I shall therefore divide this survey of the optical properties of the sources into two parts, dealing first with individual non-cluster sources in Section 2 and then with the clusters which contain extended sources in Section 3. The main theme will be the diversity of the objects in each category. Since some of the identifications are doubtful I shall only go into details where identification of the X-ray source is certain.

2. Individual Galaxies With Active Nuclei

A. NGC 5128

As Table I shows NGC 5128 (Centaurus A) is the most powerful extragalactic radio source in the sky. At a distance of 5 Mpc it is the nearest strong radio galaxy, defined as those objects with intrinsic powers in excess of 40^{41} erg s^{-1} between 10^8 and 10^{10} Hz. NGC 5128 appears to be a member of a loose group or chain of galaxies. The chain extends about 20° or roughly 1.5 Mpc across the southern sky (de Vaucouleurs, 1968; Arp 1968). Optically NGC 5128 is a unique object; it is shown in Figure 1. Although classified as DE 3 by Mathews *et al.* (1964), it differs from normal elliptical galaxies in two important respects. First it contains a broad dust lane across its center. The axis of rotation is perpendicular to this band (Burbidge and Burbidge,

Bradt and Giacconi (eds.), X- and Gamma-Ray Astronomy, 184–198. All Rights Reserved.
Copyright © 1973 by the IAU.

TABLE I

The strongest extragalactic radio sources

Name	Other	Radio flux*	Optical magn.	Redshift z	Optical spectrum	Type of galaxy	X-ray source
Cen A	NGC 5128	8700	7.0	0.0016	w.e.; abs.	DE3	Yes
Cygn. A	3C 405	8100	15.1	0.0570	s.e.	D3 (cl)	Yes?
Virgo A	3C 274 (NGC 4486)	970	8.7	0.0041	w.e.; abs	E2 (cl)	Yes; ext.
Pictor A	MSH 05-43	570	17.0	0.0342	s.e.	ND 1	?
Her A	3C 348	325	17.0	0.1540	w.e. (1 line)	D4 (cl)	Yes;?
Fornax A	NGC 1316	249	8.9	0.0058	w.e.; abs.	D4	?
Hydra A	3C 218	210	16.0	0.0530	w.e.	D2 dbl. (cl)	?
3C 353		203	15.3	0.0307	w.e.; abs.	D2	?

* Flux units at 178 MHz.

1959, 1962) as is the major axis of the radio source. Secondly the optical image of NGC 5128 contains faint extensions along the major axis. These faint outer parts give the impression of a spiral or even helical structure; the ends of the dust lane also curve around in a manner resembling a spiral.

Spectra of NGC 5128 show weak emission lines of [O II], [N II], [S II] and the Balmer lines in the vicinity of the dust band in addition to the absorption lines produced by cool stars normally found in the spectra of elliptical galaxies. The emission lines are sharp and have $H\alpha$ stronger than [N II]; in those cases where weak emission lines are found in the spectra of normal ellipticals $H\alpha < $ [N II]. Burbidge and Burbidge (1959, 1962) found that the mean recession velocity of the gas in the center of NGC 5128 is about 100 km s^{-1} greater than that indicated by the stellar absorption lines; furthermore the gas appears to be rotating more rapidly than the stars. It seems plausible that in NGC 5128 the radio components were ejected along the rotation axis of the galaxy. The Burbidges suggested that the dust lane is the remnant of an outburst which produced the extended radio source and that material is now falling back into the galaxy.

Recent infrared and radio observations of NGC 5128 have indicated that one of the knots in the center of the dust lane, relatively inconspicuous on optical photographs, (Figure 1) is the nucleus of the galaxy. (Kunkel and Bradt, 1971; Wade et al., 1971; Becklin et al., 1971). The spot in question has a size of about $3'' \times 5'' = 75 \times 110$ pc and has a luminosity of about 2.4×10^{41} erg s^{-1} between 0.35 and 1.0μ. Becklin et al. find that the 2.2μ and 3.5μ radiation from the supposed nucleus is more strongly concentrated towards the center than the shorter wavelength radiation; they infer from this that a small non-thermal core is superimposed on a nucleus composed of stars. They note, moreover, that the central 1$''$ of the nucleus has a surface brightness at 2.2μ 10 times that of a similar linear distance at the center of our galaxy and 25 times that of M 31. Wade et al. observed a weak radio component (2.9 flux units at 3.7 cm – negligible as compared with the radio emission from NGC 5128 as a whole) which is coincident with the infrared nucleus. This radio

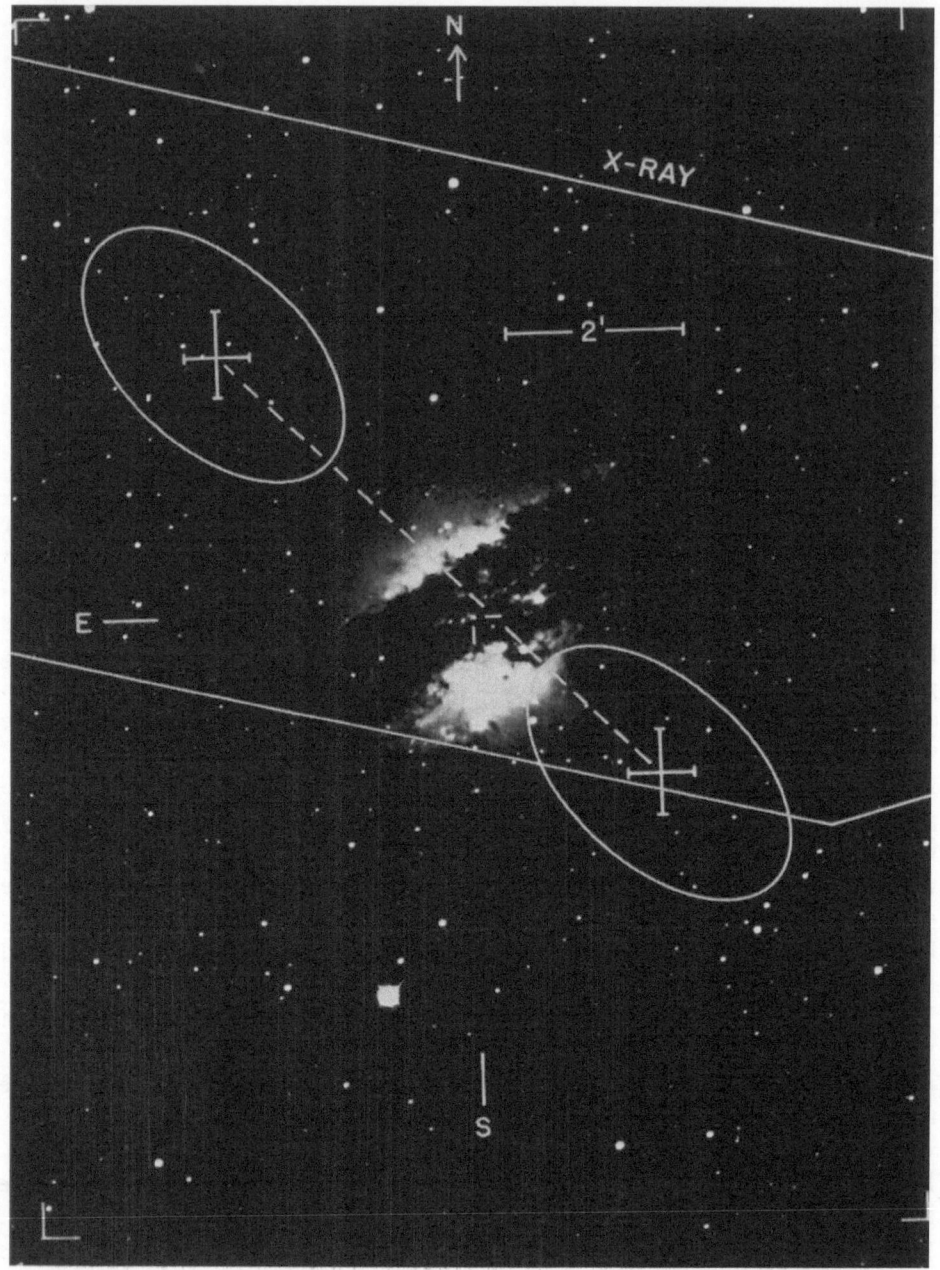

Fig. 1.   NGC 5128 – Sixty-inch, IIIa-J plate taken by Dr John Graham at CTIO. The infrared hot spot, the inner radio lobes, and the 1σ limits of a previously published UHURU X-ray error box are indicated. (From Kunkel and Bradt, 1971).

source is inferred to have a size of less than 0.25 pc and is similar to the nuclear sources found in the nuclei of M 87 and other elliptical galaxies.

There is no spectroscopic evidence in the case of NGC 5128 for a hot gas which might emit thermal X-rays – as there is in the case of the nucleus of NGC 4151. However, the UHURU measurements indicate that there is a column density of hydrogen of about 10^{23} atoms cm^{-2} in front of the X-ray source in NGC 5128. This probably indicates that the X-rays originate in the nucleus of the object because even in a spiral galaxy it would be impossible to observe such a large column density in the spiral arms. Thus the observations point to the infrared nucleus as the source of the X-ray emission; it is clearly important that further studies be made, particularly to put limits on the variability of the X-ray and infrared emission.

B. NGC 4151

This is one of the brightest Seyfert galaxies. These are mostly spiral galaxies; they contain a bright star-like nucleus which radiates a unique spectrum composed of broad emission lines (half widths from several hundred to several thousand km s^{-1}) superimposed on a smooth continuum which has both an ultraviolet and an infrared excess. The emission lines in Seyfert galaxies exhibit a wide range in ionization conditions – from [O I] to [Ne V] among the stronger lines; in this respect alone they can be distinguished spectroscopically from H II regions excited by hot stars. Some but not all of the N-type galaxies associated with strong radio sources have Seyfert spectra. Among the bright, nearby, Seyfert galaxies, some (for example NGC 1068 ≡ 3C 71 and NGC 1275 ≡ 3C 84) are strong radio galaxies while others, including NGC 4151, have only a weak nuclear radio source (Wade, 1968).

NGC 4151 is probably the best studied of the Seyfert galaxies; its spectrum is illustrated in Figure 2. Table II, taken from Oke and Sargent (1968), lists most of the emission lines observable in the region $\lambda\lambda 3300 - 11000$ with their estimated intensities at the source assuming a distance of 10 Mpc to NGC 4151. Iron is present as [Fe III], [Fe VII] and [Fe X]. The coronal line $\lambda 5303$ of [Fe XIV] is also listed in Table II; however, Weedman (1971) has recently shown that its wavelength is incorrect by a few tenths of an angstrom to be [Fe XIV].

The nucleus of NGC 4151 is unusually bright relative to the outer parts of the galaxy. The continuum between emission lines is smooth and appears to be mostly non-thermal in origin. After a contribution from hydrogen recombination has been subtracted, the remaining continuum varies as $f_v = Kv^{-1.2}$ between $\lambda 3300$ and $\lambda 11000$. The intensity of the continuous radiation varies on a time scale of months by about 1 mag.

The emission lines in the spectrum of NGC 4151 arise in a very small region, at most 2″ in diameter, which is only just resolved at the Earth's surface. The work of many authors (see, for example, Rees and Sargent (1972) for references) has led to the following picture of the nucleus.

(a) NGC 4151 is the prototype of those Seyfert galaxies which have broad wings to the permitted lines; the forbidden lines are sharper. It is supposed that the permitted

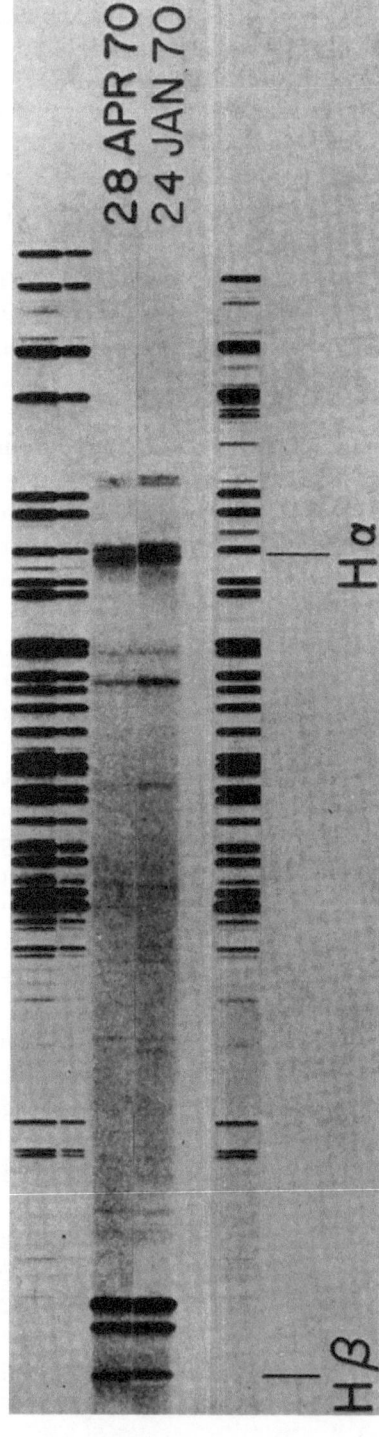

Fig. 2. Two spectra of the nucleus of the Seyfert galaxy NGC 4151, taken with the University of Arizona 90″ telescope by R. Cromwell and R. Weymann. Note the faint very broad wings on Hα and Hβ upon which are superimposed the sharp emission cores. The two strong lines to the right of Hβ are due to doubly ionized oxygen, and several of the fainter emission lines between Hα and Hβ are due to 6-times ionized iron. Of particular interest was the sudden appearance and disappearance of an absorption component of Hα and Hβ over a time scale of a few months indicating sporadic outbursts of matter in the nucleus.

TABLE II

NGC 4151: emission-line identifications and intensities

λ(Å)	Ion	Equivalent width (Å)	Flux at source (units of 10^{40} ergs s^{-1})	Strength relative to Hβ=100	
				NGC 4151	NGC 7027
10830.2	He I	163.5	5.98	81	87
10049.4	Pζ	30.0:	1.19:	16:	5
7329.9 ⎱ 7330.7 ⎰	[O II]	11.0	0.62	8	32
6731.3	[S II]	33.3	2.08	28 ⎱	6
6717.0	[S II]	27.7	1.75	24 ⎰	
6583.6	[N II]	29.0	1.83	25	90
6562.8	{ Hα wings	360.0	22.75	307 ⎱	290
	{ Hα core	34.0	2.15	29 ⎰	
6548.1	[N II]	6.2	0.39	5	30
6374.5	[Fe x]	2.8:	0.18	2:	
6363.9	[O I]	4.7	0.31	4	6
6300.2	[O I]	20.6	1.36	18	20
6085.3	[Fe VII]	7.6	0.52	7	
5875.6	He I	4.4	0.31	4	11
5754.8	[N II]	3.0:	0.22:	3:	8
5720.9	[Fe VII]	4.4	0.32	4	
5303.6	[Fe XIV]	1.0:	0.08	1:	
5006.8	[O III]	188.0	15.80	214	1460
4959.9	[O III]	62.0	5.23	70	480
4861.3	{ Hβ wings	72.0	6.07	82 ⎱	100
	{ Hβ core	16.0	1.35	18 ⎰	
4799.5	[Fe III]	1.0:	0.10:	1:	
4740.3	[A IV]	1.7	0.15	2	10
4711.4	[A IV]	1.7	0.15	2	8
4685.7	He II	21.7	1.88	25	46
4658.1	[Fe III]	5.9	0.51	7	
4471.5	He I	1.0:	0.10:	1:	4
4363.2	[O III]	5.4	0.48	7	26
4340.5	{ Hγ wings	21.4	1.92	26 ⎱	47
	{ Hγ core	5.6	0.50	7 ⎰	
4243.0	?	0.5:	0.04:	1:	
4228.0	?	0.5:	0.04:	1:	
4101.7	{ Hδ wings	5.4	0.51	7 ⎱	26
	{ Hδ core	3.7	0.35	5 ⎰	
4076.2	[S II]	2.3	0.22	3 ⎱	16
4068.6	[S II]	3.5	0.34	5 ⎰	
3970.1	Hε ⎱	10.3	1.03	14	52
3968.5	[Ne III] ⎰				
3889.1	Hζ ⎱	4.3	0.47	6	20
3888.6	He I ⎰				
3869.7	[Ne III]	19.3	2.12	29	120
3728.9	[O II] ⎱	28.7	3.75	51	35
3726.2	[O II] ⎰				
3425.8	[Ne v]	14.5	2.46	33	130

lines arise in a dense region in which $N_e \geqslant 10^8$ cm^{-3}. It is likely that the broad lines are produced by mass motions rather than by electron scattering and that this region is about 0.1 pc in diameter. The mass of dense gas is only about 50 $M_\odot$. A central mass of about 10^8 $M_\odot$ is required to retain this gas if it moves with a virial velocity of about 3000 km s^{-1}, corresponding to the widths of the Balmer line wings.

(b) most of the low ionization forbidden lines together with the cores of the forbidden lines are produced in a gas with $T_e \simeq 20\,000$ K and $N_e \simeq 5000$ cm^{-3} which extends over a region about 50 pc in diameter. This region, like the dense region cannot be completely filled, and the gas in it may be composed of clouds or filaments which occupy only about $\frac{1}{100}$ of the volume. This outer component has a mass of about 10^4 $M_\odot$ and is moving at speeds of around 400 km s^{-1}. Its total kinetic energy is about 4×10^{52} erg.

(c) The source of the non-thermal continuum was shown by measurements from Stratoscope (Danielson et al., 1968) to be less than 0".1 or 5 pc in size; the timescale of the optical variations indicate that this region is still smaller, perhaps 10^{16} cm in diameter.

The physical processes which are going on in the nucleus of NGC 4151 are still obscure and the X-ray observations are obviously an important new source of information. The main problem is how the gas in the nucleus is ionized and stirred up. Several alternative ideas have been proposed to explain the overall spectrum of NGC 4151; these are summarized by Osterbrock (1971). One idea is that the gas is ionized by the non-thermal continuum. Nüssbaumer and Osterbrock (1970) showed that the observed strengths of the various ionization stages of Fe up to [Fe x] could be explained in this way if the optical continuum $f_\nu \propto \nu^{-1.2}$ can be extrapolated below 234 eV, the ionization potential of Fe ix. A second possibility is that collisions between moving clouds or filaments produce a shock-heated gas with a temperature of several million degrees and that the thermal bremsstrahlung from this source in turn heats the remainder of the gas. This view of the nucleus of NGC 4151 as a chaotic assembly of moving and colliding gas clouds seems an attractive way of explaining the wide range of ionization conditions observed in NGC 4151 but it does not explain the ultimate source of the energy. The kinetic energy of the moving clouds would only last a few hundred years, less than the time taken for a typical cloud to cross the nucleus, at the present rate of radiative loss. Hence the problem of what stirs up the clouds is important.

The X-ray flux from the nucleus of NGC 4151 presumably arises either from the non-thermal source at the center or from the hot gas which appears to permeate the whole nucleus. In deciding between these two possibilities, the observed low energy cutoff in the X-ray spectrum is of great interest. This indicates a large column density (of order 10^{23} hydrogen atoms cm^{-2}) which, as in the case of NGC 5128, could hardly be encountered in the outer parts of NGC 4151. Using the numbers given earlier, the low density ($N_e = 5000$ cm^{-3}) gas in the nucleus should present a typical cross section of about 4×10^{21} atoms cm^{-2}, while the high density ($N_e \simeq 10^8$ cm^{-3}) should present a cross section of about 3×10^{23} cm^{-2}, very similar to that given by the X-ray absorption. Now it is likely that the high density gas is responsible for weak, variable absorption lines seen at He I λ3889 and in the blue wings of the lower Balmer lines (Anderson and Kraft, 1969). These absorption lines vary on a timescale of less than a year and this may be due to dense filaments or clouds moving across the continuum source (Rees and Sargent, 1972). In any case it seems likely that the X-rays come from the very center of NGC 4151 and not from the whole extended region that emits the forbidden line spectrum. A test of this hypothesis would be to look for variations in the X-ray flux and for variations in the amount of low energy absorption. Both should occur on timescales of several months.

c. 3C 273

This is still the brightest known quasi-stellar object and consists of a 12th magnitude starlike object which radiates a spectrum very similar to that of a Seyfert galaxy, together with a faint jet about 20″ long. The optical spectrum of 3C 273 contains very broad emission lines of Mg II, H, [O III] and Fe II. There are no emission lines indicating the presence of a very hot gas in the spectrum of 3C 273; however, the lines are much broader than those in NGC 4151 and weak features are correspondingly harder to detect. The optical continuous spectrum of 3C 273 appears to be flat and fairly satisfactory models to explain the emission line spectrum have been achieved by extrapolating a spectrum $f_v \sim v^{-0.7}$ into the far ultraviolet (Bahcall and Kozlovsky, 1969). Their model has $N_e \simeq 3 \times 10^7$ cm^{-3}, $T_e = 1.7 \times 10^4$ K, a radius $R = 14$ pc and a filling factor $f = 10^{-3}$. The mass of gas is $10^5 \, M_\odot$. Osterbrock (1971) used this model to calculate the strengths of a number of emission lines and found that these agreed well with the observed line strengths in a typical QSO spectrum (see Table III). We thus have some confidence in Bahcall and Kozlovsky's model of 3C 273; according to it the column density in an average direction through the gas is about $RfN_e = 10^{24}$ hydrogen atoms cm^{-2}. This is greater than the column density of 5×10^{22} cm^{-2} inferred from the UHURU observation of a low energy

TABLE III

Comparison of 3C 273 model with mean observed quasar spectrum

Line		Relative photon rate	Observed photon rate	Line		Relative photon rate	Observed photon rate
Lα	λ1216	166.0	S	Mg II	λ2799	3.6	S
C IV	λ1549	42.0	S	N V	λ1240	2.8	–
He II	λ1640	22.0	W	[Ne V]	λ3426	2.4	W
Hβ	λ4861	20.0	M	O III]	λ1664	2.1	–
He II	λ4686	18.0	–	Si IV	λ1397	1.7	–
Hγ	λ4340	10.0	M	[O III]	λ5007	1.6	M
C III]	λ1909	9.4	M	Si III]	λ1892	1.2	–
C II]	λ2327	8.0	obs	[Mg VII]	λ2632	1.1	–
O IV]	λ1402	6.0	M	[O III]	λ4363	1.0	obs
Hδ	λ4101	6.0	obs	[Ne V]	λ3346	0.9	obs
O VI	λ1034	5.6	–				

S = strong; M = medium; W = weak; obs = observed.

cutoff in the X-ray spectrum. There are several possible reasons for the discrepancy. The elements Ne, O and C which are chiefly responsible for the X-ray absorption, are completely ionized for an appreciable fraction of the radius in these models. Moreover, the filling factor f is only a crude average way of describing what is probably a very inhomogeneous situation so that f could be appreciably smaller in the direction of the line of sight than it is on average.

3. X-Ray Sources in Clusters of Galaxies

There is preliminary evidence that the X-ray sources in clusters are all extended; the X-ray clusters may all contain radio galaxies in which case their X-ray emission is more likely to be non-thermal rather than thermal bremsstrahlung from a hot gas in the cluster. We shall concentrate our discussion on the nearby clusters – Virgo, Perseus, Coma and Centaurus – that are known to be extended at the time of this Symposium.

A. THE VIRGO CLUSTER

This is the nearest reasonably populous cluster of galaxies. According to de Vaucouleurs (1961), The Virgo cluster consists of a diffuse elliptical cloud of spiral and irregular galaxies centered at $12^h 27^m.5, +13°.9$ and a concentrated cluster of elliptical and SO galaxies centered at $12^h 26^m.5, +13°.2$. The distribution of bright galaxies over the sky near the Virgo cluster according to morphological type is shown in Figure 3, which is taken from de Vaucouleurs' paper. The mean redshift of the E cloud is $+950$ km s^{-1} and that of the S cloud $+1450$ km s^{-1}.

The X-ray source in the Virgo cluster is centered on M 87 ≡ NGC 4486, the famous radio galaxy with the remarkable blue jet. This galaxy lies at $12^h 28^m.3, +12°40'$ and is not quite at the center of the E cloud. The dense core of the E cloud contains 19 bright ellipticals within a radius of $1°.25$ from the center; around this core the elliptical galaxies are distributed very unevenly. The E cloud in the Virgo cluster has a radial velocity dispersion of 550 km s^{-1}; this leads to an estimated mass for the system of about $4 \times 10^{14} M_\odot$ and a mean mass per galaxy of around $4 \times 10^{12} M_\odot$. This exceeds by at least an order of magnitude the mass estimated to be in the galaxies.

B. THE PERSEUS CLUSTER

The Perseus cluster, shown in Figure 4, is one of the nearest clusters of galaxies to qualify for entry in Abell's (1958) catalogue of rich clusters of galaxies. (Table IV lists all clusters of distance class 2 or less in Abell's catalogue). The Perseus cluster is very irregular in shape. The densest concentration of galaxies is around NGC 1275, the radio galaxy 3C 84 which is also a Seyfert galaxy. From NGC 1275 a chain of galaxies stretches to the west, terminating at IC 310 which is also a weak radio galaxy. Most of the galaxies in the Perseus cluster are ellipticals and SO galaxies. Recent studies of the dynamics of the Perseus cluster by Chincarini and Rood (1971) and by Gunn and Sargent (1972) have shown several peculiar features. The radial

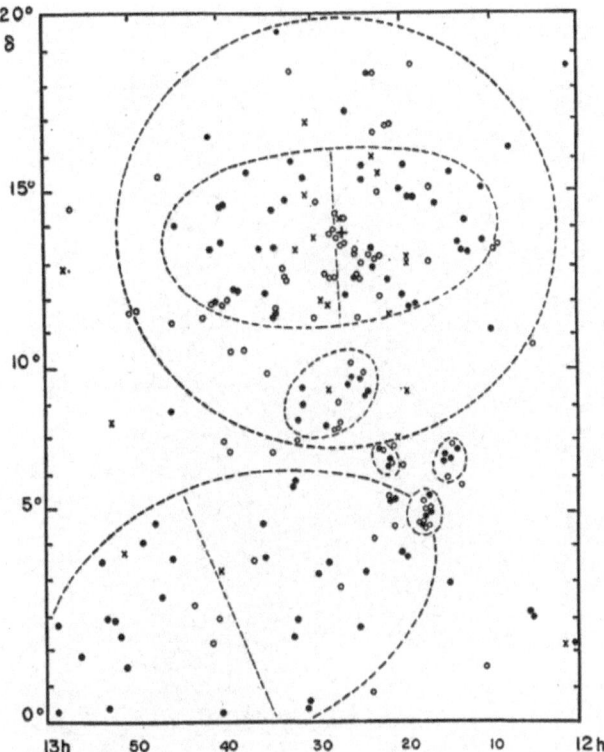

Fig. 3.   Distribution of 212 galaxies in the region of the Virgo Cluster. Dots: spirals and irregulars; circles: ellipticals and lenticulars; crosses: type unknown. (From de Vaucouleurs, 1961).

velocity dispersion of the cluster is 1420 km s^{-1} about its mean redshift of 5460 km s^{-1}. This is the largest velocity dispersion of any known cluster and leads to a mass estimate from the virial theorem of about 10^{15} $M_\odot$. Since the cluster contains at most 50 large galaxies, either most of this mass is in dark matter or the cluster is unstable on a timescale of about 3×10^8 yr. Despite the existence of an obvious chain of galaxies in the Perseus cluster, there is no evidence that the system is rotating; moreover, there is no obvious correlation of redshift with position in the cluster. Although Chincarini and Rood (1971) fitted a single Gaussian curve to the histogram of a number of galaxies versus redshift, there is some evidence in their data, as well as in that of Gunn and Sargent, (1972) that the distribution might be bimodal. This might indicate that Perseus, like Virgo, is in reality two clusters superimposed on the line of sight; more redshifts are being obtained in order to study this possibility further.

NGC 1275, whose redshift is close to that of cluster mean, is remarkable in that, as well as containing a Seyfert nucleus, it has outer emission patches which are receding from it at 3000 km s^{-1}. Lynds (1970) has published a beautiful photograph showing that this receding nebulosity has a filamentary structure, similar to that in the Crab nebula.

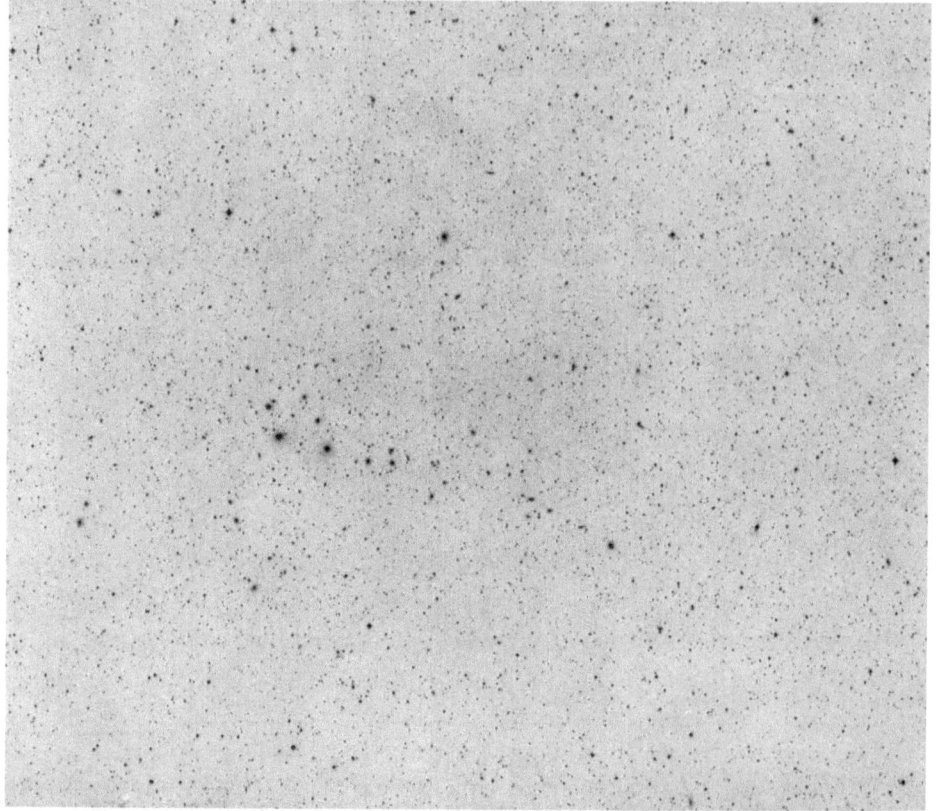

Fig. 4.   Perseus Cluster of Galaxies (Hale Observatories photo). North is up and east is to the left.

Radio observations by Ryle and Windram (1968) show possible evidence for a physical interaction between NGC 1275 and IC 310 to the W and NGC 1265, the radio galaxy 3C 83.1, which lies 27′ to the NW of NGC 1275. The radio contours around IC 310 and NGC 1265 are in each case bent away from the direction to NGC 1275 in the form of a tail; moreover, the radio center is in each case displaced from the center of the optical galaxy away from NGC 1275. This evidence for interactions between remote galaxies is particularly remarkable because NGC 1265 has a redshift which is 2370 km s^{-1} greater than that of NGC 1275 and the center of the Perseus cluster. A mass of 10^{15} $M_{\odot}$ for the cluster is again required to gravitationally bind NGC 1265. The observations by Ryle and Windram (1968) at 408 MHz show that the Perseus cluster contains an extended source 3C 84B which is centered between NGC 1275 and NGC 1265, unlike the extended source of X-ray emission which appears to be centered on NGC 1275.

C. THE COMA CLUSTER

This cluster, shown in Figure 5, is the prototype example of a spherical, centrally

TABLE IV

Abell clusters nearer than distance class 2

Abell No. (1)	D (2)	R (3)	Magn. (4)	Class (5)	Radio flux (6)	Remarks (7)	X-ray source (8)
194	1	0	13.9	II	5.5	3C 40	
262	1	0	13.3	III	0.15		
347	1	0	13.3	II–III	9.2	3C 66	
400	1	1	13.9	II–III	6.3	3C 75	
407	2	0	14.7	--	0.7	4C 35.6	
426	0	2	12.5	II–III	13.0	*Perseus*; 3C 83.1, 3C 84	Yes
539	2	1	14.4	III	0.13		
548	1	1	13.7	III	0.19		
569	1	0	13.8	–	1.05	Radio size ~ 1′4	
576	2	1	14.4	III	<0.1		
779	1	0	13.8	II	0.20		
1060	0	1	12.7	III	<0.1		
1185	2	1	14.3	–	<0.1		
1213	2	1	14.5	III	1.9	CTD 72; radio size ~ 1′	
1228	1	1	13.8	III	<0.1		
1314	1	0	13.9	–	0.93		
1367	1	2	13.5	II–III	5.90	3C 264; radio size ~ 1′5	Yes
1656	1	2	13.5	II	0.74	*Coma*; (Coma C)	Yes
1736	2	0	14.8	III	0.64		
2147	1	1	13.8	–	0.92	Radio size ~ 1′3	
2151	1	2	13.8	III	1.2	*Hercules*	
2152	1	1	13.8	III	<0.1		
2162	1	0	13.7	–	<0.1		
2197	1	1	13.9	II	<0.1		
2199	1	2	13.9	I	<3.55	NGC 6166; 3C 338	
2634	1	1	13.8	I–II	3.0	NGC 7720; 3C 465	
2666	1	0	13.8	–	<0.1		

Notes: Column (2): distance class;
Column (3): richness class;
Column (4): magnitude of 10th brightest galaxy;
Column (5): morphological class of cluster from Bautz and Morgan (1970);
Column (6): radio power in flux units at 1415 MHz from Fomalont and Rogstad (1966).

condensed rich cluster. Such clusters almost invariably contain predominantly E and SO galaxies. The Coma cluster is dominated by two very luminous galaxies, NGC 4889 to the E and NGC 4874 to the W. Counts show that the center of the cluster lies close to NGC 4874. A dynamical study of the Coma cluster has recently been made by Rood *et al.* (1972) on the basis of many new redshift measurements. The radial velocity dispersion of the cluster is 1060 km s^{-1} and the mean redshift is 6888 km s^{-1}. The autors find that the cluster is not measurably rotating and that the bright and faint galaxies have the same dispersion in radial velocity within the accuracy of the data. They conclude that the highly condensed, spherical, shape of the cluster must indicate that it is stable. The derived mass from the virial theorem is $3.4 \times 10^{15} \, M_\odot$ and the mass to light ratio is $M/L_p \sim 250$. This is higher than that obtained for individual elliptical galaxies by a factor of around 5.

Fig. 5.   Coma Cluster of Galaxies (Hale Observatories photo). North is up and east is to the left.

Weak radio sources have been observed at 408 and 1407 MHz in the Coma cluster by Willson (1970). Thirteen are certainly or probably associated with individual Coma galaxies, including NGC 4874 but not its luminous companion NGC 4889. NGC 4869 which lies about 6′ to the SW of NGC 4874 is a weak source and Willson suggests from the shape of the radio contours that the two galaxies may be interacting. In addition to discrete sources, the Coma cluster has an extended, low frequency, radio halo about 40′ arc or 80 kpc in diameter and at the center of the cluster. Willson supposes that this weak halo is associated in some way with NGC 4874. He estimates that to maintain the halo would require the ejection of 10^{51} erg yr^{-1} for the past 3.5×10^9 yr from NGC 4874.

Although the Coma cluster is usually considered in the literature to be symmetric, there is a conspicuous chain or fan of galaxies extending to the SW from NGC 4874. This is obvious to the eye and persists on contours made from deep counts (Dam and Waddington 1972), as can be seen on Figure 6.

D. THE CENTAURUS CLUSTER

An extended X-ray source about 30′ in diameter and centered roughly on NGC 4696 has been discovered by UHURU. NGC 4696 is the radio source Pks 1245-41; it is

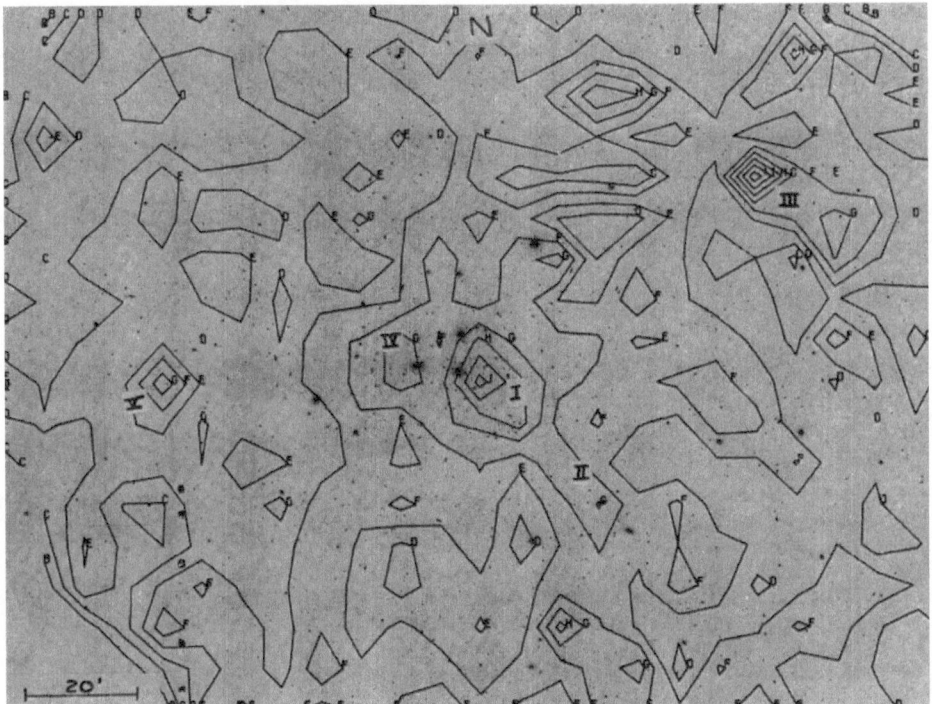

Fig. 6. Contour map of the Coma Cluster with spur extending from NGC 4875 (II). North is up and east is to the left.

an E galaxy in a rich cluster. Klemola (1969) states that the cluster contains 100 galaxies over an area $180' \times 120'$. Most of the galaxies in the cluster appear to be spirals and lenticulars. The redshift of NGC 4696 is 2790 km s^{-1} (Burbidge and Burbidge, 1972); its spectrum contains the normal absorption lines plus emission lines of [N II] and Hα. As is common in the nuclei of normal galaxies, [N II] > Hα.

4. Conclusions

A. CLUSTER SOURCES

So far all the extended X-ray sources are associated with clusters of galaxies which contain a radio galaxy. In several cases the radio galaxy is surrounded by a weak radio halo. The clusters concerned have diverse shapes, degrees of central condensation and galactic populations. These observations point to a non-thermal origin – inverse Compton effect on the microwave background photons of the relativistic electrons responsible for the radio halos – rather than thermal bremsstrahlung resulting from infalling intergalactic matter (Gott and Gunn, 1971).

B. INDIVIDUAL SOURCES

These appear to be produced in the centers of active galactic nuclei which are in turn diverse in their optical properties. The observations of the X-rays, particularly of the absorption cutoff at low energies, should lead to important new constraints on detailed models of Seyfert nuclei and quasi-stellar objects.

Acknowledgements

I am very much indebted to Dr E. Kellogg for generously informing me of the new results from UHURU in advance of publication.

References

Abell, G. O.: 1958, *Astrophys. J. Suppl.* 3, 211.
Anderson, K. S. and Kraft, R. P.: 1969, *Astrophys. J.* 158, 859.
Arp, H. C.: 1968, *Publ. Astron. Soc. Pac.* 80, 129.
Bahcall, J. N. and Kozlovsky, B. Z.: 1969, *Astrophys. J.* 155, 1077.
Bautz, L. P. and Morgan, W. W.: 1970, *Astrophys. J. Letters* 162, L149.
Becklin, E. E., Frogel, J. A., Kleinman, D. E., Neugebauer, G., Ney, E. P., and Strecker, D. W.: 1971, *Astrophys. J. Letters* 170, L15.
Burbidge, E. M. and Burbidge, G. R.: 1959, *Astrophys. J.* 129, 271.
Burbidge, E. M. and Burbidge, G. R.: 1972, *Astrophys. J.* 172, 37.
Burbidge, G. R. and Burbidge, E. M.: 1962, *Nature* 194, 367.
Chincarini, G. and Rood, H. J.: 1971, *Astrophys. J.* 16., 321.
Dam, R. and Waddington, B.: 1972, private communication.
Danielson, R., Savage, B. D., and Schwarzschild, M.: 1968, *Astrophys. J. Letters* 154, L117.
de Vaucouleurs, G.: 1961, *Astrophys. J Suppl.* 6, 213.
de Vaucouleurs, G.: 1968, *Nearby Groups of Galaxies*, preprint.
Fomalont, E. B. and Rogstad, D. H.: 1966, *Astrophys. J.* 146, 528.
Gott, J. Richard, III and Gunn, James E.: 1971, *Astrophys. J. Letters* 169, L13.
Gunn, J. E. and Sargent, W. L. W.: 1972, in preparation.
Klemola, A. R.: 1969, *Astron. J.* 74, 809.
Kunkel, W. E. and Bradt, H. V.: 1971, *Astrophys. J. Letters* 170, L7.
Lynds, C. R.: 1970, *Astrophys. J. Letters* 159, L151.
Mathews, T. A., Morgan, W. W., and Schmidt, M.: 1964, *Astrophys. J.* 140, 35.
Nüssbaumer, H. and Osterbrock, D. E.: 1970, *Astrophys. J.* 161, 811.
Oke, J. B. and Sargent, W. L. W.: 1968, *Astrophys. J.* 151, 807.
Osterbrock, D. E.: 1971, in D. J. K. O'Connell (ed.), *The Nuclei of Galaxies*, North-Holland Publ. Co, Amsterdam, p. 151.
Rees, M. J. and Sargent, W. L. W.: 1972, *Comm. Astrophys. Space Sci.* 4, 7.
Rood, H. J., Page, T. L., Kintner, E. C., and King, I. R.: 1972, *Astrophys. J.* 175, 627.
Ryle, M. and Windram, M. D.: 1968, *Monthly Notices Roy. Astron. Soc.* 138, 1.
Wade, C. M.: 1968, *Astron. J.* 73, 876.
Wade, C. M., Hjellming, R. M., Kellermann, K. I., and Wardle, J. F. C.: 1971, *Astrophys. J. Letters* 170, L11.
Weedman, D. W.: 1971, *Astrophys. J. Letters* 167, L23.
Willson, M. A. G.: 1970, *Monthly Notices Roy. Astron. Soc.* 151, 1.

17. EXTRAGALACTIC X-RAY SOURCES

G. R. BURBIDGE

*University of California, San Diego, Dept. of Physics,
La Jolla, Calif. 92037, U.S.A.*

Abstract. A review is given of the physical properties of extragalactic X-ray sources. The generation of X-rays as thermal bremsstrahlung, by the synchrotron mechanism or by Compton Scattering, is discussed. It is shown that each may be important depending on the circumstances. Only more detailed observations will enable us to decide which process dominates in any given source.

1. Introduction

Balloon and rocket observations of celestial X-ray sources have been made for the last decade, and in the last two years satellite observations have been under way. The most extensive catalogue of X-ray sources so far available is that made by the American Science and Engineering group based on observations with the UHURU satellite which was launched in December 1970. While the sky has not been completely surveyed by this satellite, the results obtained, combined with those which had been made previously, give a fairly good preliminary idea of the level of activity in the X-ray Universe.

In this paper I shall discuss only those X-ray sources which have either been tentatively identified with known extragalactic optical or radio sources.

The X-ray astronomers have begun their identifications in a period when many extragalactic objects with remarkable energetic properties in radio, optical and infrared wavelengths have been found. It is therefore natural that they have attempted to see if some of these objects can be identified as X-ray sources, as well as attempting to identify some of the bright nearby galaxies and the rich clusters of galaxies as X-ray emitters. The lists of optical objects which have been used are therefore:

  (i) Catalogues of radio sources;
  (ii) Lists of quasi-stellar objects;
  (iii) Seyfert galaxies and other galaxies with active nuclei;
  (iv) Catalogues of bright galaxies and of clusters and groups of galaxies.

The positional uncertainties in X-ray astronomy are, for the majority of sources, very large compared with those in optical and radio astronomy. In the UHURU catalogue the sizes of the 90% confidence error boxes cover a wide range of values depending on the intensity of the source and the number of times the region has been scanned. The sizes range from several square degrees to ~ 0.01 sq deg. The limit on the flux which is detectable in the 2–6 keV range is about 5×10^{-11} erg cm^{-2} s^{-1}.

This means that the weakest sources that can currently be detected at typical extragalactic distances will have the following luminosities:

Bradt and Giacconi (eds.), X- and Gamma-Ray Astronomy, 199–207. All Rights Reserved.
Copyright © 1973 by the IAU.

Object	Distance	L_x
In Magellanic Clouds	60 kpc	2.0×10^{37} erg s^{-1}
M 31	630 kpc	2.2×10^{39} erg s^{-1}
Virgo Cluster	15 Mpc	1.2×10^{42} erg s^{-1}
Source at	100 Mpc	5.3×10^{43} erg s^{-1}
Source at	1000 Mpc	5.3×10^{45} erg s^{-1}

Very little information is available so far on the X-ray spectra of the extragalactic sources. In a few cases they have been shown not to be point sources. In Table I, I give details of the extragalactic objects of various kinds which have been identified as X-ray sources.

Among the objects identified are nearby normal galaxies, two of the classical Seyfert galaxies, NGC 4151 and NGC 1275 which is associated with an extended source, one N-system 3C 390.3 and one QSO 3C 273. The remainder are identified with radio galaxies or cluster of galaxies which contain radio sources.

The types of object identified are to some extent a reflection of the classes of optical or radio objects which have been looked for. In many sources for which no identifications have been made, the uncertainty in the position is so great that many faint

TABLE I

Some identified extragalactic X-ray sources

Object	Type	X-ray flux (1 unit $= 1.7 \times 10^{-11}$ erg cm^{-2} s^{-1})
Large magellanic cloud	Dwarf galaxy	20.7, 9.3, 19.3, 14.9 (discrete sources)
Small magellanic cloud	Dwarf galaxy	28
M 31	Sb	1.9 ± 0.3
NGC 5128	Radio galaxy	7.4 ± 0.4
NGC 4151	Seyfert galaxy	3.5 ± 0.4
M 87 – Virgo cluster – 3C 274	Radio galaxy with halo	21.7 ± 0.3 (extended source)
NGC 4696 – PKS 1245-41	Radio galaxy in cluster	5.9 ± 0.4 (extended source)
NGC 1275 – Perseus cluster – 3C 84	Seyfert galaxy in cluster	43.1 ± 0.3 (extended source)
NGC 3862 – 3C 264 – Abell 1367	Radio source in cluster	3.6 ± 0.4
3C 66 – Abell 347	Radio source in cluster	4.2 ± 0.6
NGC 4874 and 4869 and Coma cluster, 5C 4.85, 5C 4.81	Extended radio source in cluster	14.9 ± 0.3 (extended source)
3C 390.3[a]	N system	3.6 ± 0.4
Cygnus A – 3C 405[b]	Radio galaxy	5.1 ± 1.4
Hercules A – 3C 348	Radio galaxy	6.7 ± 1.0
3C 273	QSO	4.2 ± 0.5
3C 129-129.1	Radio source	5.5 ± 0.9
NB 78.26 – Abell 2256	Radio source in cluster	2.9 ± 0.3 (extended source)

[a] Identification very uncertain.

[b] Identification very doubtful since sources lie in a region containing many galactic X-ray sources. Much greater positional accuracy is required before this identification can be made convincing.

galaxies lie inside the error box. Thus the possibility that X-ray observations will turn up a new class of extragalactic objects which emit most of their power in the X-ray region and are not detectable as radio sources, or as objects which have very active nuclei, cannot be tested until X-ray source positions can be measured with much higher accuracy than is presently available. We shall briefly discuss predictions along these lines later in this paper.

2. Processes Which Generate X-Rays

There are three processes which may give rise to powerful X-ray sources on the galactic scale. They are:

(1) Thermal bremsstrahlung,
(2) Synchrotron radiation,
(3) Compton scattering.

In the galactic sources it has been shown that at least two of these processes are operating. In objects like Sco X-1 a significant part of the flux is simply thermal radiation from a hot gas cloud (bremsstrahlung), while in the Crab nebula the recent polarization observations by Novick *et al.* (1972) have shown conclusively that the synchrotron process is operating in the X-ray band. However, in the extragalactic sources no definite evidence has yet been found which demonstrates that any one of the three processes dominates. Several of the arguments which will be made in what follows can be made plausible but none is conclusive.

What are the ingredients which are required for each of these processes to work:

(1) In this case what is required is a mass of gas and an energetic source which is capable of maintaining it at temperatures of $\sim 10^7$–10^8 deg.

(2) In this case we need large fluxes of highly relativistic electrons and fairly strong magnetic fields. If $\nu = 10^{19}$ Hz, we require that

$$E_x \approx 2.5 \times 10^{12} \, H_\perp^{-1/2} \text{ eV}$$

while the half-lives of the electrons,

$$\tau_{1/2} \approx 300 \, H_\perp^{-3/2} \text{ s}.$$

Thus for a range of values of $H_\perp$ ranging from 10^{-6} to 10^3 G we find the values shown in Table II.

TABLE II

Energies and half-lives of energies giving rise to X-rays by the synchrotron process

$H_\perp$ (gauss)	E_e (ev)	$\tau_{1/2}$ (seconds)
10^{-6}	2.5×10^{15}	3×10^{11}
10^{-3}	7.5×10^{13}	10^7
1	2.5×10^{12}	300
10^3	7.5×10^{10}	10^{-2}

The magnetic fields between galaxies must be very weak indeed ($\ll 10^{-6}$ G), and even in extended radio sources the 'equipartition' calculations often require fields no greater than 10^{-5} G. Thus it is clear that the generation of synchrotron X-rays in extended sources containing very weak magnetic fields is not likely unless there is an excessively large flux of very high-energy electrons present. It is only reasonable to argue that synchrotron X-rays are generated in the vicinity of a compact object (as is the case in the Crab nebula) where much stronger magnetic fields are thought to exist. Thus it is possible that X-ray sources arising in the nuclei of galaxies and QSO's are generated at least in part by the synchrotron process (Burbidge, 1970a). If it can be established that a source has a large angular size, a synchrotron origin for the X-rays is only reasonable if it can be shown that the source is made up of a large number of compact objects. When high-resolution X-ray observations are made, fine structure associated with such objects should eventually show up. If it does not, this would be evidence against a synchrotron origin. Conclusive evidence in favor of a synchrotron origin would be the detection of linear polarization.

(3) The ingredients required to generate an X-ray source by Compton scattering are a flux of relativistic electrons and a field of low-frequency irradiation. Since it has been known for nearly twenty years that extragalactic radio sources radiate by the synchrotron process, the existence of relativistic electrons in both small and extended regions centered on galaxies and QSO's has been established. We also know that violent activity in the nuclei of galaxies frequently gives rise both to relativistic particles and a high density of optical, radio, infrared and microwave radiation (cf. Burbidge, 1970b). Thus Compton scattering may generate X-ray sources confined to the regions of high lower frequency radiation in galactic nuclei (Burbidge, 1970a).

Also the existence of a universal field of microwave radiation has been established by the observations of Penzias and Wilson (1965) and many others following them. This may or may not be primordial radiation generated in a big bang. But independent of its origin, Compton scattering of relativistic electrons in this radiation field can be expected to give rise to extended X-ray sources (Brecher and Burbidge, 1972a).

3. Application of Theory to Identified Sources

For the majority of the extragalactic objects listed in Table I we do not have four critical pieces of information which might enable us to narrow down the choice of emission mechanisms. These are (1) measurements of the angular extents of the sources; (2) measurements over a wide enough energy range to give a good idea of the shape of the spectrum of the source; (3) polarization measurements; (4) detection of variability which would show that the source is very small. We do know that a few of the sources are extended (Gursky et al., 1972; Kellogg, 1972), but nothing is known about the angular sizes of the remainder.

As far as (2) is concerned the very simple-minded argument which has frequently been used and will be mentioned here is that a spectrum or exponential type suggests that the radiation is emitted by an optically thin hot gas cloud, while an inverse power law form would indicate a synchrotron or Compton scattering origin. As is well known, these arguments are far from conclusive. If the temperature varies through a hot gas cloud, the X-ray spectrum will not be a simple exponential, while in many of the compact radio or infrared sources, the spectra which arise from the synchrotron process, do not have a simple power-law form with a single negative index.

Among the sources listed in Table I we can make the following guesses concerning the mechanism which is operating.

A. MAGELLANIC CLOUDS, OUR GALAXY AND M 31

These are presumably prototypes of normal galaxies whose X-ray emission is coming from a number of discrete sources associated with stars, probably binary systems, with a minor contribution coming from supernova remnants. The emission mechanism in the binary systems is not really established in most cases, but thermal bremsstrahlung is likely to be important. At least for our own Galaxy, the nucleus is not an important X-ray source. Since only a comparatively small number of discrete sources in our Galaxy and the Small Magellanic Cloud emit at power levels $\sim 10^{38}$ erg s^{-1}, and they apparently make up the bulk of the total X-ray emission from such galaxies, there may be very considerable ranges in X-ray luminosity between galaxies of similar types since the number of powerful sources may be a function of the number of binaries at the critical emitting phase, etc.

B. NGC 5128, NGC 4151, ANY COMPACT SOURCE ASSOCIATED WITH M 87, NGC 4696, NGC 1275, 3C 390.3 AND 3C 273

All of these extragalactic sources are known to have very active nuclei in optical, infrared, microwave or radio wavelengths. The data on the Seyfert nuclei in NGC 4151 and NGC 1275 and the measurements of M 87, the N-system 3C 390.3 and the QSO 3C 273 have been discussed in detail elsewhere (Burbidge, 1970b). Recently a small bright infrared nucleus has been discovered in NGC 5128 (Kunkel and Bradt, 1971), while NGC 4696 is a southern radio galaxy containing a radio source of small angular size (Ekers, 1968).

If it can be shown that the X-ray sources arise in these active nuclei, then in principle any one of the three mechanisms previously described can be operating. Evidence for the existence of high-speed mass motions of gas in the nuclei of NGC 4151, NGC 1275 and in 3C 390.3 is available from the optical observations. Thus the X-ray emission may arise from the dissipation of kinetic energy contained in these motions. Alternatively the Compton or synchrotron processes taking place in a tiny nucleus in which the nonthermal optical and infrared flux is generated may be responsible as was originally proposed for the nucleus in M 87 (Burbidge, 1970a).

C. EXTENDED X-RAY SOURCES ASSOCIATED WITH CLUSTERS OF GALAXIES AND RADIO SOURCES

At present we have evidence that four of the sources listed in Table I, namely the sources in the Virgo, Perseus, Coma and Abell 2256 clusters (Kellogg, 1972), and the source in the cluster of which the brightest member is NGC 4696, are extended, with sizes ranging from $\sim 30'$ to $\sim 50'$ corresponding to linear dimensions ~ 200–800 kpc. Moreover a number of the other proposed identifications are with clusters of galaxies containing radio sources, though we have no size measurements for the X-ray sources and in some cases none for the radio sources either. While Gursky *et al.* (1972) have stressed the correlation between extended X-ray sources and rich clusters of galaxies, studies of correlations between (a) powerful X-ray sources and clusters of galaxies, and (b) powerful X-ray sources and extended radio sources (Brecher and Burbidge, 1972b) have shown that both correlations are present and that therefore neither hypothesis can be rejected. If (a) alone were important and if a further correlation could be found between the X-ray luminosity and the dynamical state of the cluster or some other physical property of the galaxies in the cluster, it would be reasonable to conclude that the radiation is thermal bremsstrahlung from a hot gas whose energy ultimately derives from the galaxies in the cluster or it might be generated in the galaxies directly. While the sample of sources is very small, Solinger and Tucker (1972) have claimed to have found such a correlation. Namely, for three clusters, Virgo, Perseus and Coma, they find the data to be consistent with $L_x \propto (\Delta v)^4$, where Δv is the velocity dispersion of the galaxies in the cluster. It is hard to decide how much significance should be attached to such a relation. The problem is that while the velocity dispersions in the three clusters only vary by a rather small factor, the clusters are rather different in their galactic populations and probably in their states of evolution. A discussion of the optical properties of the clusters has been given in this Symposium by Sargent (p. 184). In the Virgo cluster two subclusters of spirals and ellipticals, respectively, have been identified, the Perseus cluster contains only a comparatively small number of galaxies arranged in two chain-like structures, while the Coma cluster is a very rich cluster with a more symmetrical distribution of galaxies, and it apparently only contains ellipticals and SO galaxies. In each cluster the visible galaxies are not able to bind the system, and if it is argued that they are stable configurations very large amounts of dark material, either in the form of uncondensed gas or of condensed objects with very high mass-to-light ratios are required to stabilize them. While many investigators have taken this as evidence that much dark matter must be present, others have taken the view that the clusters are unstable and are expanding (cf. the discussion of Burbidge and Sargent, 1969). Thus it is not possible from these arguments based on the virial theorem to draw any certain conclusions about the amount of hot gas that may be present. However, in each case it is easily shown that only a comparatively small mass of hot gas, perhaps only a few percent of the mass of the galaxies if optimum values are taken, is required to account for the X-rays as thermal bremsstrahlung. Such gas could have been ejected from the galaxies and be heated by the motions of the galaxies.

If the correlation between extended radio sources and X-ray sources were shown to be all important, this would be a strong argument in favor of the suggestion that the X-rays are generated by Compton collisions between photons of the microwave background radiation and the relativistic electrons in the radio sources. This argument looks very plausible for the Coma cluster (Brecher and Burbidge, 1972a) and for other sources listed in Table I (Brecher and Burbidge, 1972b; Bridle and Feldman, 1972; Perola and Reinhardt, 1972; Costain et al., 1972).

The relationship between the power emitted by the synchrotron process and that radiated by Compton scattering is given by (Felten and Morrison, 1966)

$$\frac{L_r}{L_x} = \frac{H^2}{8\pi\varrho} \left\{ \frac{20000 \, T_b}{H} \right\}^{(3-m/2)}$$

where H is the effective magnetic field in the source, ϱ is the energy density of the radiation field and T_b is its brightness temperature, and m is the electron spectral index $(N(E) = kE^{-m})$. For $m = 3$ corresponding to a synchrotron or Compton radiation spectrum $I \propto \nu^{-1}$, this relation simply reduces to the fact that the ratio $L_r/L_x = (H^2/8\pi)/\varrho$.

Except in the case of Centaurus A, the extended radio source surrounding NGC 5128 from which no X-ray flux has been detected, the sensitivity limits of the X-ray detectors presently in use lead to the result that where an extended X-ray source has been identified with a known radio source, $L_x \gg L_r$.

In the simplest case, with $m = 3$, and assuming that the microwave background radiation is of black body form with $T_b \simeq 3\text{K}$,

$$L_r = L_x (2.5 \times 10^5 \, H)^2.$$

Thus, if $L_x \gg L_r$, $H \ll 4 \times 10^{-6}$ G. If the microwave background radiation has a much higher radiation density, as has been suggested by some of the rocket observations, then the lower limit on H will be raised accordingly. Calculations of the energetics of the well-known extended radio sources (Burbidge, 1959; Maltby et al., 1963) show that if the so-called equipartition condition applies, though there is no physical basis for this, the magnetic field strengths required are in general $\sim 10^{-5}$ G or greater. Thus, provided that the same regions give rise to the radio emission and the X-ray emission, this argument would suggest that the magnetic field is considerably weaker than the equipartition value (perhaps $\sim 10^{-7}$ G) and the total energy in particles in the sources must be increased over the equipartition value by a factor $(H_{eq}/H)^{3/2}$. However, to generate X-ray photons with energies ~ 5 keV from photons of the microwave background $(\lambda_{max} \simeq 1 \text{ mm})$ requires electrons with energies $\sim 1-2$ GeV, and in a field as weak as 10^{-7} G such electrons will radiate synchrotron radiation of rather low frequencies, < 10 MHz. Consequently, if the magnetic fields are very weak, the electrons largely responsible for the X-rays will have lower energies than those responsible for the radio flux in the meter and centimeter wavelength range. Thus a more detailed investigation is required to apply relation (1) precisely. We require to know the shapes of the radio and X-ray spectra over a wide spectral range and also to determine from observation whether the regions of radio and optical emission coexist. A fairly detailed model which will fit the observations has been made for the Coma cluster (Brecher and Burbidge, 1972a).

It is well known that many of the strong radio sources are double with two fairly well defined regions of emission symmetrically placed about the optical galaxy, which may be the brightest member of a cluster of galaxies. It is possible in such cases that the X-ray emission comes from a much more extended volume than the double radio source and that the magnetic field in this more extended region is much weaker than that in the strongly radio-emitting source.

From this discussion it is clear that it is too early for us to decide whether the extended X-ray sources are due to Compton scattering or thermal radiation from a hot gas. The third possibility, that such extended emitting regions are due to X-ray emission from many powerful galaxies or other discrete objects, also cannot be excluded.

D. NEW TYPES OF EXTRAGALACTIC OBJECTS

In this paper we have only discussed so far extragalactic X-ray sources which have been identified with special types of optical or radio sources. We conclude by pointing out that it is clearly possible that some of the X-ray sources are of a completely new type which are not known to radiate powerfully in optical, infrared or radio wavelengths. Can we predict what kinds of objects might be found?

One possibility is that they are highly compact massive objects which generate nonthermal radiation in ultraviolet wavelengths by the synchrotron process. The prediction of the existence of such objects is a natural extension of our present knowledge of QSO's or of the nuclei of galaxies. If an intense flux of far UV radiation, say in the wavelength range 500–1500 Å, is generated in a small volume, Compton scattering of comparatively low-energy electrons ($E \gtrsim 100$ MeV) would lead to the generation of very powerful fluxes of X-rays in the 1-100 keV energy range. Estimates of the flux can be made in a way similar to that calculated for the efficiency of the Compton process involving microwave radiation in the nucleus of M 87 (Burbidge, 1970b). Such objects need not be powerful UV emitters, and they would not have been discovered by other techniques so far. Since the nuclei of galaxies which are powerful infrared emitters may be radiating by a process involving nonthermally generated UV photons heating dust, such sources might be condensed objects in the nuclei of faint galaxies in which dust is absent.

Acknowledgement

Research in extragalactic astronomy at the University of California, San Diego, is supported by the National Science Foundation and by NASA through Grant Number NGL 05-005-004.

References

Brecher, K. and Burbidge, G. R.: 1972a, *Astrophys. J.* **174**, 253.
Brecher, K. and Burbidge, G. R.: 1972b, *Nature* **237**, 440.
Bridle, A. H. and Feldman, P. A.: 1972, *Nature Phys. Sci.* **235**, 168.
Burbidge, G. R.: 1959, in R. N. Bracewell (ed.), 'Paris Symposium on Radio Astronomy', *IAU Symp.* **9**, 541, Stanford University Press.

Burbidge, G. R.: 1970a, *Astrophys. J. Letters* **159**, L105.

Burbidge, G. R.: 1970b, *Ann. Rev. Astron. Astrophys.* **8**, 369.

Burbidge, G. R. and Sargent, W. L. W.: 1969, *Comm. Astrophys. Space Phys.* **1**, 220.

Costain, C. H., Bridle, A. H., and Feldman, P. A.: 1972, preprint.

Ekers, R. D.: 1968, *Austral. J. Phys. Astrophys. Suppl.*, No. 6.

Felten, J. and Morrison, P.: 1966, *Astrophys. J.* **146**, 686.

Gursky, H., Solinger, A., Kellogg, E. M., Murray, S., Tananbaum, H., Giacconi, R., and Cavaliere, A.: 1972, *Astrophys. J. Letters* **173**, L99.

Kellogg, E. M.: 1972, this volume, p. 171.

Kunkel, W. E. and Bradt, H. V.: 1971, *Astrophys. J. Letters* **170**, L7.

Maltby, P., Matthews, T. A., and Moffet, A. T.: 1963, *Astrophys. J.* **137**, 153.

Novick, R., Weisskopf, M. C., Berthelsdort, R., Linke, R., and Wolff, R. S.: 1972, *Astrophys. J. Letters* **174**, L1.

Perola, G. and Reinhardt, M.: 1972, *Astron. Astrophys.* **17**, 432.

Penzias, A. A. and Wilson, R.: 1965, *Astrophys. J.* **142**, 419.

Sargent, W. L. W.: 1972, this volume, p. 184.

Solinger, A. and Tucker, W.: 1972, preprint.

Weisskopf, M. C., Novick, R., Berthelsdort, R., Linke, R., and Wolff, R. S.: 1972, *Bull. Am. Phys. Soc.* **17**, 501;

18. EXTRAGALACTIC X-RAY SOURCES AND THEIR CONTRIBUTION TO THE DIFFUSE BACKGROUND

Invited Paper

G. SETTI

Laboratorio di Radioastronomia CNR, University of Bologna, Italy

and

L. WOLTJER

Dept. of Astronomy, Columbia University, New York, N.Y., U.S.A.

Abstract. We present estimates of the integrated contribution of extragalactic sources to the diffuse X-ray background in the 2–10 keV energy interval. It appears that classes of objects already detected as X-ray sources easily account for at least 10% of the background. Quasistellar objects, and possibly Seyfert and radio galaxies might contribute a larger fraction of the background intensity.

We also suggest that several of the unidentified X-ray sources in the UHURU catalogue at high galactic latitudes may be bright QSO's as yet unknown. The detection of a larger number of QSO's in the X-ray band may allow a decisive test for the cosmological interpretation of the redshifts of QSO's.

1. Introduction

Since the basic mechanisms which might be responsible for the X-ray emission from extragalactic sources have already been reviewed by Burbidge (paper in this volume, p. 199), we shall here mainly concentrate on estimating the possible contribution of various classes of objects to the isotropic component of the diffuse X-ray background in the 2–10 keV energy range, whose measured intensity is ~ 30 keV cm^{-2} s^{-1} ster^{-1} (Gorenstein *et al.*, 1969). Also we shall not discuss the diffuse processes which might take place in intergalactic space. These processes and their possible role in explaining the diffuse X-ray background for energies $\gtrsim 1$ keV have been extensively reviewed by Setti and Rees (1970), and since then the situation has not basically changed either theoretically or experimentally. Because we shall find that the extragalactic sources are likely to make a substantial contribution to the background, if not explaining all of it, a rediscussion of the importance of the diffuse processes must necessarily await further observations on a larger number of such sources.

An earlier paper on this topic was published about two years ago (Setti and Woltjer, 1970a) and the present report is intended to update the estimates given there in the light of new observational evidence, in particular that from UHURU catalogue (Giacconi *et al.*, 1972).

2. Normal Galaxies

Assuming that the optical and X-ray luminosities of normal galaxies are proportional and deriving the constant of proportionality from the observed X-ray emission from M 31, the integrated background from all normal galaxies is ~ 0.6 keV cm^{-2} s^{-1} ster^{-1}

Bradt and Giacconi (eds.), X- and Gamma-Ray Astronomy, 208–211. All Rights Reserved.
Copyright © 1973 by the IAU.

in the energy interval 2–10 keV. Therefore, unless cosmological evolutionary effects of the type discussed by Silk (1968, 1969) are present, normal galaxies may account for about 2% of the diffuse background.

On the present assumption, the predicted flux from the Large Magellanic Cloud would be 0.7 keV cm^{-2} s^{-1}, compared to the observed 0.6 keV cm^{-2} s^{-1}. For the Small Magellanic Cloud, in which most of the emission is due to one very strong X-ray source, the predicted flux is about half of that, following from the above assumption. The next bright galaxy is NGC 598 (M 33) for which we predict a flux of $\sim 6 \times 10^{-3}$ keV cm^{-2} s^{-1}, about 8 times weaker than the flux from M 31.

For our own galaxy, assuming that it has the same mass to luminosity ratio as M 31, the X-ray luminosity would be $\sim 1 \times 10^{39}$ erg^{-1} consistent with the observed emission from the galactic disk.

3. Seyfert Galaxies

To account for the whole X-ray background by Seyfert galaxies, assumed to be 1% of all bright galaxies, representative objects like NGC 1068, or NGC 4151, would have to have an X-ray emission of about 6×10^{51} keV s^{-1} in the energy interval 2–10 keV. In the case of NGC 4151 this would correspond to an X-ray flux of ~ 0.5 keV cm^{-2} s^{-1}, while the observed flux is only 0.04 keV cm^{-2} s^{-1}. Moreover the upper limits to the X-ray fluxes from two other Seyfert galaxies, NGC 1068 and NGC 4051, indicate X-ray emission $\lesssim \frac{1}{3}$ the emission for NGC 4151 (Kellogg, paper in this volume, p. 171). As a result Seyfert galaxies may perhaps account for about 5% of the background flux, but probably not for all of it unless evolutionary effects of a cosmological nature are present (Setti and Woltjer, 1970b). It can be easily shown that luminosity and/or density evolutions proportional to $(1+z)^4$ out to a redshift $z \approx 2.5$ would increase the resulting background by a factor of ten.

In the previous estimate we have not included an object like NGC 1275 both because it probably cannot be classified as a typical Seyfert and because its X-ray luminosity, about two order of magnitude greater than that of NGC 4151 (Gursky et al., 1971), appears to be extended over a large volume with angular size $\sim 35'$. Therefore at the moment it is not clear if the X-ray emission is mainly associated with the Perseus cluster of galaxies or with the galaxy itself.

4. Quasistellar Objects

To date only 3C 273 has been observed with a flux of 0.04 keV cm^{-2} s^{-1} in the 2–10 keV interval. Its X-ray spectrum corresponds to an energy flux $\propto E^{-0.9 \pm 0.3}$ which appears to be compatible with the spectrum of the background. Adopting a constant ratio of optical to X-ray luminosity for all quasistellar objects *including the radioquiet ones*, the predicted background due to QSO's down to a blue magnitude 19.4 would be ~ 4 keV cm^{-2} s^{-1} ster^{-1}, or 13%.* However including the fainter

---

* This estimate is based on the detailed computations made in our previous paper (Setti and Woltjer, 1970a) where, however, a slightly different spectrum was adopted for the average QSO.

QSO's in Schmidt's (1971) extrapolation to the 23rd magnitude the contribution would become 20 keV cm^{-2} s^{-1} ster^{-1}. Consequently QSO's may well suffice to account for most of the observed background, but a definite conclusion has to await data on a larger sample of QSO's. It should be noted that 3C 273 appears as a relatively weak source in the UHURU catalogue, and therefore if our assumption is correct the next brightest known QSO would not have been detected.

We note that in Schmidt's table of numbers of QSO's as a function of magnitude there are estimated to be 5 QSO's of the 13th magnitude and 25 of the 14th. As a consequence the interesting possibility arises that several of the unidentified X-ray sources in the UHURU catalogue at higher galactic latitudes (21 sources $\geqslant 20°$ off the galactic plane) may well be bright QSO's. We also note that if the QSO's were 'local' objects, say nearer than 30 Mpc, the integrated background from all QSO's in the universe would become at least ten times the observed value, and therefore a 'local' theory is possible only if 3C 273 is an anomalously bright object in X-rays.

With regard to the uniformity of the background we note that with Schmidt's extrapolation to the 23rd magnitude there would be of the order of 10^4 QSO's per 25 deg^2 of the sky, and consequently the fluctuations observed with this type of resolution would be no more than 1–2%, in agreement with present observational evidence.

5. Radio Galaxies and Rich Clusters of Galaxies

X-ray emission has been observed from several radio galaxies, but in the majority of cases there is doubt whether the X-rays are due to the radio galaxy itself or to an associated cluster of galaxies. If we assume, following the suggestion of Gursky *et al.*, (1972), that all rich clusters of galaxies are strong X-ray sources, then on the basis of the UHURU data we find that their contribution to the diffuse background is only about 3% of the total. As there should be about one richness 2 cluster per square degree of the sky out to the Hubble distance, the fluctuations observed with a resolution of 25 sq deg would be about 1%, which in turn excludes the possibility that the same clusters can contribute a substantial fraction of the background (say 30% or more, as has been suggested by some) because the fluctuations then expected in the X-ray background would exceed existing upper limits.

Since about half of the radio galaxies occur in clusters, we conclude that neither the clusters nor the radio galaxies are likely to make a significant contribution to the background. Of course this does not exclude that radio galaxies at large z may be of importance if cosmological effects of the kind considered by Rees and Setti (1968) are present.

6. Concluding Remarks

The above estimates show that classes of extragalactic objects already detected as X-ray sources are likely to account altogether for at the very least 10% of the background intensity in the 2–10 keV energy interval. If all the background is due to sources the most likely candidates are QSO's, and perhaps Seyfert galaxies and radio galaxies at large z.

Various authors (Apparao, 1968; Tucker, 1970) have proposed that the X-ray emission produced during the early phases of a supernova outburst might be sufficient to account for the whole X-ray background in terms of the integrated flux from supernova explosions in normal galaxies. Up to now there is no direct evidence that supernovae are powerful X-ray sources during their early phases, and there are only rather large upper limits several days after the optical outbursts. However Cavallo and Messina (1972) have shown that broad beam observations in the direction of the Virgo Cluster already demonstrate that the average X-ray emission from such super-nova is at least a factor of 20 below the required output, unless the emission took place in rather short time ($\lesssim 2$ days) after the explosion; however in this case the resulting fluctuations would be incompatible with the observed uniformity of the X-ray background.

In the present paper we have only analysed the contribution of sources to the intensity of the background in a limited energy interval. Of course when more data become available about the spectra of the various kinds of sources, a comparison between the source spectra and the background spectra should provide a much more conclusive test. Finally the graininess of the background may contain useful information with regard to the sources but here the effect of X-rays from our own galaxy may well prevent a unique interpretation, as is actually the case with regard to the 100 MHz radio background.

References

Apparao, M.: 1968, *Nature* **219**, 145.

Cavallo, G. and Messina, A.: 1972, *Astrophys. Letters* **10**, 61.

Giacconi, R., Gursky, H., Kellogg, E. M., Murray, S., Schreier, E., and Tananbaum, H.: 1972, 'The UHURU Catalogue of X-Ray Sources, Submitted to *Astrophys. J.*

Gorenstein, P., Kellogg, E. M., and Gursky, H.: 1969, *Astrophys. J.* **156**, 315.

Gursky, H., Kellogg, E. M., Leong, O., Tananbaum, H., and Giacconi, R.: 1971, *Astrophys. J.* **165**, L43.

Gursky, H. Solinger, A., Kellogg, E. M., Murray, S., Tananbaum, H., Giacconi, R., and Cavaliere, A.: 1972, 'X-Ray Emission from Rich Clusters of Galaxies', submitted to *Astrophys. J. Letters*.

Rees, M. J. and Setti, G.: 1968, *Nature* **219**, 127.

Schmidt, M.: 1971, in 'Semaine d'étude sur les noyaux des galaxies' Pontificiae Academiae Scientiarum Scripta Varia, No. 35, p. 387.

Setti, G. and Rees, M. J.: 1970, in L. Gratton (ed.), 'Non-Solar X- and Gamma-Ray Astronomy', *IAU Symp.* **37**, 352.

Setti, G. and Woltjer, L.: 1970a, *Astrophys. Space Sci.* **9**, 185.

Setti, G. and Woltjer, L.: 1970b, *Nature* **227**, 586.

Silk, J.: 1968, *Astrophys. J.* **151**, L19.

Silk, J.: 1969, *Nature* **221**, 347.

Tucker, W. H.: 1970, *Astrophys. J.* **161**, 1161.

PART IV

INTERSTELLAR MEDIUM
AND SOFT X-RAY BACKGROUND

19. THE SOFT X-RAY BACKGROUND

H. FRIEDMAN, G. FRITZ, and S. D. SHULMAN

*E. O. Hulburt Center for Space Research, Naval Research Laboratory,
Washington, D.C. 20390, U.S.A.*

and

R. C. HENRY

*E. O. Hulburt Center for Space Research, and The Johns Hopkins University,
Baltimore, Md. 21218, U.S.A.*

Abstract. A survey of soft X-ray background observations in the 0.1–10 keV range is presented. In the region above 1 keV, recent results on point X-ray sources are discussed and their integrated contribution to the diffuse background is estimated. However, the average luminosity of various classes of extragalactic X-ray sources is still not sufficiently well known to permit this estimate to be made with any certainty. A discussion is given of recent observations at energies below 1 keV where the effects of interstellar absorption are important. It is argued that although some fraction of the background radiation in the 0.1–1 keV range must be galactic in origin, there is still substantial evidence for an extragalactic component. Proposed theories for generating both the galactic and extragalactic X-ray background are briefly reviewed.

1. Introduction

Over the past decade the X-ray background radiation has been studied almost as intensively as have discrete sources. From 1 keV to 1 MeV, the spectrum is fairly well defined, and general isotropy in the 1–20 keV region has been established to a considerable degree. Above 20 keV the spectrum approximates a power law with an index of about −2.3 for the differential photon flux. At lower energies the spectrum flattens, and the index, from 2 to 8 keV, falls to about −1.4. For the lowest observed energies (100 eV to 1 keV), the measurement techniques introduce large uncertainties. Nevertheless, the body of data accumulated in the past three or four years indicates a significant excess over any simple extrapolation of the higher energy spectrum. In this review we confine our attention to the range 0.1 to 10 keV. The work has been carried out with rocket- and satellite-borne proportional counters. As an Appendix, we include a discussion of the instrumental problems specific to soft X-ray measurements with proportional counters.

The background may be composed of the integrated contributions of discrete sources and of truly diffuse X-rays generated in interstellar or intergalactic space. From the wealth of data on discrete sources that is now available in the soft X-ray range, we shall assess the contribution of discrete sources. At the very lowest energies (100 eV to 1 keV), it is still debatable whether the observed background is partly extragalactic or entirely local in origin. Our most recent results favor an extragalactic component which can be distinguished from soft X-rays generated within the galaxy. Intrinsic to any understanding of the X-ray background is the possible role of intergalactic gas; we shall discuss X-ray observations of clusters of galaxies which offer important clues in this regard.

Bradt and Giacconi (eds.), X- and Gamma-Ray Astronomy, 215–234. All Rights Reserved.
Copyright © 1973 by the IAU.

1.1. THE X-RAY LUMINOSITY OF NORMAL GALAXIES

A few years back, estimates of the luminosity of our galaxy (Friedman *et al.*, 1967), based on a sampling of some 30 sources, led to a value of about 7×10^{39} erg s^{-1}. In arriving at this figure, it was assumed that the observed sources are grouped in the nearest spiral arms and represent only about 2% of the volume of the galactic disk. Subsequent observations of spectral extinction below 1 keV implied that the observed sources were distributed over a larger range of distances and represented a larger sample of the volume of the galaxy. Although fewer in number, the deduced average luminosity of the sources was higher. The new estimate of L_{gal} became $2-3 \times 10^{39}$ erg s^{-1}. Andromeda is a giant spiral similar to the Milky Way. The flux (2 to 7 keV) observed by UHURU leads to a luminosity of about 2×10^{39} erg s^{-1}, in good agreement with earlier estimates for our galaxy. From an observation of the Large Magellanic Cloud by Mark *et al.* (1969), a luminosity of about 4×10^{38} erg s^{-1} was deduced, consistent with the roughly tenfold lower stellar content than that of the Milky Way.

The UHURU catalog now lists 125 sources, about two-thirds of which lie close to the galactic plane and the remaining one-third at latitudes $> \pm 20°$. Amongst the galactic sources we can identify supernova remnants such as the Crab Nebula, the Supernova of 1572, and Cas A. For all of these relatively younger supernovae, the X-ray luminosities are in the range $10^{36}-10^{37}$ erg s^{-1} (1 to 10 keV). A second, perhaps larger, class of sources are binary systems in which a compact object – white dwarf, neutron star, or, perhaps, a black hole – accretes gas from a large stellar companion. Cyg X-1, Cen X-3, and 2U 1702+35 in Hercules are representative of this group. From a combination of 21-cm emission and X-ray absorption studies, the estimated distance to Cyg X-1 is about 1000 parsec and its luminosity is, therefore, also about 10^{36} erg s^{-1}. It appears that this luminosity may represent close to the average value for most of the roughly 100 sources detected thus far in the galaxy. Their sum would amount to about 10^{38} erg s^{-1}, about an order of magnitude less than our previous estimates of the galactic luminosity. However, there is evidence of discrete sources in the Magellanic Clouds with luminosities of 10^{38} erg s^{-1} and one example of a luminosity as great as 10^{39} erg s^{-1}. In the region of the galactic bulge of our galaxy there appear to be several sources as powerful as 10^{38} erg s^{-1}. We may speculate that the bulk of the galactic luminosity could be made up of just a few sources with luminosities in the range of 10^{38} erg s^{-1}, perhaps a few hundred sources in the $10^{36}-10^{38}$ erg s^{-1} range and, possibly, thousands of sources with average luminosities between 10^{35} and 10^{36} erg s^{-1}.

In any case, the direct observation of Andromeda supports the estimate that the X-ray flux from normal galaxies is about 10^{39} erg s^{-1}. The contribution of all such galaxies, assuming a space density of 0.03 Mpc^{-3}, amounts to between 10^{-1} and 10^{-2} of the observed background of 2×10^{-7} erg cm^2 s sterad. It is possible that evolutionary effects could provide a much higher X-ray luminosity in earlier epochs than at present, but we are a long way from the observational capability to test such a hypothesis.

1.2. Pulsars

The pulsar NP 0532 in the Crab Nebula radiates about 10^{36} erg s^{-1} from 1 to 10 keV. Several attempts have been made to estimate the contribtuion of supernovae to the diffuse background on the assumption that the Crab is a typical supernova remnant and that the average rate of appearance of supernovae is about one per 100 yr per normal galaxy. If a new pulsar is endowed with 10^{52} erg of rotational kinetic energy and if the conversion efficiency to X-ray emission is about one percent, 10^{50} erg would be available in the pulsar lifetime. Averaged over time, the contribution from pulsars alone would yield a galactic luminosity of 10^{41} erg s^{-1} and the combined X-ray background emission from all galaxies would be equal to the observed diffuse background.

However, the estimate of the initial energy of rotation is very uncertain. Some authors prefer as high a value as 1.5×10^{53} erg (Gunn and Ostriker, 1969) and others as little as 5×10^{49} (Woltjer, 1969). To avoid exceeding the observed upper limit on anisotropy, an average emission lifetime of 4 days per pulsar would be required. Thus far, only upper limit flux results are available from X-ray observations directed toward known supernova events in distant galaxies. Bradt et al. (1968) set an upper limit of 2×10^{42} erg s^{-1} for a supernova in NGC 4254. Since the observation was made 40 days after the phase of maximum light, they could only deduce that the emission lifetime would need to be shorter than 0.1 yr for the event to have decayed before observation, or greater than 5 yr for the average emission to be less than the detector sensitivity. For four supernova events observed from the OSO III satellite in 1967 and 1968, Ulmer et al. (1972) found only upper limits on the fluxes which give upper limits on total energies of from 10^{50} to 10^{51} erg. The observations, therefore, do not yet answer the supernova question in any useful way.

1.3. X-rays from Radio Galaxies, Seyfert Galaxies, Quasars, and Clusters

If normal galaxies cannot account for the X-ray background, can a much smaller number of powerful radio galaxies, Seyferts, or quasars be responsible? Among the distant galaxies that have been observed as X-ray sources are M 87, Cen A, NGC 1275, NGC 4151, and the quasar 3C 273. The claims for identification are, of course, based on the uniqueness of these sources as radio galaxies and Seyferts rather than on any precise location of the X-ray source. At its apparent red shift distance of 630 Mpc, 3C 273 radiates $\sim 10^{46}$ erg s^{-1} and is the most powerful X-ray source detected thus far. Estimates of integrated background, based on the assumption that some of these galaxies are typical members of their class, can lead to embarrassingly high values of the background. This is true for quasars, if 3C 273 is typical, and for Seyferts, if NGC 1275 is representative. However, NGC 4151 is far weaker than NGC 1275, and NGC 1068, a strong infrared Seyfert galaxy which was predicted to be a powerful X-ray emitting object was undetected by UHURU and is probably one to two orders of magnitude weaker than NGC 1275. Clearly, our sampling to date does not permit any quantitative assessment of the integral contribution of quasars, Seyferts and strong radio galaxies such as M 87. Perhaps we have detected thus far only the most unusual specimens of these classes of unusual galaxies.

Observations of clusters of galaxies with rockets and by the UHURU satellite now include several prominent clusters such as Virgo, Coma, and Perseus. It is not entirely clear how much of the observed flux is attributable to one or two relatively powerful radio galaxies in any particular cluster, for example, M 87 in Virgo, or to the larger number of galaxies that make up each cluster. In both Coma and Virgo (minus the contribution of M 87) the luminosity per galaxy appears to be about 20 times that of our own galaxy. It is possible that the X-rays originate largely in hot intracluster gas rather than in the individual galaxies, but the observations are still too primitive to make any clear determination of the nature of the source.

1.4. ISOTROPY

Any model which attributes the diffuse background (1–10 keV) to discrete sources must be consistent with the limits on anisotropy. Schwartz *et al.* (1970) obtained X-ray background data from OSO III which covered about 50% of the sky and showed no rms intensity fluctuations larger than 3% on the scale of their beam width ($\sim 30°$). Wolfe and Burbidge (1970) compared these observations with theoretically expected fluctuations corresponding to clusters or super-clusters of galaxies. Their results argued against any hierarchical universe models in which the distribution of X-ray sources corresponded to the cosmological mass distribution.

Fabian and Sanford (1971) have observed nine regions of sky and conclude from the observed isotropy that there must be at least 10^7 sources of equal apparent brightness over the entire sky. The corresponding number for sources of equal absolute intensity is $\sim 10^9$, which implies a number density exceeding that for Seyfert galaxies. Schwartz *et al.* (1971) believe that the Fabian and Sanford experimental limit should be lowered considerably and suggest a reduction to about 10^4 sources over the entire sky. Furthermore, from an autocorrelation of their count rate as a function of angle, they find results consistent with random statistical scatter and inconsistent with a supercluster model and some cluster models.

2. Recent Observations of the (0.1–1 keV) X-Ray Background

In the 1-10 keV region there is very good agreement between various groups (Boldt *et al.*, 1969; Ducros *et al.*, 1970; Bunner *et al.*, 1971; Shukla and Wilson, 1971; Shulman *et al.*, 1971; and Toor *et al.*, 1972) as to the spectrum of the diffuse background. A power law representation, $A E^{-\gamma}$ photons (cm² s ster keV)⁻¹, is fit by these groups to their data. A is found to be in the range 8–11, and γ is between 1.3 and 1.4 .In the soft X-ray region, 10–100 Å (0.1–1 keV) there is general agreement that the flux near 44 Å exceeds that expected from an extrapolation of the spectrum in the 1–10 keV region. The original interpretation attributed most of the excess at the galactic pole to an extragalactic orgin (Bowyer *et al.*, 1968; Henry *et al.*, 1968). However, Henry *et al.* (1968) also observed a significant flux of 44 Å X-rays from the galactic plane. The soft X–ray emission from the galactic plane is now well established (Henry *et al.*, 1971; Bunner *et al.*, 1971; Palmieri *et al.*, 1971, Hayakawa *et al.*, 1971 and Yentis *et al.*, 1972) and this emission almost certainly requires a galactic origin since the mean free path for 44 Å X–rays in the interstellar medium is on the order of only 100 pc. In a recent observation, the Wisconsin group (McCammon *et al.*, 1971) found no evidence for absorption of the diffuse background by the Small Magellanic Cloud (SMC), and placed an upper limit of 25% on the fraction of 44 Å

TABLE I

Soft X-ray background intensities

Wavelength [Å]	Energy [keV]	Pole	Plane	Ratio pole/plane	Extrapolation of 1–10 keV Power law (no interstellar absorption) [photons (cm² s ster keV)⁻¹]
		[photons (cm² s ster keV)⁻¹]			
18–24[a]	0.69–0.52	50	35	1.4	18
44–60[a]	0.28–0.21	300	120	2.5	65
66–90	0.19–0.14	∼500	∼500[b]	∼1[b]	116

[a] Intensities are from Davidsen *et al.* (1972).
[b] Estimated using the flux at the pole observed by Yentis *et al.* (1972) and the observation of Gorenstein and Tucker (1972) that the pole/plane ratio is 1.5 in the 44–80 Å band. Since the 44–60 Å ratio is about 2.5, the 66–90 Å ratio must be about 1 to give the 44–80 Å result.

X–rays which originate beyond the SMC. As seen in Table I, this upper limit corresponds roughly to the ratio of the 44 Å flux calculated from an extrapolation of the 1–10 keV. power law to the total flux observed toward the poles. This raises the question of whether or not there is any extragalactic contribution to the soft X–ray background beyond that expected from an extrapolation of the power law; we discuss this below.

2.1. OBSERVATIONAL DATA

The currently available data can be summarized for three separate ranges where broad-band measurements have been made within the 0.1–1 keV energy region. Table I gives average values for the flux in the 66–90 Å, 44–60 Å, and 18–24 Å regions. In the 66–90 Å range, only two preliminary results are available (Yentis *et al.*, 1972; Gorenstein and Tucker, 1972). Most of the recent soft X–ray observations have been made in the 44–60 Å region, and a large fraction of the sky has now been surveyed. Figure 1 shows the Naval Research Laboratory (NRL) results (Davidsen *et al.*, 1972) for the galactic anti-center direction, and Figure 2 shows the Wisconsin results (Bunner *et al.*, 1972) for the galactic center. These two groups find very similar results in regions where they have both observed. The Lawrence Radiation Laboratory

(LRL) group (Palmieri *et al.*, 1971) using a somewhat different proportional counter dectector with a considerably thinner window of Formvar, rather than the Kimfol used by the other two groups, find a flux near $b^{II} = 50°$ of 270 photons (cm^2 s ster keV)$^{-1}$ which is in reasonable agreement with the other results. However, they also

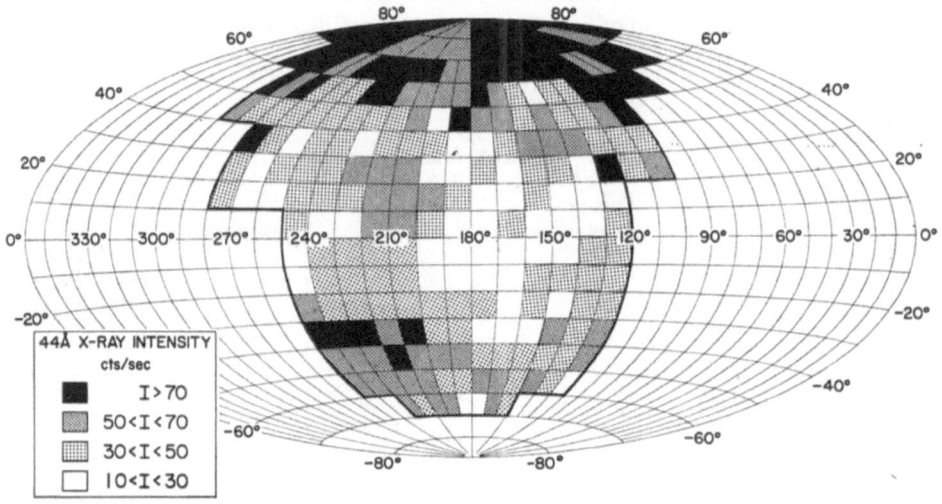

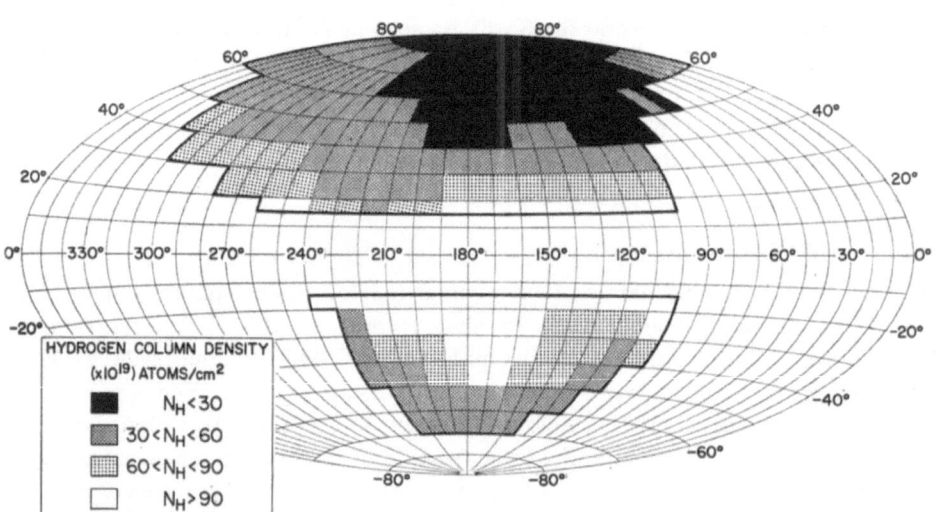

Fig. 1.   Top: Map of the sky in the direction of the galactic anti-center in the 44–60 Å X-ray band. A count rate of 80 s^{-1} corresponds to a flux of 300 photons (cm^2 s ster keV)$^{-1}$ (Davidsen *et al.* 1972). Bottom: Hydrogen (21 cm) column density of Tolbert(1971). Both maps are in New Galactic Coordinates.

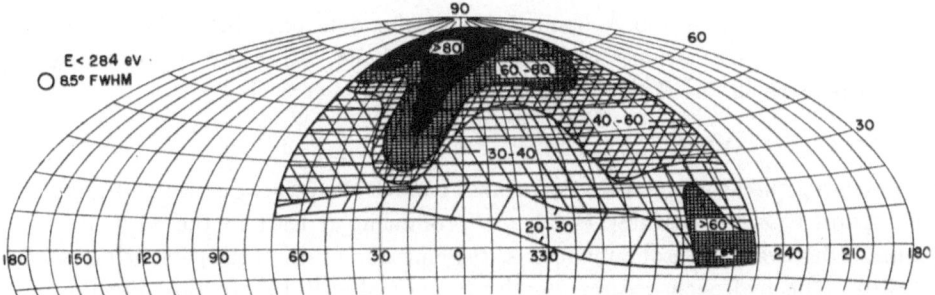

Fig. 2.   Map of the sky in the galactic center direction in the 44–60 Å X-ray band. A count-rate of 20 s^{-1} corresponds to a flux of 120 photons (cm^2 s ster keV)$^{-1}$ (Bunner *et al.*, 1972).

find a large flux of 630 photons (cm^2 s ster keV)$^{-1}$ in a band $l^{II} = 300$–$340°$ near $b^{II} = 20°$ which is absent in the Wisconsin results (Figure 2). This discrepancy may be due to low energy particle or UV contamination in the thinner, Formvar window detector.

There is also a considerable amount of data in the region from 0.5–1 keV. The NRL group (Henry *et al.*, 1971; Davidsen *et al.*, 1972) has used a Teflon window proportional counter which provides a fluorine K-edge filter with a transmission band from 18–24 Å. The peak transmission is only 20% at 18 Å, as opposed to 45% for Kimfol at 44 Å, so that the statistics do not permit a detailed mapping. However, if the counts are put in 10° galactic latitude bins and summed over all longitudes, the results are statistically significant. These data are shown in Figure 3 (right) along with

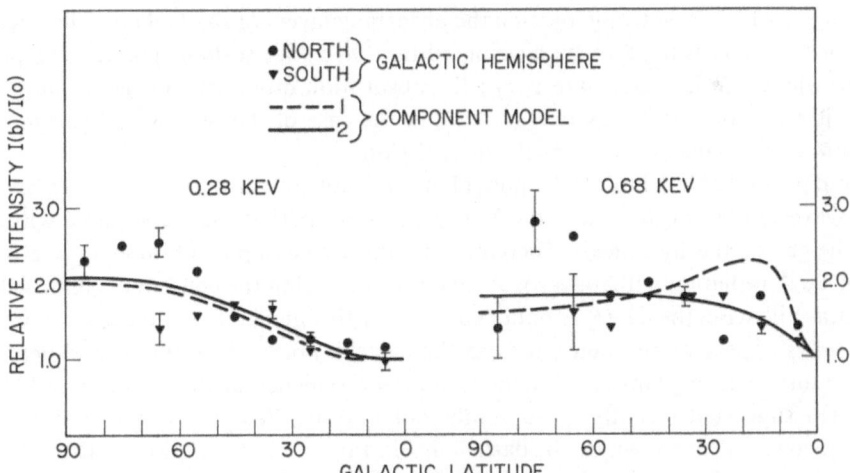

Fig. 3.   Latitude dependence of the 44–60 Å, 0.28 keV (left), and 18–24 Å, 0.68 keV (right), X-ray data. The intensity is normalized to the values in the galactic plane. The dashed and solid curves represent the latitude dependence predicted by the one and two-component models discussed in the text, respectively (Davidsen *et al.*, 1972).

similar results in the 44–60 Å band (left). Similar data have been obtained by the Wisconsin and LRL groups using the broader band, 0.5–1 keV, provided by the pulse–height resolution of the proportional counter.

2.2. DISCRETE FEATURES

In presenting the average soft X–ray fluxes toward the galactic poles and plane, we have ignored several features which are prominent at least in the 44 Å data. In Figure 1, there is an area of enhanced X–ray emission north of the galactic plane near $l^{II} = 210°$. The hydrogen column density is anomalously low in this direction and may account for a high transmission of either a galactic or extragalactic flux. A clustering of strong sources could also produce the high intensity, but no strong sources have been reported. In Figure 2, there is a prominent ridge of high intensity which extends from near the North Galactic Pole down towards the plane at $l^{II} = 0$–30°. Bunner *et al.* (1972) have pointed out that this coincides fairly closely with the North Polar Spur. There is also a high intensity feature in the plane near $l^{II} = 260°$ which is associated with the Vela supernova remnant. In spite of these features, however, the soft X–ray background in a large part of the sky has a general dependence on galactic latitude from which it seems possible to deduce fluxes near the pole and in the plane which are more or less independent of galactic longitude.

2.3. MODELS

We can use these data to examine two simple models for the origin of the soft X–ray background. The first is a one-component model in which the origin is entirely galactic, with the emission occurring in a disk of scale-height z_s. The second is a two-component model in which there is both a galactic component and an extragalactic component which is isotropic outside the absorbing layers of the Galaxy. The neutral hydrogen is also assumed to be distributed in a disk with scale-height z_H. The parameters which can be varied are z_s/z_H; the absorption cross–section per H–atom, σ, which is a function of X–ray energy; and, in the case of the two-component model, the ratio of the galactic to extragalactic emission.

The data in Table I indicate a pole/plane ratio of ~ 1 in the 66–90 Å region. At these wavelengths, σ is large and the X–ray mean free path is small compared with the scale-height of the hydrogen. Therefore, in the one-component model, if $z_s \geqslant z_H$, the 66–90 Å radiation will appear isotropic thus satisfying the constraints imposed by the data. Likewise, the 44–60 Å data can be fit by this model if we take $z_s > z_H$ so that the X–rays appear more intense toward the galactic poles where the optical depth is about unity. This explanation for the latitude dependence of the 44–60 Å radiation has been suggested recently by a number of authors. A detailed calculation with $z_s/z_H \simeq 5$ gives the fit shown by the dashed line in Figure 3 (left). However, the 18–24 Å data cannot be fit any simple way with this model. The optical depth toward the galactic poles is only about 0.15 so that the X–ray intensity should first increase with decreasing galactic latitude before it decreases near the plane where the optical depth becomes larger than unity. A detailed calculation for the one-component model gives the dashed curve in Figure 3 (right). The data do not seem to indicate a peak intensity at low to intermediate latitudes. This problem in fitting the latitude dependence at wavelengths shorter than 44 Å was also noted by the Wisconsin group (Bunner *et al.*, 1969) in attempting to fit their data in an energy region from 0.5–1.0 keV.

The two-component model avoids this problem with the 18–24 Å data by allowing us to keep $z_s \simeq z_H$. With this constraint, the sum of the galactic and extragalactic components can be made to decrease monotonically with decreasing latitude in the 18–24 Å region. The results of detailed calculations for both the 44–60 Å and the 18-24 Å data are shown by the solid curves in Figure 3. The model predicts that the 44–60 Å radiation observed toward the galactic poles is approximately $\frac{2}{3}$ extragalactic and $\frac{1}{3}$ galactic. The 66–90 Å radiation will still appear isotropic, however, since the mean free path is still less than the neutral hydrogen scale-height, and all but a tiny fraction of the extragalactic component will be absorbed out.

It appears, therefore, that while the 44–60 Å and 66–90 Å observations could be explained by a totally galactic origin for the soft X–rays, the 18–24 Å flux is more naturally explained by a combination of galactic and extragalactic emission. This same model can also provide the flux observed in the two longer wavelength bands as it must if we wish to keep the same mechanisms and source distributions for all the soft X–ray emission. The hypothesized extragalactic component must be examined, however, in the light of the negative result for absorption of the background radiation by the Small Magellanic Cloud (McCammon *et al.*, 1971). As mentioned previously, an upper limit of 25% was set on the fraction of the radiation in the 44–60 Å band which originates beyond the SMC. From our two-component model we find that 67% of the observed X–rays should be extragalactic in origin in the direction of the galactic poles where the optical depth is about unity.

There are several ways in which these observations might be reconciled. The SMC is at galactic latitude, $b^{II} \simeq -40°$, considerably away from the South Galactic Pole At this latitude, the two-component model then predicts that only 50% of the radiation is extragalactic in origin. This value is somewhat closer to the upper limit quoted by the Wisconsin group (McCammon *et al.*, 1971). Another possibility is that there is soft X–ray emission in the SMC itself which fills in for the extragalactic X–rays which have been absorbed. The Wisconsin group points out that the only known source in the SMC which is observed in the 2–6 keV energy range does not appear to provide enough soft X–ray photons to compensate for the absorption. However, in our own galaxy we know that there is soft X–ray emission which is not associated with any known sources of X–rays with energies above 1 keV. In fact, most of the galactic component in our two-component model is exactly such soft X–ray emission. if we attribute a similar source of soft X–rays to the SMC, we find that the emission could compensate for about one-half of the absorbed X–rays so that the fraction of extra-galactic X–rays in these observations will be reduced to 50%, or less, of the total flux.

These considerations introduce some uncertainty into the interpretation of the SMC observation, and allow an origin beyond the SMC for a sizeable fraction of the soft X–ray flux. Even if none of the soft X–ray flux originates beyond the SMC, a two-component model could still be a good approximation with a large spherical galactic halo source replacing the extragalactic component. The emission would still be isotropic, and the calculational details of the model would remain unchanged. Using data from the same flight as the SMC observation, McCammon *et al.* (1971) also rule out the existence of a substantial background contribution from such a hot galactic halo because they see no correlation between changes in the hydrogen column density and the 44 Å X–ray intensity. However, the interpretation of this observation relies heavily on the accuracy of the 21-cm data. The survey data of McGee *et al.* (1966), which was used by the Wisconsin group, may not be reliable at the low hydrogen column densities $(1-2 \times 10^{20} \text{ cm}^{-2})$ in the observed direction. In contrast to the Wisconsin observation, Davidsen *et al.* (1972) do find a correlation at high galactic latitudes over a large area of the sky between the hydrogen column density and the 44 Å intensity.

3. Theories of the Origin of the Soft X-Ray Background

The low energy X–ray background (0.1–10 keV) may be the integrated effect of large numbers of discrete X–ray emitters, as discussed in Section 1, or it may be due to processes occuring in the reaches of intergalactic, interstellar, or even interplanetary space. The X–ray and γ–ray background spectrum is considerably more complex than that of any single discrete source that has been observed, suggesting that many processes may contribute to the observed total background. We first summarize those theories that involve a diffuse emission mechanism, and then consider models for the 0.1–1 keV galactic emission. An extensive review of extragalactic mechanisms has been published by Silk (1970).

3.1. DIFFUSE EMISSION SOURCES

Diffuse emission sources involve either non-thermal or thermal mechanisms for the production of the radiation. Non-thermal mechanisms produce X–rays through the interaction of high energy protons or electrons with electromagnetic radiation, magnetic fields, or other particles. Sources of the high energy particles which have been proposed include radio galaxies, 'normal' galaxies, galactic nuclei, and neutron stars in galaxies. Difficulties with such mechanisms center on inadequate primary particles, insufficient target material, and inefficient energy conversion. On the other hand, theories of thermal emission have difficulty in accounting for the heating of the proposed gas to the high temperatures required, and difficulty in ionizing the gas in the time available.

The most carefully studied non-thermal mechanism is the inverse Compton effect, in which high energy electrons from strong radio galaxies (Felten and Morrison, 1966) or from 'normal' radio galaxies (Brecher and Morrison, 1969) scatter from photons of the 3 K microwave background radiation and produce X–rays. Clearly, this interaction must occur; the only question is whether the bulk of the X–ray background could be so produced. Cowsik (1971) has argued that the spectral form of the 2–1000 keV background is not compatible with this theory for its origin; he suggests, however, that the radiation observed below 2 keV and above 1000 keV might be (at least partly) produced by the inverse Compton mechanism.

3.2. THERMAL MECHANISMS

The thermal mechanism of radiation from a hot intergalactic gas was described by Field and Henry (1964) and proposed by Henry *et al.* (1968) as the source of the background radiation observed at 44 Å. In the 'big bang' cosmology, however, non-thermal components such as high energy particles are also required to ionize and heat the intergalactic gas. These particles (protons and/or electrons) will lose part of their energy through non-thermal bremsstrahlung in collisions with thermal electrons and, therefore, the spectrum produced will be modified from a pure thermal spectrum. The inefficiency of the non-thermal bremsstrahlung, however, probably ensures that this effect will be relatively small.

Arons (1971) has examined the ionization process on the assumption that the flux of ionizing radiation is produced by quasars, and that the quasars are at their cosmological distance and had a very high space density in early times. He concludes that in the case of a high–density intergalactic medium, the zones ionized by the quasars never join up. This would mean that broad Lyman-α absorption troughs should appear in the red shifted ultraviolet spectra of quasars, contrary to observation. However, a lower–density medium (up to $\frac{1}{3}$ closure density) can become fully ionized.

It should eventually be possible to obtain more information on the efficacy of quasars as ionizers of the intergalactic gas by observing indirectly the particle emissions from nearby (recent) quasars, on the assumption that these are not very different from the earlier quasars. Byram *et al.* (1970), Bowyer *et al.* (1970), and Kellogg *et al.* (1971a) have detected X–rays from the quasar 3C 273. If the X–rays are produced by relativistic particles, it might be possible to determine the rate of emission and spectrum of the particles through careful study of the spectrum and spatial distribution of the X–ray emission.

If an ionized intergalactic gas does exist, it may be at a temperature of $\sim 300 \times 10^6$ K (Cowsik, 1971) or at $\sim 3 \times 10^6$ K (Henry *et al.*, 1968). The higher temperature gas would be used to explain the greater part of the X–ray spectrum, while the lower temperature gas would be used only to explain the excess observed below 1 keV. These alternatives permitted by the observations result in very different relationships between the diffuse and condensed matter in the universe. The condensed matter (galaxies) is distributed in a distinctly non-uniform manner (Limber, 1954). If the ionized matter is at only 3×10^6 K, it must respond gravitationally to the presence of these irregularities. In particular, it will tend to fall into clusters of galaxies (Gott and Gunn, 1971). It will, therefore, be clumped at the present time, and a smaller total quantity will be required to account for the observed 44 Å X–ray emission. If, on the other hand, the gas is at 300×10^6 K it would escape freely from clusters of galaxies and presumably would be uniformly distributed. A density of 3×10^{-6} particles cm^{-3} would be needed to account for the emission, probably sufficient to close the universe. The 2–100 keV background radiation is isotropic to within a few percent (Schwartz *et al.*, 1970) arguing consistency with a 300×10^{-6} K plasma. Angular variations at a level of a few percent, reported recently (Murray *et al.*, 1971), could indicate that an extremely hot gas is not the origin of the radiation, or could simply indicate that a second, highly irregular X–ray component is involved.

3.3. CLUSTERS OF GALAXIES

Observations of the relationship of the X–ray background to clusters of galaxies are clearly important. Meekins *et al.* (1971) detected a flux of X–rays from the direction of the Coma cluster of galaxies. Gursky *et al.* (1971) showed that the X–ray emission comes from nearly the position of the optical cluster center, and that the source is extended. It is not possible to distinguish between a thermal and a power-law spectrum, but if the radiation is interpreted as thermal, a temperature of $\sim 70 \times 10^6$ K is implied. The intensity of the radiation is so low that there cannot be enough matter present in the form of gas at this temperature to gravitationally bind the cluster. Can the additional 'missing mass' be present in the form of gas at lower temperature? The upper limit of Meekins *et al.* at 44 Å (0.28 keV) rules out

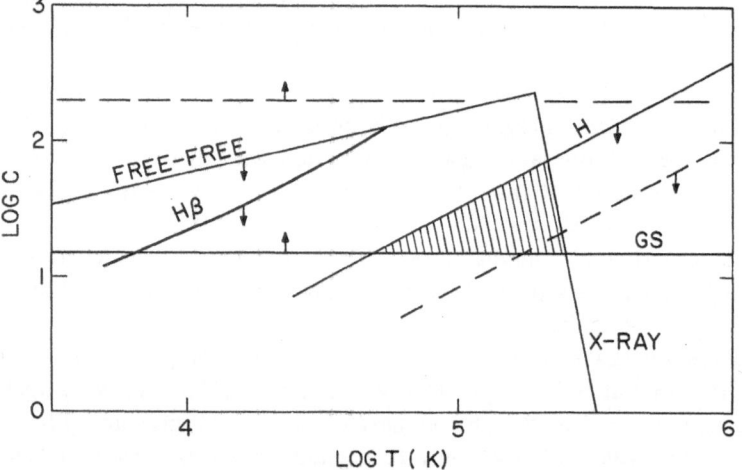

Fig. 4.   The logarithm of $C(=1/f)$, where f is the fraction of the Coma cluster of galaxies occupied by hot clouds of gas, is plotted against the logarithm of the temperature. Observations and theoretical considerations limit the regions of the plane that might contain the point representing the actual condition of the Coma cluster, if the Hubble constant is 50 km s^{-1} Mpc^{-1} and if the cluster contains a 'binding' mass of ionized gas. In particular, Woolf (1967) set the free-free emission and Hβ limits shown, while Meekins *et al.* (1971) set the X-ray limit. Henry (1972) set the Lα emission limit shown (line labelled H, the result of a more detailed calculation of his original (parallel dashed line) model), while the limit due to Goldsmith and Silk (1972) is the horizontal solid (GS) line (revision of their original, 1971, limit). The hatched area is the surviving region on the plane that could represent the actual state of a gravitationally-binding mass of ionized gas.

the possibility of sufficient quantities of gas at $\sim 5 \times 10^5$ K, but if a general inter-galactic medium exists at the latter temperature, one could consider a situation in which matter falls gravitationally into the Coma cluster over a time period of 10^9 yr, and then collapses into clouds in the temperature range 10^4 to 10^6 K. (Cooler clouds are ruled out from the lack of 21-cm absorption or emission.) Motion of these clouds, and of the galaxies, heats a small amount of residual gas to 70×10^6 K. Goldsmith and Silk (1972) have studied such a cloud model, and suggested its Lα recombination radiation as a means of detecting it. Henry (1972) set an upper limit to the red-shifted Lα flux from the Coma cluster of galaxies. His result is summarized in Figure 4. That limit, with a theoretical limit on the degree of clumping of the gas in the cluster by Goldsmith and Silk (1971), led to the virtual exclusion of the possibility of the presence of a binding mass of hot gas in the cluster. However, Goldsmith and Silk (1972) have revised their theoretical limit. Also, we have done a detailed numerical integration of the Coma cluster density model of Omer *et al.* (1965), resulting in the revised meaning of the Lα limit shown in Figure 4. The result suggests that if a binding mass of gas exists, it is in the form of clouds at about 2×10^5 K which occupy about 6% of the volume of the cluster. Clearly, further intensive examination of the Coma cluster of galaxies in all wavelength ranges is desirable.

The problem of heating a dense intergalactic medium to 3×10^6 K is severe, and the problem is clearly more difficult in the case of the higher temperature model. In the latter case, the energy for heating could not come from nuclear energy sources. Gravitational collapse can provide arbitrarily large amounts of energy in theory, but putting that energy in the right place – the low–density regions of intergalactic space – is very difficult. The problem may be avoided by adopting a steady-state origin, but this is not a very satisfying solution.

3.4. LOW ENERGY GALACTIC X-RAYS

No matter how the problem of the origin of the extragalactic component is resolved, it is clear that some galactic component is present at low energies (0.1–1 keV). This galactic emission is rather smoothly distributed and the production mechanism (or mechanisms) is not understood. Many origins have been proposed, most of which involve discrete sources. These models can be divided into those in which the sources have a larger scale-height than the neutral hydrogen and those in which the scale-height is the same as the hydrogen. These correspond to the galactic components required by the one- and two-component models, respectively, which were discussed in the previous section.

The large scale-height models include proposals for soft X–ray emission from a hot corona surrounding white dwarfs (Strittmatter *et al.*, 1972) and from flare stars (Cavallo and Horstman, 1972) where the emission is presumably due to a hot stellar atmosphere which also produces the observed optical and radio flares. The models which have scale-heights comparable to the neutral hydrogen include soft X–rays from supernova remnants (Ilovaisky and Ryter, 1971) and accretion of interstellar matter by defunct pulsars (Ostriker *et al.*, 1970). The supernova remnant model suffers from the problem that the number of such sources is small and, therefore the average distance is great. The supernova remnants only contribute at galactic latitudes within a few degrees of the plane and must have very large soft X–ray luminosities to overcome interstellar absorption. The smoothness of the soft X–ray distribution may already rule out such a low source density model. The defunct pulsar model, on the other hand, produces a high density of sources if one accepts the exponential decay of the neutron star birth rate assumed by the authors.

Another class of models which could produce the galactic soft X–ray component is nearby diffuse sources. In particular, Shklovski and Sheffer (1971) have proposed that the galactic spurs, which they assume are old nearby supernova remnants, are such diffuse soft X–ray sources. They propose that the spurs, because they are nearby, provide the source of emission in the direction of the galactic poles while the integral effect of old type II supernova remnants accounts for the emission in the plane. Recently, the North Polar Spur has been observed to be a source of soft X–rays (Bunner et al., 1972). However, other spur structures such as Loop III and parts of other loop structures were observed by Davidsen et al. (1972) and not found to be X-ray emitters with an upper limit significantly below that of the North Polar Spur. It appears, then, that the spurs are not generally soft X-ray sources of sufficient intensity to account for a large fraction of the background.

One other possible galactic source of soft X-rays that was mentioned in the preceding section is the galactic halo proposed by Spitzer (1956). However, even in his highest temperature model (10^6 K), where the electron density is 5×10^{-4} cm^{-3} and the radius is 8 kpc, there is insufficient soft X-ray emission to explain the observed background. The crude spectral observations seem to require a slightly higher temperature of 3×10^6 K and an emission measure, $\int n_e^2 \, dl$, of about 10^{-1} pc cm^{-6}. If one is to maintain Spitzer's assumption of static pressure equilibrium at the interface between the halo gas and the neutral hydrogen disk, then the temperature and emission measure can be used to determine the radius of the halo. For the parameters which best fit the observations, the radius is 10^6 pc. Therefore, the halo is not really associated with the Galaxy alone and, in fact, would not be gravitationally bound. Instead, it is really an intracluster gas, and the distinction between galactic and extragalactic becomes a semantic one. Burke and Hartwick (1972) have recently suggested, however, that one could give up the assumption of static pressure equilibrium and still support a local halo with the required temperature and emission measure by means of a galactic wind mechanism. These authors propose that an extended X-ray source recently observed near the galactic center (Kellogg et al., 1971b) may be evidence for a galactic nucleus with a temperature of more than 10^7 K and densities of 0.1 cm^{-3} which would drive the galactic wind.

Appendix

OBSERVING TECHNIQUES

Observations of the soft X-ray background radiation require the detection of a small flux of photons of rather low energy in the presence of substantial interference from non-X-ray events in the detector. Further difficulties are encountered when the weak photon flux is subdivided in terms of energy to form a spectrum. In addition, conventional astronomical techniques are difficult to apply in the X-ray region: polished surfaces reflect X-rays only at small angles, solid materails transmit soft X-rays only for very short distances ($\gtrsim 0.1$ mm), photographic film is relatively insensitive, and so on. In fact, detection of the soft X-ray background has been achieved almost exclusively by means of the photoionization of gases.

1. PROPORTIONAL COUNTER DETECTORS

In a typical gas-filled X-ray detector, a window serves both to transmit the X-rays and to contain the gas; a high electric field in the gas volume causes avalanche multiplication of the original photoelectrons. The multiplication is necessary because, for example, a 1 keV photon can produce only about 36 ion pairs in the detector gas, and such a small signal would not be electronically detectable. With a gas multiplication of 10^4 to 10^5, each photon event is easily detectable, even down to energies ~ 0.1 keV, since the equivalent noise in the amplifier can be kept below 10^4 electrons. In the earlier days of X-ray astronomy, the gas-filled detectors were operated at much higher gains ($\sim 10^8$), in the saturated avalanche or Geiger counter mode, but it was soon realized that the lower gain detectors (proportional counters) were almost as simple to build and offered the great advantage of some spectral resolution (by analyzing the heights of the pulses).

Since the gas multiplication in the detector is nearly constant for all events, the spectral resolution can be approximately determined by considering the Poisson statistics of the number of original ion pairs. For monoenergetic photons at 1 keV, the average of ~ 36 ion pairs produced has $\sigma \simeq \sqrt{36} = 6$; the pulse height analysis of such 1 keV line emission would show an approximately Gaussian distribution whose full width at half maximum (FWHM) would be $\sim 2.35\sigma = 14$ (in units of ion pairs). Then the percentage FWHM for 1 keV would be $\sim \frac{14}{35} = 40\%$; this corresponds closely to the measured value. Figure 5 illustrates the efficiency and resolving power of a typical proportional counter. The great advantage of the gas-filled proportional counter over other possible photoelectric detectors for soft X-rays is that it provides a crude photon spectrum, and it can be scaled up to almost any size (currently in the thousands of cm^2) with virtually no degradation of the spectral resolution.

2. FIELD OF VIEW

For studies of the soft X-ray background, the small flux of X-rays necessitates a large field of view (as well as a large collecting area) to improve the statistical precision of the measurements. However, the large and increasing number of known point sources, as well as the spatial variations in the background flux itself that are now becoming apparent, require a small field of view. The majority of observations to date have used mechanical baffles to define the field of view, which is generally in the range 10–200 deg^2; collimation in one direction has been as fine as $\sim 1°$ and as large as $\sim 45°$.

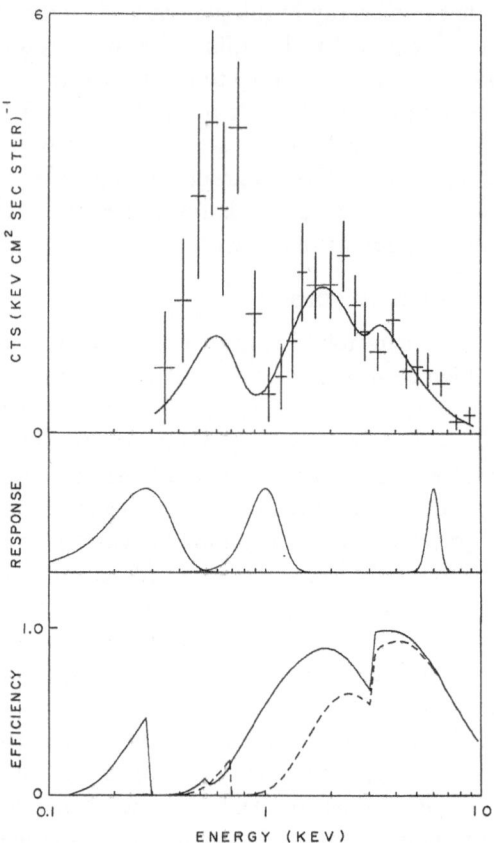

Fig. 5.   Spectral resolution of a gas-filled proportional counter. The abscissa for all curves is energy in units of keV. At the bottom of the figure is shown the detection efficiency of argon-filled counters with 2μ polycarbonate window (solid curve) and 3.2μ Teflon window (dashed curve). The center section shows the response these detectors would have to line emission at 0.28, 1.0 and 6.0 keV. In the upper part of the figure, data obtained near the galactic pole with the Teflon window counter are plotted together with the theoretical response of the counter to a simple power-law spectrum (solid curve) which fits the data above 1 keV; the low-energy excess in the data is clearly apparent.

X-ray telescopes are useful only below 1 keV, because the reflectance drops sharply at higher energies; even below 1 keV, the field of view is restricted by the reflection angles to $\sim 5°$. A further complication with the telescopes is the fact that the efficiency of the overall instrument varies both with wavelength and with angle of arrival of the incoming photons. Proportional counters with simple collimators may have some small reflection effects themselves at the longest wavelengths (0.1–0.3 keV), but generally the efficiency at different angles and wavelengths is easily calculable from the geometry of the collimator and the mass absorption coefficients of the window and gas materials.

3. THE NON-X-RAY BACKGROUND

Proportional counters are sensitive to any source of ionization that deposits an energy in the counter gas equivalent to the X-ray energy being observed (0.1–10 keV). The main components of this non-X-ray background in the detector are cosmic rays, soft electrons trapped by the Earth's magnetic field, γ-rays that scatter in the detector, the natural radioactivity of materials in the detector assembly, and other minor sources of particles (such as those associated with solar activity). It is worth noting that other types of photoionization or photoelectric detectors for X-rays, such as X-ray photo-cathodes, are considerably more susceptible to these interfering effects than the proportional counter because even events outside the range of X-ray energies are counted in such devices.

The techniques used to eliminate the non-X-ray background include the use of coincidence counters (also called guard or veto detectors) surrounding the X-ray detector, discrimination based on the risetimes of the electrical pulses in the proportional counters, and the use of X-ray telescopes to collect the X-rays on a small area detector. Risetime analysis depends upon the fact that the electrons liberated in the counter gas by the incoming X-ray are all collected nearly simultaneously, while the ionization produced by a particle track in the gas is collected over a longer time. Unfortunately, this difference disappears at energies below a few keV, and the technique is efficient only above 3 keV. Focusing X-ray collectors reduce the non-X-ray background in two ways: first, by reducing the size of the detector relative to the effective area of the total instrument and, second, by shielding the detector against direct paths of soft electrons that could penetrate the counter window.

Coincidence counters around the central X-ray detector have taken almost every conceivable form. Some improvement can be made by placing the coincidence counters in the same gas volume as the X-ray counter, separated only by wires or screens. This arrangement has the advantages that (1) low energy particles are not stopped by the walls of the X-ray counter before reaching the veto counters, and (2) photon scattering effects in the walls are minimized. While this technique is quite effective for penetrating particles, it suffers from the drawback that only 5 sides of the X-ray counter can be protected. In the soft X-ray region the detector windows are already so thin that there is no possibility of placing a veto counter in front of the window.

Another source of non-X-ray background is the ultraviolet sensitivity of the X-ray detectors. The problem exists because the inner surfaces (generally metallic) of the gas-filled detectors do not have a negligible photoelectric yield; the small pulses from UV photons pile up so as to mimic low energy X-ray events. Most groups have taken steps to desensitize the counter walls and have made measurements to verify that the UV response is sufficiently low.

4. SPECTROSCOPY OF THE SOFT X-RAY BACKGROUND

Although several well-established techniques exist for producing high resolution X-ray spectra (notably use of crystals and gratings), the difficulty of applying them to the soft X-ray background is the requirement for a nearly parallel beam, which thus limits the field of view, and therefore the flux, to quite low values. This factor, combined with the small collecting areas typical of such devices, means that such spectroscopy of the X-ray background is not likely to be done in the near future.

In the meantime, a considerable effort has gone into a crude form of spectroscopy using pulse height analysis in proportional counters and 'filter photometry' based upon the absorption edges in the detector window material (or occasionally in a separate filter). In fact, the strongest evidence that the background fluxes below 1 keV are true X-rays is that the pulse height spectra show the characteristic peaks of the window elements (usually carbon or fluorine). Of particular importance is the 'peak-to-valley' ratio at these low energies, since the interfering effects of UV response, soft electron fluxes, and general system noise tend to be smoothly varying functions of photon energy; the peaking of the spectrum is thus characteristic of true X-rays transmitted through the window. This effect is shown clearly in Figure 5 (top) where the peak near 0.6 keV is due to the fluorine edge of the Teflon window detector.

Above 1 keV the detector response is generally much smoother (also shown in Figure 5), and here the problem is to obtain accurate wavelength and efficiency calibrations so that the X-ray background spectrum can be unfolded with some precision. In practice, the intrinsic background spectrum cannot be easily unfolded from the pulse height spectrum of the detector. A reverse process is normally used in which a trial source spectrum is folded with the known parameters of the detector (such as efficiency, resolution, non-linearity, etc.); the resulting theoretical curve is then compared with the observational data (as shown in Figure 5, top), and the trial spectrum is modified until a satisfactory fit is obtained.

The difficulty in obtaining accurate spectral information with coarse resolution devices and such fitting techniques is illustrated by recent efforts to study spectrum details in the 1–10 keV energy range. Ducros et al. (1970) and Shulman et al. (1971) have reported possible emission features in their diffuse X-ray background data at 5 keV and 7 keV, respectively. The 7 keV peak appears weakly in the data of Figure 5 (top), which shows only part of the counts actually obtained. Boldt et al. (1971) and Toor et al. (1972), however, find no evidence for such features and place upper limits that are an order of magnitude smaller than the reported features which were equal to the integrated continuum in the same energy region. Toor et al. (1972) obtained their negative results by using detectors with different fields of view and extracting that part which scales with the solid angle. The fact that no line feature was observed indicates that the original results may have been due to particle interference, which is independent of the collimator field of view. The problem with this interpretation is that it requires equal amounts of X-ray and non-X-ray background in the detectors, whereas earth-looking data and data taken before the rocket doors were opened suggest that the non-X-ray background was much lower. If particle contamination should turn out to be the cause of these anomalies, it may also affect the interpretation of spectral index changes in the 1–20 keV energy range. A greater effort is needed to understand the origin of these discrepancies.

References

Arons, J.: 1971, 136th AAS Meeting (San Juan).

Boldt, E. A., Desai, U. D., and Holt, S. S.: 1969, *Nature* **224**, 677.

Boldt, E. A., Desai, U. D., Holt, S. S., and Serlemitsos, P. J.: 1971, *Astrophys. J. Letters* **167**, L1.

Bowyer, C. S., Field, G. B., and Mack, J. E.: 1968, *Nature* **217**, 32.

Bowyer, C. S., Lampton, M., Mack, J., and de Mendonca, F.: 1970, *Astrophys. J. Letters* **161**, L1.

Bradt, H., Naranan, S., Rappaport, S., Zwicky, F., Ogelman, H., and Boldt, E.: 1968, *Nature* **218**, 856.

Brecher, K. and Morrison, P.: 1969, *Phys. Rev. Letters* **23**, 802.

Bunner, A. N., Coleman, P. L., Kraushaar, W. L., McCammon, D., Palmieri, T. M., Shilepsky, A., and Ulmer, M.: 1969, *Nature* **223**, 1222.

Bunner, A. N., Coleman, P. L., Kraushaar, W. L., and McCammon, D.: 1971, *Astrophys. J. Letters* **167**, L3.

Bunner, A. N., Coleman, P. L., Kraushaar, W. L., and McCammon, D.: 1972, *Astrophys. J. Letters* **172**, L67.

Burke, J. A. and Hartwick, F. D. A.: 1972, *Nature Phys. Sci.* **236**, 4.

Byram, E. T., Chubb, T. A., and Friedman, H.: 1970, *Science* **169**, 366.

Cavallo, G. and Horstman, H.: 1972, *Nature Phys. Sci.* **235**, 110.

Cowsik, R.: 1971, 136th AAS Meeting (San Juan).

Davidsen, A., Shulman, S., Fritz, G., Meekins, J. F., Henry, R. C., and Friedman, H.: 1972, *Astrophys. J.* **177** (in press).

Ducros, G., Ducros, R., Rocchia, R., and Tarrius, A.: 1970, *Astron. Astrophys.* **7**, 162.

Fabian, A. C. and Sanford, P. W.: 1971, *Nature Phys. Sci.* **231**, 52.

Felten, J. E. and Morrison, P.: 1966, *Astrophys. J.* **146** 686.

Field, G. B. and Henry, R. C.: 1964, *Astrophys. J.* **140**, 1002.

Friedman, H., Byram, E. T., and Chubb, T. A.: 1967, *Science* **156**, 374.

Goldsmith, D. and Silk, J.: 1971, 136th AAS Meeting (San Juan).

Goldsmith, D. and Silk, J.: 1972, *Astrophys. J.* **172**, 563.

Gorenstein, P. and Tucker, W. H.: 1972, *Astrophys. J.* **176**, 333.

Gott, J. R., III and Gunn, J. E.: 1971, *Astrophys. J. Letters* **169**, L13.

Gunn, J. E. and Ostriker, J. P.: 1969, *Astrophys. J.* **157**, 1395.

Gursky, H., Kellogg, E., Murray, S., Leong, C., Tananbaum, H., and Giacconi, R.: 1971, *Astrophys. J. Letters* **167**, L81.

Hayakawa, S., Kato, T., Makino, F., Ogawa, H., Tanaka, Y., Yamashita, K., Matsuoka, M., Miyamoto, S., Oda, M., and Ogawara, Y.: 1971, *Astrophys. Space Sci.* **12**, 104.

Henry, R. C., Fritz, G., Meekins, J. F., Friedman, H., and Byram, E. T.: 1968, *Astrophys. J. Letters* **153**, L11.

Henry, R. C., Fritz, G., Meekins, J. F., Chubb, T., and Friedman, H.: 1971, *Astrophys. J. Letters* **163**, L73.

Henry, R. C.: 1972, *Astrophys. J. Letters* **172**, L97.

Ilovaisky, S. A. and Ryter, Ch.: 1971, *Astron. Astrophys.* **15**, 224.

Kellogg, E., Gursky, H., Leong, C., Schreier, E., Tananbaum, H., and Giacconi, R.: 1971a, *Astrophys. J. Letters* **165**, L49.

Kellogg, E., Gursky, H., Murray, S., Tananbaum, H., and Giacconi, R.: 1971b, *Astrophys. J. Letters* **169**, L99.

Limber, N.: 1954, *Astrophys. J.* **119**, 655.

Mark, H., Price, R. E., Rodrigues, R. M., Seward, F. D., and Swift, C. D.: 1969, *Astrophys. J. Letters* **155**, L143.

McCammon, D., Bunner, A. N., Coleman, P. L., and Kraushaar, W. L.: 1971, *Astrophys. J. Letters* **168**, L33.

McGee, R. X., Milton, J. A., and Wolfe, W.: 1966, *Australian J. Phys. Suppl.* No. 1.

Meekins, J. F., Fritz, G., Chubb, T. A., Friedman, H., and Henry, R. C.: 1971, *Nature* **231**, 107.

Murray, S., Tananbaum, H., Matilsky, T., Kellogg, E., Gursky, H., and Giacconi, R.: 1971, 136th AAS Meeting (San Juan).

Omer, G. C., Page, T. L., and Wilson, A. G.: 1965, *Astron. J.* **70**, 440.

Ostriker, J. P., Rees, M. J., and Silk, J.: 1970, *Astrophys. Letters* **6**, 179.
Palmieri, T. M., Burginyon, G. A., Grader, R. J., Hill, R. W., Seward, F. D., and Stoering, J. P.: 1971, *Astrophys. J.* **169**, 33.
Schwartz, D., Hudson, H. S., and Peterson, L. E.: 1970, *Astrophys. J.* **162**, 431.
Schwartz, D. A., Boldt, E. A., Holt, S. S., Serlemitsos, P. J., and Bleach, R. D.: 1971, *Nature Phys. Sci.* **233**, 110.
Shklovski, I. S. and Sheffer, E. K.: 1971, *Nature* **231**, 173.
Shukla, P. G. and Wilson, B. G.: 1971, *Astrophys. J.* **164**, 265.
Shulman, S., Fritz, G., Meekins, J. F., Chubb, T. A., Friedman, H., and Henry, R. C.: 1971, *Astrophys. J. Letters* **166**, L9.
Silk, J.: 1970, *Space Sci. Rev.* **11**, 671.
Spitzer, L.: 1956, *Astrophys. J.* **124**, 20.
Strittmatter, P. A., Brecher, K., and Burbidge, G. R.: 1972, *Astrophys. J.* **174**, 91.
Tolbert. C. R.: 1971, *Astron. Astrophys. (Suppl.)* **3**, No. 5.
Toor, A., Price, R., and Seward, F.: 1972, *Astrophys. J. Letters* **172**, L73.
Ulmer, M., Grace, V., Hudson, H. S., and Schwartz, D. A.: 1972, *Astrophys. J.* **173**, 205.
Wolfe, A. M. and Burbidge, G. R.: 1970, *Nature* **228**, 1170.
Woltjer, L.: 1969, *Problems of the Crab Nebula*, Columbia Astrophys. Lab. Contra. 280–4501, No. 15.
Woolf, N.: 1967, *Astrophys. J.* **148**, 287.
Yentis, D. J., Novick, R., and Vanden Bout, P.: 1972, *Astrophys. J.* **177** (in press).

20. ABSORPTION AND PRODUCTION OF SOFT X-RAYS
IN THE GALAXY

S. HAYAKAWA

Nagoya University, Nagoya, Japan

Abstract. The column densities of interstellar hydrogen to X-ray sources derived from their spectra are compared with those obtained from 21 cm radio observations. Referring to several observed results on Cyg X-2, Cygnus Loop etc., the interpretation of the low energy cut-off of the spectrum in terms of the interstellar absorption is subject to ambiguities due to a modification of the emission spectrum by Compton scattering in the sources and the contribution of emission lines.

The result of soft X-ray sky surveys indicates that the diffuse component of soft X-rays consists of the extragalactic and the galactic components. The former has a hard component with a power law spectrum and a soft component which may be represented by an exponential spectrum. The galactic component is so soft that its spectrum may also be explained by thermal bremsstrahlung of temperature of about 0.1 keV. Its generation rate may account for the heating and ionization of interstellar matter. It is suggested that galactic diffuse soft X-rays are produced by active stars of a rather high number density.

1. Introduction

The observation of cosmic X-rays has been extended towards low energy to see the distances of X-ray sources through the modification of their spectra by interstellar absorption and the density of interstellar matter through the distribution of the diffuse component over the celestial sphere. However, results of soft X-ray observations thus far obtained are more complicated than one would have thought.

For some X-ray sources such as Cygnus Loop, the distances derived from the turn-over of their spectra seem to be significantly different from those obtained from the radio observations of the 21 cm line and other astronomical information. As for the diffuse component of soft X-rays, their intensity at low galactic latitudes is much stronger than that expected from the interstellar absorption of the extragalactic component, thus suggesting the contribution of the galactic component.

2. Interstellar Absorption and Scattering of X-Rays

The X-ray spectrum of a source is expressed by the product of the emission spectrum $Q(E)$ and the attenuation coefficient $A(E)$ as

$$J(E) = A(E) Q(E), \ A(E) = \exp\left[-N_H \sigma(E)\right],
\tag{1}$$

where N_H is the column density of hydrogen atoms and $\sigma(E)$ is the effective attenuation cross section per hydrogen atom for X-rays of energy E.

In most works the cross section given by Brown and Gould (1970) is adopted. This is based on the photoelectric absorption by interstellar atoms and may be approximately expressed, below and above the oxygen K-edge, by

$$\sigma(E) = \begin{cases} 6.0 \times 10^{-20} \left(0.1 \text{ keV}/E\right)^3 \text{ cm}^2 & \text{for } 0.1 \text{ keV} \leqslant E \leqslant 0.53 \text{ keV}, \\ 2.0 \times 10^{-22} \left(1 \text{ keV}/E\right)^{2.5} \text{ cm}^2 & \text{for } 0.53 \text{ keV} \leqslant E \leqslant 10 \text{ keV}, \end{cases}
\tag{2}$$

Bradt and Giacconi (eds.), X- and Gamma-Ray Astronomy, 235–249. All Rights Reserved.
Copyright © 1973 by the IAU.

It may be worth mentioning the role of dust grains. A considerable fraction of carbon, oxygen and heavier nonvolatile elements may form dust grains whose density is of the order of 10^{-12} cm^{-3}. For X-rays of energies higher than 1 keV the optical depth of a grain is so small that the interstellar absorption is independent of whether these heavy elements form grains or not. As energy decreases, however, the effective absorption coefficient may be reduced, since the optical depth of a single grain approaches unity, if the fraction of the elements forming grains is considerable. On account of that oxygen is mainly responsible for the photoelectric absorption at energies above 0.53 keV, the K-edge of oxygen, and that it is about 30 times more abundant than silicon, with which one of the major constituents of dust grain, SiO_2 is formed, the photoelectric absorption of dust grains is considered to be of minor importance, unless grains adsorb a great amount of H_2O.

Nevertheless, the effect of dust grains is not always negligible because they scatter soft X-rays. The half width of the scattering angle for a spherical grain of radius a in μ is as large as (Hayakawa, 1970)

$$\theta_s = 10 \left(1 \text{ keV}/E\right) \left(0.1 \; \mu/a\right) \text{ arc min} \tag{3}$$

and the scattering coefficient at $\frac{1}{4}$ keV may be as large as $\frac{1}{4}$ of the photoelectric absorption coefficient predicted by (2). If, therefore, a compact source is measured with a good angular resolution, the attenuation effect for the source is greater than expected from the photoelectric absorption.

The effect of dust grains will be more clearly observed through time variations. Since photons scattered are considerably delayed, even if the scattered angle is very small, pulsations of short periods may be washed out by scattering (Slysh, 1969; Naranan and Shah, 1970). Such an effect could be observed for Cen X-3 and Cyg X-1 at about 0.8 keV and lower energies.

Since detailed properties of interstellar dust grains are not yet established, their effects as discussed above are not quantitatively predictable but will have to be revealed by future observations.

3. Distances of X-Ray Sources

The hydrogen column density to an X-ray source has been derived by assuming an emission spectrum

$$Q(E) \propto E^{-1} \exp\left(-E/kT\right) \tag{4a}$$

or

$$Q(E) \propto E^{-\alpha}. \tag{4b}$$

It should be noted that the value of N_H thus derived, which is denoted as N_{Hx}, depends rather sensitively on the shape of the spectrum assumed. The expression (4a) implies that X-rays are emitted by thermal bremsstrahlung. Some authors use a more exact expression applicable to thin plasmas

$$Q(E) \propto E^{-1} \exp(-E/2\,kT)\, K_0(E/2\,kT),\tag{4c}$$

$$K_0(x) \simeq \begin{cases} \ln(2/x) - 0.577 & \text{for} \quad x \ll 1 \\ (\pi/2x)^{1/2} \exp(-x) & \text{for} \quad x \gg 1, \end{cases}$$

in place of (4a). This gives a larger value of N_{Hx} if the turn-over energy is much smaller than kT. For optically thick sources such as Sco X-1 the emission spectrum has a complicated shape and may turn over without the interstellar absorption.

The value of N_{Hx} for Sco X-1 obtained with (4a) is $(30–40) \times 10^{20}$ cm^{-2} (Grader et al., 1970a; Burginyon et al., 1970), whereas the 21 cm emission line intensity gives about 10×10^{20} cm^{-2} (Goldstein and MacDonald, 1969). The difference has been interpreted as due to circumstellar cold matter. It is not easy to find such thick cold matter that is left unionized under a high X-ray flux, that is held against a high radiation pressure and that does not produce absorption lines. Since the turn-over point may change from one experiment to the other, it is likely that the spectrum at low energies is partly modified by Compton scattering in the X-ray emitting region, and the modification changes with the optical depth for the Compton scattering.

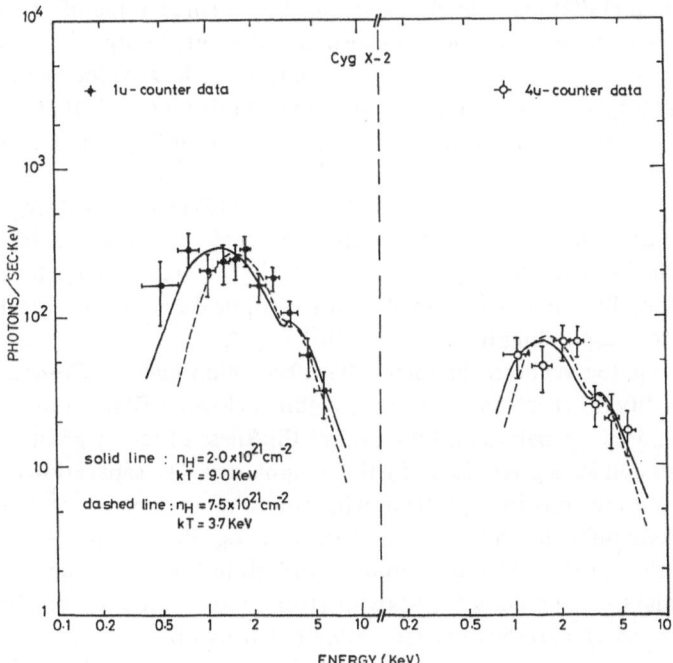

Fig. 1.   Energy spectrum of Cyg X-2 (Bleeker et al., 1972). The pulse height distributions obtained with two kinds of counters of different window thicknesses are compared with those expected from the thermal bremsstrahlung spectra modified by the interstellar absorption. The solid lines represent the best fit to all experimental points, whereas the dashed lines fit to the points for $E \geqslant 0.9$ keV.

The variation of the spectrum in the low energy part may be of experimental origin rather than of genuine variations in the source. The spectrum of Cyg X-2 provides an instructive example. An observation with counters of poor efficiency at low energies gave N_{Hx} of about 200×10^{20} cm^{-2} although the quoted errors were rather large (Gorenstein et al., 1967), whereas another with high efficiency resulted in a value as small as 20×10^{20} cm^{-2} (Hayakawa et al., 1971). It has been demonstrated by Bleeker et al. (1972) that the value of N_{Hx} depends on the lower cut-off energy of the spectrum using a single observation with high efficiency counters. $N_{Hx} \simeq 75 \times 10^{20}$ cm^{-2} is derived on the basis of the spectrum for $E \geqslant 0.9$ keV, whereas $N_{Hx} \simeq 20 \times 10^{20}$ cm^{-2} if lower energy data ($E \geqslant 0.16$ keV) are included, as shown in Figure 1. A reason for giving such considerably different values is due, at least in part, to the flatness of the spectrum around 1 keV. It is likely that the emission spectrum cannot be represented by any one of the simple spectra given in (4a–c).

Another example is provided by Cygnus Loop. Grader et al. (1970b) obtained $N_{Hx} = 2.5 \times 10^{20}$ cm^{-2}, assuming $\alpha = 3.2 \pm 0.3$ for (4b) or $kT = 0.4$ keV for (4c). Gorenstein et al. (1971) also gave a small value, $N_{Hx} = 2.6 \times 10^{20}$ cm^{-2}, assuming $kT = 0.37$ keV for (4a) plus a contribution of a 0.65 keV line of O VIII. The average hydrogen density over the distance to the Cygnus Loop would then be as small as 0.1 cm^{-3} if 770 pc is adopted as the distance of Cygnus Loop. The spectrum observed by Bleeker et al. (1972) may again be explained by a small value of N_{Hx} with a weak contribution of emission lines, but may also be represented only by a continuum with a greater value of N_{Hx}, $(5–7) \times 10^{20}$ cm^{-2}. Even if such a large value of N_{Hx} is adopted, the average hydrogen density of 0.2–0.3 cm^{-3} over the distance of 770 pc is smaller than that currently adopted for the density in the direction of Cygnus Loop (Daltabuit, 1970).

The third example is Cyg X-1. Gursky et al. (1971) obtained $N_{Hx} = (16 \pm 4) \times 10^{20}$ cm^{-2} assuming a power spectrum of $E^{-1.7 \pm 0.1}$, whereas a larger value of N_{Hx} is not ruled out for a steeper spectrum $E^{-2.6 \pm 0.3}$ as observed by Bleach et al. (1972). Incidentally, the distance of its possible optical counter part, HD 226868, is about 2 kpc, corresponding to $N_{Hx} \simeq 40 \times 10^{20}$ cm^{-2}.

It is unfortunate that both the interstellar absorption and the Compton scattering in sources modify the spectrum at energies slightly below 1 keV for sources at distances of several hundreds of parsec and the optical thickness of ten or greater. The contribution of emission lines gives rise to further complexity. The separation of the absorption effect from the emission spectrum will require more experimental effort.

Keeping these difficulties in mind, the values of N_{Hx} obtained for X-ray sources are compared with N_H (21 cm) from radio observations in Figure 2, extending a compilation by Ilovaisky (1971). Only for SNR of strong radio emission the absorption line of 21 cm is observed, and the hydrogen column density obtained therefrom is denoted as N_{Hab}. For other sources comparison can be made only with the emission line intensity in the direction of each X-ray source, and the hydrogen column density for the emission line is denoted as N_{Hem}. The values of N_{Hab} and N_{Hem} derived from radio observations depend on the spin temperature, so that the reliability of these values becomes poorer as the galactic latitude decreases. For $|b| \gtrsim 10°$ the optical depth at 21 cm becomes so small that N_{Hem} depends only weakly on the spin temperature. Moreover, X-ray sources at such latitudes are likely to lie towards the edge of the galactic disk, so that N_{Hem} may be close to the hydrogen column densities to the sources. Hence the comparison of N_{Hx} with N_{Hem} is of greater significance for sources with $|b| \gtrsim 10°$.

The relation between N_{Hx} and N_H (21 cm), shown in Figure 2, should be considered as illustrative, on account of the discussions given above. Most values of N_{Hx} are taken from Hill *et al.* (1972) and are subject to considerable statistical errors and to dependences on the spectral shape assumed. The values of N_{Hem} are based on a compilation by Daltabuit (1970), in which different 21 cm observations do not always give mutually consistent results, in particular for $b = 0°$.

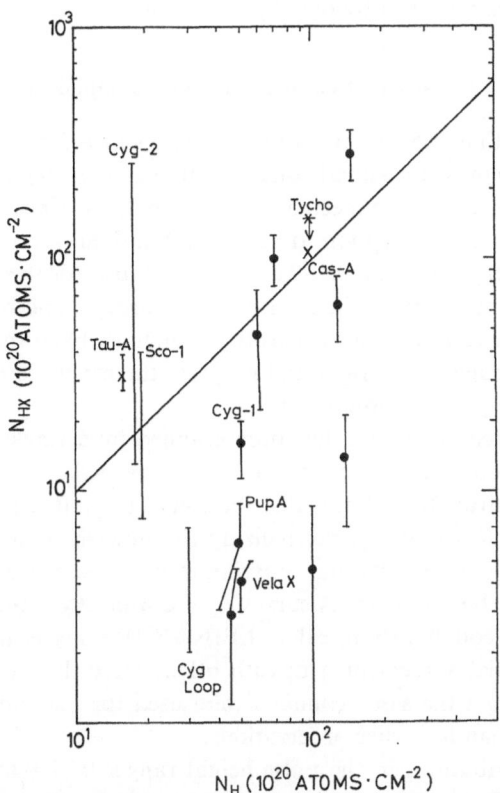

Fig. 2. The hydrogen column densities obtained from X-ray spectra (N_{Hx}) and 21 cm radio intensities (N_H). In the latter the crosses ($\times$) represent N_{Hab}, whereas others represent N_{Hem}. The uncertainties in the values of N_{Hx} due to fitting of the observed spectra to expressions (1) with (4) are indicated by vertical bars with solid circles. For Sco X-1, Cyg X-2 and the Cygnus Loop the experimental values are so widely scattered that the ranges of N_{Hx} are indicated by vertical bars. References to the values of N_{Hx} for the named sources can be found in the text, whereas the unnamed ones are based on Hill *et al.* (1972).

Within the uncertainties in the numerical values of N_H, a general trend indicates $N_{Hx} \lesssim N_{Hem}$. For Sco X-1 and Cyg X-3 there hold $N_{Hx} > N_{Hem}$, thus suggesting the presence of circumstellar matter or emission spectra with maxima at corresponding energies.

There are several sources for which $N_{Hx} << N_{Hem}$. For SNR's, Cygnus Loop, Vela X and Pup A, the distances are estimated from the expansion velocities and the angular diameters (Milne, 1970). In all three cases the hydrogen densities averaged over the distances are as small as 0.2–0.3 cm^{-3}, which is compared with about 0.7 cm^{-3} obtained from 21 cm observations. If, however, these SNR's are surrounded by H II regions with radii depending on their ages, as in the case of the Gum nebula around Vela X (Brandt et al., 1971), the hydrogen density in the H I region estimated from N_{Hx} may have to be increased by a factor of two or so.

Other sources with small N_{Hx} are not fully studied yet. Some of them may be variables, since those observed by one flight are not detected by others.

4. Distribution of the Diffuse Component

At energies greater than 1.5 keV the diffuse component is isotropic within a few per cent (Cooke and Pounds, 1971) and consequently has been regarded as extragalactic origin. Anisotropy should be expected at lower energies, since extragalactic X-rays are subject to interstellar absorption. It has been found, however, that the intensity of diffuse X-rays of energies around 0.25 keV decreases towards the low galactic latitude more slowly than expected from the interstellar absorption and remains to be considerable at the galactic equator. This is hardly understandable if the diffuse component is exclusively of extragalactic origin and suggests the existence of the galactic component in the energy range below 1 keV.

Since only a limited sky region has been scanned by a single rocket observation, and since different observations have not always given the same absolute intensity, it is not easy to construct the celestial distribution of the soft X-ray intensity. This has been attempted by Kato (1972), normalizing the intensities in the direction of the lowest hydrogen column density and referring to five observations with high counter efficiencies at the carbon window. A survey over a wide sky region has been made by the Leiden-Nagoya collaboration, called LEINAX (Bleeker et al., 1972). These two results are in essential agreement with eath other. Here the LEINAX result is presented, in view of that the same counters were used for scanning and the statistical accuracy is better than in earlier observations.

The celestial distributions in the pulse height ranges 0.37–0.65 keV and 0.65–0.90 keV are shown in Figure 3a and Figure 3b, respectively. Since the counts obtained with the 1μ counters are averaged over the spin angle of 10°, the spatial resolution is 25° × 13° (FWHM) with the greater angular extension perpendicular to the scan path. In the range 0.37–0.65 keV the counter efficiency has a dip, so that incident X-rays in the energy range 0.23–0.8 keV contribute to this pulse height range. Because of the softness of the X-ray spectrum the distrubition in Figure 3a is dictated by that in the

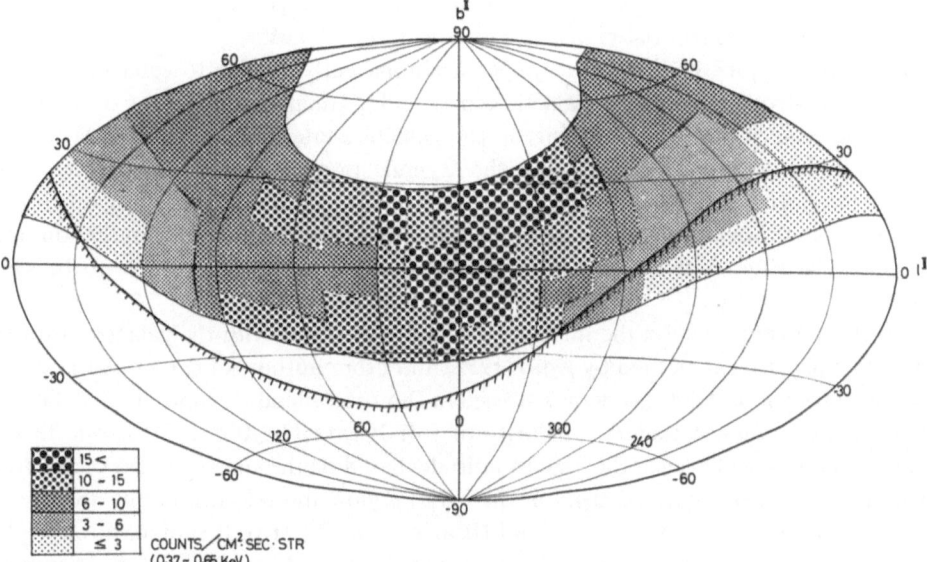

Fig. 3a.

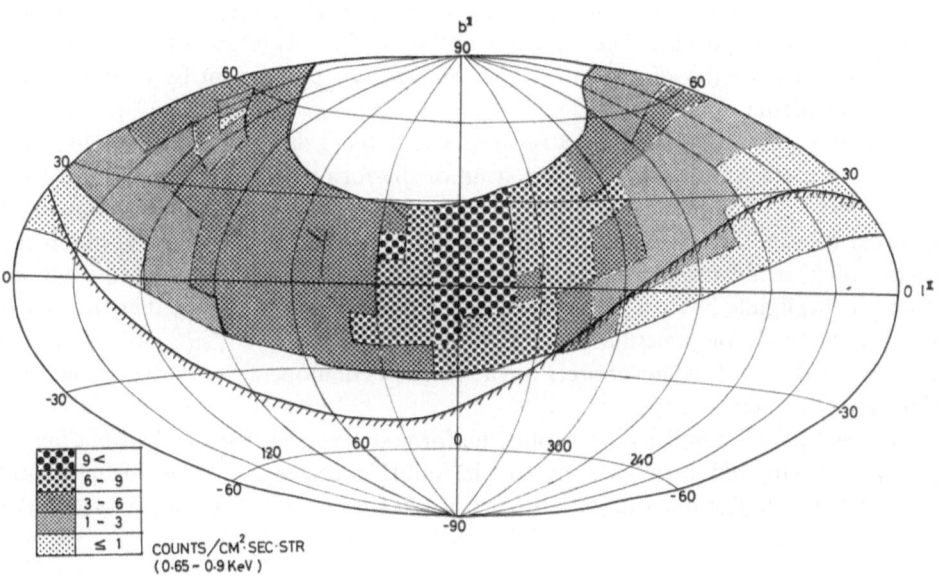

Fig. 3b.

Figs. 3a–b. Celestial distributions of the soft X-ray intensity. (a) $E=0.37$–0.65 keV, (b) $E=0.65$–0.90 keV. The position at which the counting rate is given corresponds to the direction of the counter axis.

X-ray energy range 0.23–0.28 keV. In the pulse height range 0.65–0.90 keV, the main contribution comes from X-rays of energies around 0.8 keV.

The distributions in these two ranges are apparently similar to each other. The intensity is strong towards the galactic center and gradually decreases as one recedes from the center. The strong intensity in the galactic center region is due partly to the contribution of X-ray sources; only the Cygnus Loop is avoided in mapping the distribution. In a still lower energy range the enhancement in the center region almost disappears, although in this range spurious counts due to ultraviolet radiation and the noise arising from the electrostatic field for electron rejection have to be subtracted.

In the anticenter region the intensity gradually increases with the galactic latitude. This fact has been attributed by Kato (1972) and other authors of earlier observations to the absorption of extragalactic X-rays. On the other hand, Gorenstein and Tucker (1972) have assumed that the majority of soft X-rays ($E \lesssim 0.25$ keV) is due to the galactic component even at the highest latitude, based on the absence of the correlation with the hydrogen column density in the Virgo region and referring to the absence of absorption by Small Magellanic Cloud (Bunner *et al.*, 1971b). It should, however, be remarked that the comparison of the intensities in two selected regions, as attempted by Gorenstein and Tucker (1972), may not always be an acceptable procedure, in view of irregularities in the intensity distribution.

The correlation between the X-ray intensity and the hydrogen column density is shown in Figure 4 in the anticenter region ($90° \leqslant l \leqslant 270°$), in which irregularities seem to be less important. The N_H dependence observed is in good agreement with that expected from the absorption of the extragalactic component by homogeneous interstellar matter.

The separation between the extragalactic and the galactic components depends also on their energy spectra. The spectrum of the former has been represented by a power law, $E^{-1.4}$, between 1 and 10 keV (Bunner *et al.*, 1971a). If this is extrapolated to lower energy, there remains an excess intensity below 1 keV. At low latitudes the excess soft X-rays cannot be due to the extragalactic component, since its contribution, if any, is negligible because of the interstellar absorption. Assuming that the excess part observed in the celestial region, $90° \lesssim l \lesssim 180°$ and $-10° \lesssim b \lesssim 30°$, in which $N_H \gtrsim 30 \times 10^{20}$ cm^{-2}, is due entirely to the galactic component, we derive the generation rate of galactic X-rays.

For the sake of simplicity we assume uniform slab models for the distributions of X-ray emissivity and interstellar matter, the half thicknesses of the slabs being β_x and β_H, respectively. The intensity at latitude b is related to the X-ray emissivity $g(E)$ as

$$j_g(E, b) = \frac{1}{4\pi} g(E) \left[\frac{1 - \exp(-\sigma(E) N_H)}{\sigma(E) n_H} + \frac{\beta_x - \beta_H}{\sin b} e^{-\sigma(E) N_H} \right], \qquad (5)$$

where $\beta_x \geqslant \beta_H$ is assumed. In the low latitude region, $\exp(-\sigma N_H)$ is negligibly small

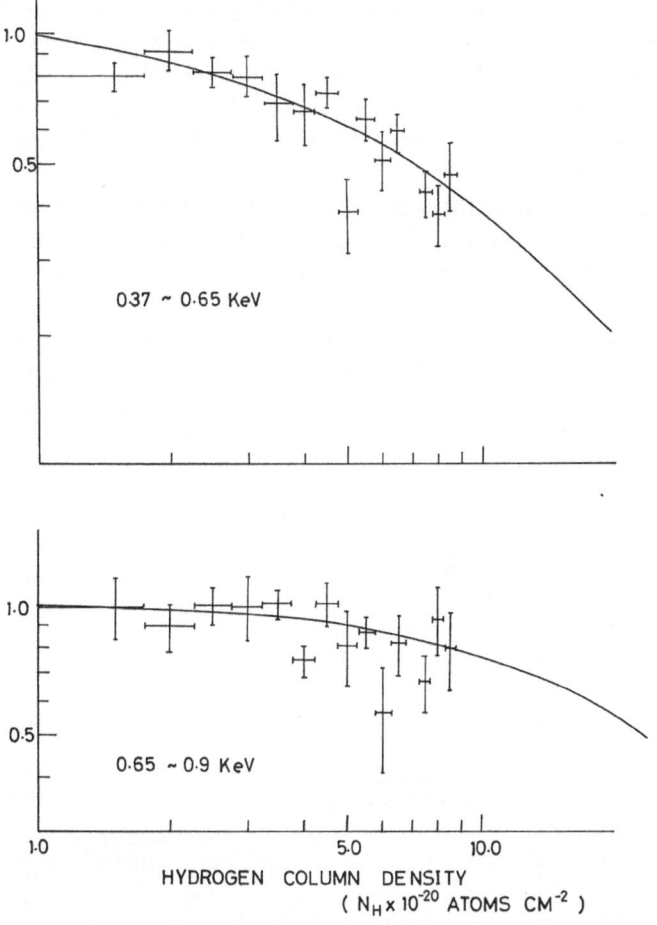

Fig. 4. The soft X-ray intensities versus the hydrogen column density. The intensities are adopted from the observations in the anticenter side and the curves represent the absorption of extragalactic X-rays expected in the respective energy ranges.

in the energy range of interest. Hence the X-ray emissivity is obtained as

$$g(E) \simeq g_0 n_H E^{-1} \exp(-E/kT) \text{ photons cm}^{-3} \text{ s}^{-1} \text{ keV}^{-1}, \tag{6}$$

with $g_0 = 3 \times 10^{-17}$ photons s^{-1}, $kT = 0.1$ keV.

This gives the energy generation rate

$$G = \int Eg(E)\,dE \simeq 5 \times 10^{-27} \, n_H \text{ erg cm}^{-3} \text{ s}^{-1}. \tag{7}$$

In the high latitude region the contribution of the galactic component can be evaluated from the expression (5) with (6), provided that the value of β_x is known.

However, the contribution of the term containing β_x can be distinguished from that of the extragalactic component only through its $\mathrm{cosec}\,b$ dependence. The counting rates around $l=120°$, $b=25°$ and around $l=240°$, $b=55°$ are nearly the same, and the values of N_H in these regions are about equal, 5×10^{20} cm^{-2}. This results in an upper limit of $(\beta_x/\beta_H)-1 \lesssim 0.2$, provided that the diffuse soft X-rays are exclusively of the galactic origin.

Since the contribution of the second term in (5) seems to be of secondary importance, we neglect this term for the time being. In the low latitude region a superposition of the galactic component with emissivity given in (6) and the extragalactic component of the $E^{-1.4}$ spectrum modified by the interstellar absorption reproduced the oberved spectrum, as shown in Figure 5. A superposed spectrum in the high latitude region, $90° \lesssim l \lesssim 270°$ and $b \gtrsim 40°$, in which $N_H \lesssim 5 \times 10^{20}$ cm^{-2}, is compared with the observed spectrum. The latter is found to be about twice greater than the

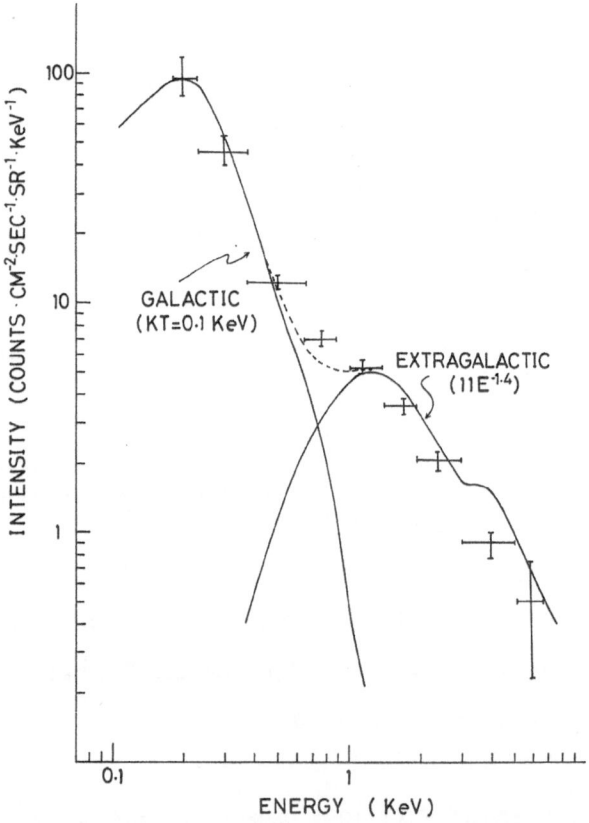

Fig. 5.   X-ray spectrum in the low latitude region. The solid curves represent the pulse height spectra of the galactic and the extragalactic components, taking into account the interstellar absorption for $\langle N_H \rangle = 30 \times 10^{20}$ cm^{-2}, the counter efficiency and the energy resolution. The dashed curve represents their sum. The observed points are based on the results obtained with the 1 μ LEINAX counters.

former in the pulse height range below 0.37 keV. The excess is accounted for in terms
of the exponential spectrum of $kT=0.1$ keV modified by the absorption of $N_H=3$
$\times 10^{20}$ cm^{-2}. A superposition of the galactic component and the extragalactic
component, the latter consisting of the power law and the exponential law spectra,
is compared with the observed one in Figure 6.

Thus the general behaviour of the diffuse soft X-rays may be understood as a
mixture of the galactic and the extragalactic components. However, the conclusion
will be reserved for future investigation, since the LEINAX data are contaminated
with ultraviolet radiation and the angular resolution is not good enough for unambiguous separation between the N_H and the b dependences.

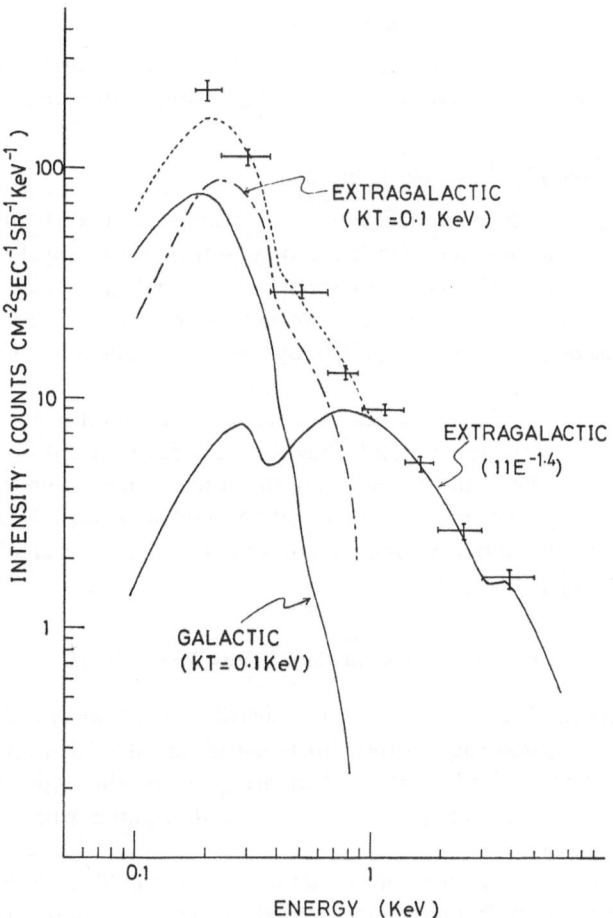

Fig. 6.   X-ray spectrum in the high latitude region. The solid curves are the same as those in Figure 5
except for the interstellar absorption, for which $\langle N_H \rangle = 3 \times 10^{20}$ cm^{-2} is assumed. The dot-dashed
curve represents the pulse height distribution of the soft extragalactic component. The dashed curve
represents the sum of these three components. The observed points are based on the results obtained
with the 1 μ LEINAX counters.

5. Ionization of Interstellar Medium

The energy generation rate of soft X-rays is equal to the energy dissipation rate in the interstellar medium. Since the energy per ionization is about 6×10^{-11} erg, the rate of ionization per hydrogen atom caused by X-rays, ζ_x, is given by

$$\zeta_x \simeq G/6 \times 10^{-11} \, n_H \simeq 1 \times 10^{-16} \, s^{-1} \text{ per H atom}, \tag{8}$$

where the value of G in (7) is substituted to obtain the value of ζ_x. On the other hand, the rate of ionization by cosmic rays is limited by the abundance of Be, which results from the spallation of cosmic ray nuclei colliding with interstellar matter, to (Reeves *et al.*, 1970; Meneguzzi *et al.*, 1971)

$$\zeta_{CR} \simeq 10^{-17} \, s^{-1} \text{ per H atom}. \tag{9}$$

These are compared to the ionization rate required for ionization and thermal equilibria for H I regions (Spitzer and Scott, 1969; Goldsmith *et al.*, 1969; Hjellming *et al.*, 1969)

$$\zeta_H \simeq 5 \times 10^{-16} \, s^{-1} \text{ per H atom}. \tag{10}$$

In view of the fact that the numerical value of ζ_x is subject to uncertainties, such as the ambiguity in distinguishing the galactic component from the extragalactic one and an unknown flux below 0.1 keV, the difference between ζ_x and ζ_H given above may not be serious. It is, therefore, likely that the soft X-rays are a main cause of interstellar ionization, as discussed by Sunyaev (1969), by Silk and Werner (1969) and by Werner *et al.* (1970).

The X-rays from individual sources also produce ionization in the vicinity of the sources. The radius of the H II region formed by an X-ray source is limited by its age. In the case of Cygnus Loop the dimension of the H II region is estimated to be smaller than the size of the X-ray source. In view of the fact that most X-ray sources are short-lived, an estimate of the H II region radius under the asumption of the stationary state is not always justified.

6. Mechanisms of Soft X-Ray Production

There are two alternative cases for the spatial distribution of galactic X-ray emissivity, a continuous distribution such as the diffuse matter in interstellar space or a superposition of as yet unresolved sources. There are also two alternative mechanisms of X-ray production, thermal or non-thermal. We shall examine which of four combinations is most likely.

Non-thermal processes in the diffuse matter are most unlikely to be the origin of galactic soft X-rays. If X-rays were emitted by bremsstrahlung of non-thermal particles, the amount of energy spent by these particles for ionization would be several orders of magnitude greater than that for X-rays. As shown in the preceding section, the X-ray energy generation rate is nearly as large as the heating rate required for interstellar matter. The heating rate by the ionization would greatly exceed the latter. In similar arguments other non-thermal processes in interstellar space are ruled out.

The thermal bremsstrahlung could account for the X-ray emissivity if $n_e^2 f = = 10^{-2} - 10^{-3}$ cm^{-6}, where f is the fraction of volume in which the electron density is n_e and the electron temperature is $T \sim 10^6$ K. Since the major part of interstellar matter is known to be of much lower temperature, such hot regions would have to be restricted to a small volume. Hence this hypothesis is not different from the case of assuming soft X-ray sources.

A granularity observed by Gorenstein and Tucker (1972) provides evidence against the diffuse distribution of emissivity. They have concluded from the granularity that the source density is about 10^{-2} (pc)$^{-3}$ or greater and the average luminosity of the sources is about 3×10^{30} erg s^{-1} in the energy range 0.16–0.28 keV.

The above considerations are in favour of interpreting the diffuse galactic soft X-rays in terms of a superposition of discrete sources which may be stellar and/or nebulous. A number of possible candidates of such sources have been proposed. Since there are too many theories, we here mention only a few of them.

Supernova explosions can supply as much energy as to account for the generation of soft X-rays (Ilovaisky and Ryter, 1971). However, the number density of active SNR's is as small as 10^{-7} (pc)$^{-3}$, far smaller than that required from the granularity. The angular resolutions in most of the observations available are good enough to resolve such SNR's, which could otherwise contribute to the diffuse component. SNR's could not be responsible for the diffuse soft X-rays, unless their size were as large as or larger than the size of the Gum nebula (Ramaty et al., 1971), so that sky could be covered by a small number of extended SNR's.

The number density increases if old SNR's are included. They may maintain activities due to the matter accretion on to neutron stars. The activities may produce such a hot region around the neutron star that is heated to about 10^6 K (Ostriker et al., 1970; Shvartsman, 1971). This may remain to be a possible candidate, if the galactic emission takes place within the gas disk, in which most of pulsars are found.

Radio spurs may also be responsible for the diffuse soft X-rays, in view of the fact that a strong X-ray emission has been observed in the North Polar Spur (Bunner et al., 1972). Although the enhancement is not so large in nearly the same region scanned by other observations and is not always detectable in other radio spurs (Kato, 1972), the positive result indicates that at least a part, if not all, of diffuse soft X-rays are likely to be associated with radio spurs.

Due to the fact of that the number density of the soft X-ray sources seems to be rather high, the sources may be main sequence stars. Gorenstein and Tucker (1972) have argued that the matter accretion onto a main sequence star results in a temperature of 10^6 K or lower and a matter ring formed in a closed binary system supplies sufficient power to explain the generation of soft X-rays.

Late type main sequence stars are known to produce strong optical and radio flares at a high frequency. They are highly convective, so that their activities are regarded as similar to those of T-Tauri type stars.

Ryter *et al.* (1970) considered thermal bremsstrahlung emitted from the atmosphere of T-Tauri type stars due to the heating by energetic particles that may be responsible for the production of light elements. Grindlay (1970) and Gurzadyan (1971) estimated the intensities of bremsstrahlung X-rays caused at flare stars indirectly and directly by energetic electrons responsible for radio emission, respectively. According to their theories, X-rays of energies greater than 1 keV are also produced and they would result in anisotropies of the diffuse component in the higher energy range. On the other hand, Hayakawa *et al.* (1970) considered the thermal balance of lower corona between the heating by energetic particles and thermal bremsstrahlung loss. Hence a lower corona with the electron density 10^{11} cm^{-3} and the thickness 10^{10} cm can be heated to about 10^6 K and can emit X-rays at a rate of about 10^{30} erg s^{-1}. Taking the flare frequency of once in every hour observed for YZ CMi as representative and the number density of red dwarf stars as 0.1 (pc)$^{-3}$, the average energy output is estimated to be about 3×10^{-28} erg cm^{-3} s^{-1}. Although this is smaller than that required in (7) by an order of magnitude, slight changes in the numerical values of the parameters can increase the theoretically estimated value of the energy generation rate.

Although the sources of the galactic soft X-rays cannot yet be identified in an undisputable way, since too many parameters are left undetermined, it is plausible that the galactic soft X-rays are generated at discrete sources whose density is comparable to the density of stars and that they are emitted by thermal bremsstrahlung of temperature about 0.1 keV.

Acknowledgements

The present review could not be prepared without prompt information from the soft X-ray groups of AS&E, GSFC, LRL at Berkeley, LRL at Livermore, MIT, NRL, Saclay and University of Wisconsin. The author would also like to thank the members of LEINAX who have made unpublished data available. Among them Drs. T. Kato and Y. Tanaka directly helped the author in the preparation of Section 4.

Addendum

After completion of the manuscript, the author received a preprint by A. Davidsen, S. Shulman, G. Fritz, J. F. Meekins, R. C. Henry, and H. Friedman. Their result and its interpretation are in essential agreement with those given in Section 4. Also, Dr Henry made the following comment at the symposium: "E. B. Jenkins (*Bull. Am. Astron. Soc.* **4**, 226, 1972), has observed 8 hot stars in the far ultraviolet and has found, from the strength of interstellar absorption lines, that oxygen and silicon are of canonical abundance, but carbon is apparently 10 × overabundant. H. Friedman and the NRL X-ray astronomy group (*Astrophys. J.* **174**, 389) have found a stronger absorption of soft X-rays from the Crab than the 21-cm column density and canonical abundances predict; the discrepancy would be substantially moderated if carbon were actually 10 × overabundant in the interstellar medium."

References

Bleach, E. A., Boldt, E. A., Holt, S. S., Schwartz, D. A., and Serlemitsos, P. J.: 1972, *Astrophys. J.* **171**, 51.

Bleeker, J. A. M., Deerenberg, A. J. M., Yamashita, K., Hayakawa, S., and Tanaka, Y.: 1972, paper submitted to this Symposium.

Brandt, J. C., Stecher, T. P., Crawford, D. L., and Maran, S. P.: 1971, *Astrophys. J. Letters* **163**, L99.

Brown, R. L. and Gould, R. J.: 1970, *Phys. Rev.* **D1**, 2252.

Bunner, A. N., Coleman, P. L., Kraushaar, W. L., and McCammon, D.: 1971a, *Astrophys. J. Letters* **167**, L3.

Bunner, A. N., Coleman, P. L., Kraushaar, W. L., and McCammon, D.: 1971b, *Astrophys. J. Letters* **168**, L33.

Bunner, A. N., Coleman, P. L., Kraushaar, W. L., and McCammon, D.: 1972, *Astrophys. J. Letters* **172**, L67.

Burginyon, G. A., Grader, R. J., Hill, R. W., Price, R. E., Rodrigues, R., Seward, F. D., Swift, C. D., Hiltner, W. A., and Mannery, E. J.: 1970, *Astrophys. J.* **161**, 987.

Coleman, P. L., Bunner, A. N., Kraushaar, W. L., and McCammon, D.: 1971, *Astrophys. J. Letters* **170**, L67.

Cooke, B. A. and Pounds, K. A.: 1971, *Nature* **229**, PS 144.

Daltabuit, E.: 1970, University of Wisconsin, preprint.

Goldsmith, D. W., Habing, H. J., and Field, G. B.: 1969, *Astrophys. J.* **158**, 173.

Gorenstein, P., Giacconi, R., and Gursky, H.: 1967, *Astrophys. J. Letters* **150**, L85.

Gorenstein, P., Harris, B., Gursky, H., Giacconi, R., Novick, R., and van de Bout, P.: 1971, *Science* **172**, 369.

Gorenstein, P. and Tucker, W. H.: 1972, *Astrophys. J.*, in press.

Grader, R. J., Hill, R. W., Seward, F. D., and Hiltner, W. A.: 1970a, *Astrophys. J.* **159**, 201.

Grader, R. J., Hill, R. W., and Stoering, J. P.: 1970b, *Astrophys. J. Letters* **161**, L45.

Grindlay, J. E.: 1970, *Astrophys. J.* **162**, 187.

Gursky, H., Gorenstein, P., Kerr, F. J., and Grayzeck, E. J.: 1971, *Astrophys. J. Letters* **167**, L15.

Gurzadyan, G. A.: 1971, *Astron. Astrophys.* **13**, 348.

Hayakawa, S.: 1970, *Prog. Theor. Phys.* **43**, 1224.

Hayakawa, S., Kato, T., Makino, F., Ogawa, H., Tanaka, Y., and Yamashita, K.: 1970, *Proc. VI Interamerican Seminar on Cosmic Rays*, Vol. 1, p. 106.

Hayakawa, S., Kato, T., Makino, F., Ogawa, H., Tanaka, Y., Yamashita, K., Matsuoka, M., Miyamoto, S., Oda, M., and Ogawara, Y.: 1971, *Astrophys. Space Sci.* **12**, 789.

Hill, R. W., Burginyon, G., Grader, R. J., Palmieri, T. M., Seward, F. D., and Stoering, J. P.: 1972, *Astrophys. J.* **171**, 519.

Hjellming, R. M., Gordon, C. P., and Gordon, K. J.: 1969, *Astron. Astrophys.* **2**, 202.

Ilovaisky, S. A.: 1971, *Astron. Astrophys.* **11**, 134.

Ilovaisky, S. A. and Ryter, C.: 1971, *Astron. Astrophys.* **15**, 224.

Kato, T.: 1972, *Astrophys. Space Sci.* **16**, 478.

Meneguzzi, M., Audouze, J., and Reeves, H.: 1971, *Astron. Astrophys.* **15**, 337.

Milne, D. K.: 1970, *Austral. J. Phys.* **23**, 425.

Naranan, S. and Shah, G. A.: 1970, *Nature* **225**, 836.

Ostriker, J. P., Rees, M. J., and Silk, J.: 1970, *Astrophys. Letters* **6**, 179.

Ramaty, R., Boldt, E. A., Colgate, S. A., and Silk, J.: 1971, *Astrophys. J.* **169**, 87.

Reeves, H., Fowler, W. A., and Hoyle, F.: 1970, *Nature* **226**, 727.

Ryter, C., Reeves, H., Gradsztajn, E., and Audouze, J.: 1970, *Astron. Astrophys.* **8**, 389.

Shvartsman, V. F.: 1971 *Sov. Astron. – AJ* **14**, 662.

Silk, J. and Werner, M. W.: 1969, *Astrophys. J.* **158**, 185.

Slysh, V. I.: 1969, *Nature* **224**, 159.

Spitzer, L. and Scott, E. H.: 1969, *Astrophys. J.* **158**, 161.

Sunyaev, R. A.: 1969, *Astron. Zh.* **46**, 929.

Werner, M. W., Silk, J., and Rees, M. J.: 1970, *Astrophys. J.* **161**, 965.

21. 'EVOLUTIONARY' THEORIES OF THE X-RAY BACKGROUND

M. J. REES*

Harvard College Observatory, Cambridge, Mass. 02138, U.S.A.

Abstract. It is unclear whether evolutionary effects must be invoked to account for the intensity of the X-ray background, nor whether the background could all be due to unresolved X-ray sources of the types already known. Intensity fluctuations on small angular scales are probably on the threshold of detectability. Future studies of these fluctuations – with improved sensitivity – should help to elucidate the nature of the background. The interpretation of such studies is discussed.

At the time of the 1969 IAU Symposium on X-ray astronomy, the existence of an apparently isotropic background was already well established. Its spectrum was known to follow a power law, at least over the range 1–20 keV; and the large-scale isotropy was established to a precision of around 10%. Because no class of extragalactic object seemed capable of accounting for the strength of the background (unless drastic evolutionary effects were invoked) some astrophysicists devoted much attention and ingenuity to devising emission mechanisms which might operate uniformly throughout intergalactic space, or at least in very extended regions such as clusters of galaxies. Examples of such 'diffuse' mechanisms are inverse Compton scattering of microwave background photons, and bremsstrahlung (thermal or non-thermal). The inverse Compton process, in particular, yielded a reasonable fit to the *shape* of the spectrum, but it was hard to reproduce the *strength* of the observed background without relegating its production to early epochs ($z \approx 2-5$), when more energy might have been available, and/or the emission mechanism more efficient. To account for the *hard* x-ray and γ-ray background, even larger redshifts – z up to ~ 100 in some theories – have been invoked. Though these 'evolutionary' models are not inherently implausible, nor even completely *ad hoc*, they are very speculative and difficult to test observationally. But they certainly make the X-ray background seem even more interesting to cosmologically-inclined theorists, because of the possibility that it may tell us something about even remoter epochs than the most distant known discrete sources, and about the thermal history and density of intergalactic matter. These ideas were reviewed by Setti and Rees (1970) at the Rome meeting (see also the article by Silk, 1970), and there do not seem to have been any really significant developments along these lines in the subsequent three years.

On the observational front, however, progress has been much more substantial. Two particular developments are especially important for increasing our understanding of the X-ray background: the number of identified extragalactic sources has risen from ~ 1 to ~ 10, which permits us to make a somewhat less conjectural estimate of how much of the background could come from different categories of

---

* On leave from Institute of Theoretical Astronomy, Madingley Road, Cambridge, England.

Bradt and Giacconi (eds.), X- and Gamma-Ray Astronomy, 250–257. All Rights Reserved.
Copyright © 1973 by the IAU.

discrete sources; also, the limits on possible small-scale anisotropies in the background are now more stringent. We may expect rapid accumulation of data in both these areas to continue. It therefore seems worthwhile to devote the rest of this paper primarily to considering what such information tells us – or may soon tell us – about the nature of the background.

If the mean emissivity at frequency v, per unit comoving volume, is $\mathscr{E}_z[v]$ (the suffix z indicating that $\mathscr{E}$ may vary with epoch) then the mean energy flux per unit frequency received by a detector with effective solid angle Ω is

$$I(v) = \Omega \int_{0}^{r_{max}} \frac{\mathscr{E}_z[v(1+z)]}{1+z} \, dr. \tag{1}$$

r is the coordinate distance, defined so that $dr = c \, dt (1+z)$ (t being the epoch corresponding to redshift z), and r_{max} corresponds to $z = \infty$. When $\mathscr{E}$ is independent of z, and the X-rays are emitted with a power law spectrum $\mathscr{E}[v] \propto v^{-\alpha}$, (1) can be readily integrated. For an Einstein-de Sitter universe one obtains

$$I(v) = \frac{3}{3 + 2\alpha} \tau_{\mathrm{H}} \mathscr{E}[v] \, \Omega, \tag{2}$$

where τ_{H} is the inverse Hubble constant. This means that, for $\alpha \approx 1$, we would get a correct estimate of the background – in the non-evolutionary case – by considering a static Euclidean universe of radius $\sim \frac{3}{5}$ the Hubble radius. The precise value of this factor depends on the cosmological model; but it is always around $\frac{1}{2}$ except in Lemaitre models with a long 'coasting phase', where it may be much larger. Another consequence of (2) is that about half of the integrated background comes from $z \gtrsim 0.3$.

If the X-rays are coming from discrete sources, then we can write

$$\mathscr{E}_z[v] = \int_{L} \varrho_z(L[v]) L[v] \, dL, \tag{3}$$

$\varrho(L) dL$ being the coordinate density of sources whose luminosity lies between L and $L + dL$. Setti and Woltjer (this volume, p. 208) have used (2) and (3) to estimate the contribution to the background from different types of extragalactic sources. Despite the limited observational material and the consequent difficulty of making reliable calculations, it cannot be excluded that rich clusters of galaxies, and Seyfert galaxies, may each collectively contribute up to 10% of the total background, even if there is no evolution with z. The likely contribution from quasars is more uncertain still, but could well be *even more* if the evolution inferred from the steep number-magnitude relation is allowed for.

The recent data thus leave open the possibility that *all* the background could come from discrete sources. A special mechanism for diffuse X-ray emission may therefore be quite superfluous. Even if the bulk of the X-ray background *does* have a genuinely diffuse origin, the contribution from sources is almost certainly not negligible. But there is no reason why the X-rays from sources and the diffuse X-rays (if any) should have the same spectrum – completely different radiation processes may be involved. This means that we cannot ascertain what the spectrum of the hypothetical diffuse component is; and this uncertainty is unlikely to be resolved before the luminosity function and evolutionary behavior of extragalactic discrete sources has been established. Until that time, it is perhaps premature to dwell on the virtues and defects of rival 'evolutionary' theories.

Even if the background were all due to (say) the inverse Compton effect, we might expect some spatial irregularities in the electron distribution, which would give rise to inequalities between the intensities measured in different areas of sky. But *some* of the background is due to discrete sources, and this component will obviously produce a certain amount of 'graininess' if the background is studied with high angular resolution.

Provided that there is some length scale $l \ll c\tau_H$ such that the universe is statistically homogeneous if averages are taken over regions larger than l – and this is an article of faith among most cosmologists, to which the high overall isotropy of the microwave background, radio surveys, and the X-ray background lend strong support – the spatial structure of the X-ray emissivity can be adequately described by a 3-dimensional auto-correlation function. Knowledge of this function, and its z-dependence, allows us to calculate the 2-dimensional auto-correlation function for the brightness of the X-ray sky. A calculation of this kind has been given by Wolfe and Burbidge (1970) who considered whether the anisotropy limits found by Schwartz (1970) were compatible with a model where the X-ray emissivity displayed the same spatial autocorrelation as galaxies. This and related questions have also been discussed by Silk (1970), Schwartz *et al.* (1971), Fabian (1972), Webster (1972) and Craven and Sciama (1972). The essential features of Wolfe and Burbidge's work – for instance, the way the fluctuations depend on detector beam area and on the 'clumpiness' of the emission – can be clarified by a more simple-minded analysis.

Suppose that the background intensity is measured, in a particular bandwidth, in n non-overlapping areas of sky. For simplicity, we consider a detector with circular beam and uniform sensitivity over the whole beam area, but the arguments can be straightforwardly extended to detectors with more general and more realistic properties.

Consider first a model in which the background is entirely due to *randomly distributed point sources* with a single luminosity L, and there are no evolutionary effects. I shall follow the treatment given by Craven and Sciama (1972). The expected mean-square deviation $\overline{\delta I^2}$ from the expected flux (1) is given by

$$\overline{\delta I^2} = \Omega \int_0^{r_{max}} \varrho r^2 \left(L\left[(1+z)\,v\right]\right)^2 \left\{ \frac{1+z}{r^2\,(1+z)^2} \right\}^2 \, dr. \tag{4}$$

The most important feature of (4) is that *the main contribution to $\overline{\delta I^2}$ comes from small r*. In one respect this simplifies things, because it means that the fluctuations can be adequately discussed in a Euclidean approximation. (In contrast to the *mean* integrated background which, by the familiar Olbers argument, comes mainly from 'cosmological distances'.) We observe, however, that the integral (4) actually diverges as $r \to 0$! This is because the contribution of the nearest source to the mean-square fluctuation depends on the inverse fourth power of its distance, whereas the probability that there *is* one source within a given distance varies only as $(\text{distance})^{+3}$. This divergence is symptomatic of the highly skew and non-Gaussian character of the probability distributions involved, for which rms methods of analysis are not really appropriate. In practice, however, we are concerned merely with one realization of the ensemble of possible source distributions, in which the closest source will be at some definite distance r_{min}. It is therefore legitimate to take r_{min}, instead of zero, as the lower limit of integration. The median distance of the closest source expected within a solid angle $n\Omega$ is $(3 \log_e 2/n\Omega\varrho)^{1/3}$. Taking this value for r_{min} yields

$$\overline{\delta I^2} \simeq (\varrho\Omega)^{4/3} (L[\nu])^2 (n/3 \log_e 2)^{1/3} ; \tag{5}$$

and so, defining δI as $(\overline{\delta I^2})^{1/2}$,

$$\frac{\delta I}{I} \simeq \frac{3 + 2\alpha}{3\varrho^{1/3}\tau_H} \Omega^{-1/3} \left(\frac{n}{3 \log_e 2}\right)^{1/6}. \tag{6}$$

The appearance in this expression of n (the number of areas surveyed) perhaps at first sight seems rather surprising. It reflects the fact that there will generally be one area in which the nearest source is $n^{1/3}$ times closer than the average, and it is this area alone which makes the main contribution to $\overline{\delta I^2}$. This result again indicates that the single quantity $\delta I/I$ does not characterise the fluctuations adequately – however many areas are surveyed, its numerical value remains sensitive to how close we happen to be to the nearest source which is included.

If n is *not* very large, (6) reduces, very roughly, to

$$\delta I/I \simeq R/(c\tau_H) \Omega^{-1/3} \tag{7}$$

when R denotes the mean spacing between neighboring sources. The fractional fluctuations from one beam area to another are thus roughly given by the inverse *cube* root of the number of sources within the beam. Or, in other words, $\delta I/I$ is approximately the ratio of the distance of the nearest source in the beam to the Hubble radius (i.e. the fractional contribution this source makes to I) and it is the variations in the actual distance of this nearest source that account for about half of the total $\delta I/I$.

If the sources span a range of luminosities, then (6) and (7) are modified only by an extra factor $(\overline{(L[\nu])^2})^{1/2}/\overline{L[\nu]}$ on the right hand side. The higher-luminosity sources obviously contribute proportionately more to δI than to I.

The above analysis is easily generalised to the case when the 'sources' are extended, with (say) a diameter d. This is relevant if we wish to test whether the X-ray background could be due to many galaxies in each cluster (the individual galaxies not being resolvable) or to intracluster gas. A source closer than $d/\Omega^{1/2}$ would subtend a a solid angle larger than the beam. Its contribution to I cannot exceed a definite finite value $\sim L\Omega/d^2$, however close it is, and this removes the divergence in (4). If Ω is so large that even the nearest source is smaller than the beam (as is marginally true for the experiments so far carried out, if clusters are the sources) then (6) and (7) still hold. But if Ω were smaller, so that the nearest sources *did* fill the beam, then one finds that

$$\delta I/I \simeq (3 + 2\alpha)/\tau_H (\varrho d)^{-1/2} \Omega^{-1/4}. \tag{8}$$

$\delta I/I$ then becomes $\propto \Omega^{-1/4}$ instead of $\propto \Omega^{-1/3}$. The major contribution to $\delta I/I$ is due to sources at distances $\sim d/\Omega^{1/2}$. If Ω is so small, or d so large, that even a source at a distance $\sim c\tau_H$ fills the beam, then the fluctuation amplitude is almost independent of Ω. This means that the largest value attained by $\delta I/I$ is

$$\sim (d^2 c\tau_H/R^3)^{-1/2}; \tag{9}$$

the quantity in parentheses being the mean number of sources intercepted by a line of sight extending out to the Hubble radius.

It is obviously inaccurate to approximate clusters as uniform spheres. It might be more realistic to assume that the emissivity falls off in a Gaussian fashion away from the cluster center. However the crude results given above are good enough to reproduce, within a factor two, the form of the autocorrelation function for the X-ray sky brightness calculated by Wolfe and Burbidge (1970). Note that we can always obtain an *upper limit* to $\delta I/I$ by using the point source approximation: if the sources are extended the fluctuations will always tend to be smeared out.

So far, none of the attempts to measure $\delta I/I$ has detected *any* fluctuations exceeding those attributable to the finite number of photons counted, and therefore only upper limits are available. Comparing these limits with (6), one finds that, in a non-evolutionary model, the number of sources out to the Hubble radius must be at least 10^7–10^8. This is just compatible with a mean source separation $R \approx 10$ Mpc, and thus with an origin in clusters of galaxies. But an origin exclusively in rich clusters already seems ruled out. Wolfe and Burbidge were led to an over-strong conclusion owing to an error later pointed out by Webster (1972).

Another type of model which can be tested against observational limits on $\delta I/I$ is what we might call an 'extreme evolutionary model', in which the background is attributed to sources which are *all* at cosmological distances and all have the same *apparent* brightness. The expected fluctuations then depend on the inverse *square* root (instead of the inverse cube root) of the number of sources in the beam. The present observations are thus compatible with extreme evolutionary models even if there are only 10^5–10^6 effective sources in the sky. Models in which the background comes from quasars or strong radio sources with large redshifts are thus still in the field.

If we wish to discuss evolutionary models in greater generality, it is better to approach the problem in a rather different way. This alternative viewpoint will also clarify both the limitations and the potential usefulness of studies of fluctuations in the background. Even if nothing is known about the luminosity function and z-dependence of the sources, one can in principle determine their number-versus-intensity relation – i.e. the function $N(S)$, which denotes the number of sources per steradian whose intensity exceeds S – and, if the sources are randomly distributed, $N(S)$ contains all the information relevant to the magnitude and probability distribution of $\delta I/I$. We have

$$I = \Omega \int_{\infty}^{0} S\,(dN/dS)\,dS \qquad (10)$$

and

$$\overline{\delta I^2} = \Omega \int_{\infty}^{0} S^2\,(dN/dS)\,dS. \qquad (11)$$

In the particular case when $N(S) \propto S^{-3/2}$ (with a low-S cut-off at the intensity appropriate to a source at the Hubble radius), these two equations are precisely equivalent to (2) and (4). But $\overline{\delta I^2}$ could obviously be evaluated for *any* $N(S)$. In particular, one may consider the case $N(S) \propto S^{-\beta}$ ($\beta > 1$), with a truncation at some intensity $S = S_{min}$ to ensure convergence of (10). If $\beta > 2$, the situation resembles the 'extreme evolutionary model' already mentioned: the fluctuations (as well as, of course, the integrated flux) are dominated by sources with $S \approx S_{min}$. For $1 < \beta < 2$, however, the fluctuations are due mainly to the high-S sources. One can derive a generalised version of (6) in which the dependence of $\delta I/I$ on the beam area and the number of regions surveyed is $\Omega^{1/\beta - 1} n^{1/\beta - 1/2}$.

Substantially more would be learnt about the nature of the background if one could determine not merely $\overline{\delta I^2}$, but the whole probability distribution $p(\delta I)$ of δI. This might be feasible if high sensitivity measurements could be made in a large number of areas of sky. The form of $p(\delta I)$ in point source models can be calculated in terms of $N(S)$. For $N(S) \propto S^{-\beta}$ ($1 < \beta < 2$) it is skew, with a sharp cut-off on the negative side, but a long tail with the approximate form $p(\delta I) \propto (\delta I)^{-(\beta+1)}$ on the positive side, which reflects the probability of there being one exceptionally bright source within the beam. One could calculate the modifications in $p(\delta I)$ arising from the finite extent of the sources. Thus in principle the fluctuations contain information on the linear dimensions, correlation length, *and* evolutionary behavior of the contributors to the background. This procedure is essentially the same as the so-called $P(D)$ technique of Scheuer (1957) and Hewish (1961), whereby the form of $N(S)$ could be inferred for radio sources below the confusion limit by analysing the 'noise' from a radio interferometer.

A sufficiently large positive value of δI would in practice be attributed to a resolved source. This suggests that a more realistic (and more precisely measurable) value of $\delta I/I$ could be defined if $p(\delta I)$ were truncated at a certain level, any larger-amplitude fluctuation being deemed a 'detected source'. It is then interesting to consider the following problem. Suppose one wishes to know whether a particular class of objects – e.g. quasars or clusters – can account for the X-ray background. One could *either* check whether the closest members of the population would be strong enough sources to have shown up in surveys, *or* calculate whether the number of objects in the class is large enough that the small scale fluctuations in the background would be undetectable. Which of these tests is the most sensitive, bearing in mind that similar detectors, observing for similar lengths of time, would be utilised in each case? Craven and Sciama (1972) have discussed this question. The precise answer depends on the characteristics of the detector, and on the criteria that must be fulfilled before a source is said to have been detected, but in non-evolutionary models – perhaps not surprisingly – it turns out that the two procedures are more or less equally sensitive. When evolutionary effects are important, however, the balance tilts in favor of the fluctuation technique. This is because the power-law tail of the function $p(\delta I)$ steepens with increasing β (and in fact disappears for $\beta > 2$), making it less likely that individual sources will be detected.

If we compare recent developments in X-ray astronomy with the early history of *radio* astronomy, it is striking that, whereas investigations of the background have been regarded as an important part of X-ray astronomy, the radio background never achieved such a major role. This is partly because the isotropic radio background is swamped by emission from the Galactic Disk, but it also indicates a genuine difference in the type of extragalactic source which dominates in the respective frequency bands: it was clear from the very early days of radio astronomy that some discrete sources were so powerful that they could be readily detected at cosmological distances (and there is no doubt that much of the isotropic radio background is attributable to such sources); most of the extragalactic X-ray sources so far detected are, however, at distances $\ll (c\tau_{\mathrm{H}})$, and this automatically implies that only a minority of the X-ray photons reaching us come from resolved sources.

Another way of investigating the background and elucidating the nature of the emission mechanism is by studying the spectrum. It is obvious that we would not expect to see any lines (except perhaps in a Lemaître universe), because of the range of redshifts which would contribute. In principle it might be possible to detect 'edges' in the spectrum, and to infer the evolutionary properties of the emission from the shape of the spectrum at wavelengths longward of the edge, but in practice such observations would always be very ambiguous (see, for example, Tinsley, 1972). Note also that it would be quite possible for bremsstrahlung from gas with a range of temperatures to yield an integrated spectrum that mimics a power law. Thus the basic question of whether the emission is thermal or non-thermal remains open still.

Improved study of the small-scale isotropy seem much more likely to yield useful clues as to the nature of the background. The review I have given here is obviously over-simplified, being intended merely to illustrate the nature of the problem. In practice, there are probably several quite different processes, each making a significant contribution to the total background. Since the different contributions need not have the same spectrum, one would ideally like to have measurements at different wavelengths of the same regions of sky. (Wavelength-dependent absorption might also be important, especially if the background comes from large redshifts). Another complication is that the fluctuations could be dominated by a component which makes a relatively minor contribution to the total intensity, if that component has a larger correlation-length than the others. A further possible cause of confusion is that our own Galaxy may contribute significantly to the fluctuations (especially on larger angular scales), even though the overall isotropy tells us that it cannot contribute more than a few per cent of the total background.

Acknowledgements

I am grateful for helpful discussions with many colleagues, and would especially like to thank P. Craven, A. C. Fabian, J. E. Felten, H. Gursky, C. Hazard, D. W. Sciama, and A. M. Wolfe.

References

Craven, P. and Sciama, D. W.: 1972, in preparation.
Fabian, A. C.: 1972, *Nature Phys. Sci.* **237**, 19.
Hewish, A.: 1961, *Monthly Notices Roy. Astron. Soc.* **123**, 167.
Scheuer, P. A. G.: 1957, *Proc. Cam. Phil. Soc.* **53**, 764.
Schwartz, D. A.: 1970, *Astrophys. J.* **162**, 439.
Schwartz, D. A., Foldt, E. A., Holt, S. S., Serlemitsos, P. J., and Bleach, R. D.: 1971, *Nature Phys. Sci.* **233**, 110.
Setti, G. and Rees, M. J.: 1970, in L. Gratton (ed.), 'Non-Solar X- and γ-ray Astronomy', *IAU Symp.* **37**, 352.
Silk, J.: 1970, *Space. Sci. Rev.* **11**, 671.
Tinsley, B. M.: 1972, *Astrophys. Letters* **10**, 31.
Webster, A. S.: 1972, *Nature* **238**, 20.
Wolfe, A. M. and Burbidge, G. R.: 1970, *Nature* **228**, 1170.

22. 'LOCAL' THEORIES OF THE X-RAY BACKGROUND

JAMES E. FELTEN

Steward Observatory, University of Arizona, Tucson, Ariz. 85721, U.S.A.

Abstract. Recent theories of the origins of diffuse-background X-rays are reviewed, with emphasis on theories of the soft flux in the galactic plane and at the poles. This is probably partly galactic and partly extragalactic in origin. Failure to observe absorption by the Small Magellanic Cloud and by galactic gas in neighboring directions may be due to sources in the Cloud and to statistical fluctuations in galactic emission and absorption. Several models for numerous low-luminosity sources in the Galaxy are available. True 'diffuse' emission seems unnecessary. Absorption by Galactic gas seems to agree roughly with theory. The soft extragalactic component may arise in a hot intergalactic medium.

The existence of a 'diffuse' galactic-plane excess in 1–100 keV is in some doubt. Low-luminosity sources may contribute to this as well.

For isotropic X-rays in 1 keV – 1 MeV, superposition theories involving clusters of galaxies, Seyfert galaxies, etc. over a cosmological path length are now roughly viable. Simple 'metagalactic' Compton theories seem excluded if the break at 40 keV is sharp, but this is now in doubt. A very hot intergalactic medium at $T \approx 3 \times 10^8$ K would give the possibility of a sharp break.

A recent upper limit on the line source strength of 100-MeV photons in the galactic plane may create some difficulties for cosmic-ray theory. The spectral shape of π-γ photons has become a matter of theoretical dispute.

1. Introduction

My task in this review is to deal with 'local' theories and interpretations of the background X-rays and γ-rays, where by local I refer to notions of galactic sources as well as of 'metagalactic' contributions out to a distance roughly equal to the Hubble radius. In the next paper Dr Rees will discuss models involving redshifts $\geqslant 1$, strongly evolutionary cosmologies, and the early universe. In the time available I cannot even mention all the recent papers on my subject, and I wish to apologize in advance to those authors who will get short shrift. It seemed wise to select a few important and timely topics for major comment. I have been asked to give particular attention to the soft X-rays. Figure 1 shows a recent compilation of diffuse background data. Developments at this Symposium suggest that the observed 'structure' in this curve, especially around 40 keV and 1 MeV, may be rather evanescent. I will, however, deal with some of the recent attempts to explain this structure, as befits a theorist. I will start at low photon energies and work my way up.

2. Soft X-Rays, $h\nu < 1$ keV

The flux below 1 keV appears to exceed the extrapolation from higher energies. There is great diversity of views concerning these very soft X-rays. In trying to arrive at an understanding of the data I will lean most strongly on a recent preprint by the Naval Research Laboratory group (Davidsen *et al.*, 1972). This observation employed two windows, whose effective transmission energies are about 280 and 680 eV. Figure 1 of Friedman *et al.* (this volume, p.215) compares '280-eV' fluxes in

Bradt and Giacconi (eds.), X- and Gamma-Ray Astronomy, 258–275. All Rights Reserved.
Copyright © 1973 by the IAU.

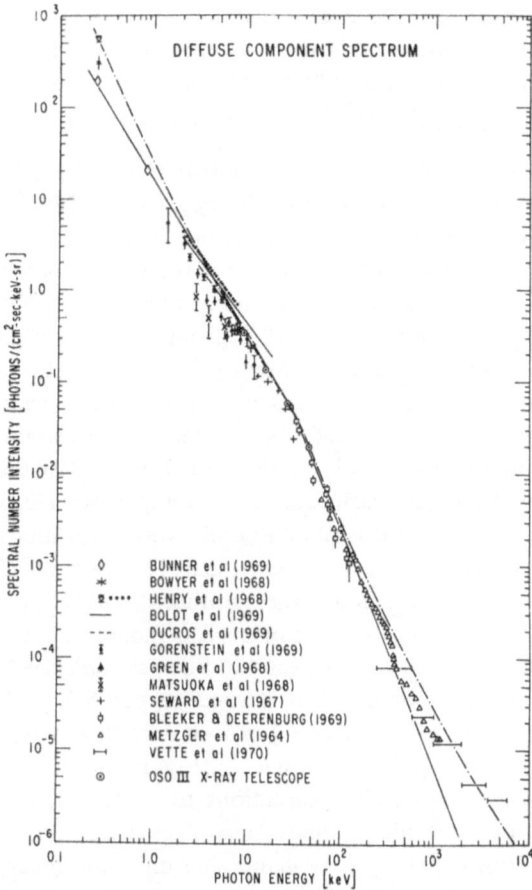

Fig. 1.   Collected data on the spectrum of diffuse-background X-rays (Peterson, 1971).

this observation, plotted in galactic coordinates, with neutral-hydrogen column densities N_H. In general the soft X-ray flux rises at the poles, where N_H is low. There are even rises in the count rate at some specific low-N_H features, e.g. $l^{II} \approx 210°$, $b^{II} \approx 20°$, though the matchup is far from exact. The rise toward the poles suggests strongly that most of this radiation comes from outside the absorbing disk of gas. This need not mean that there is an isotropic extragalactic component; we might have a disk of sources with a scale height z_s greater than that of the gas, $z_H \approx 120$ pc. Davidsen *et al.* (1972) assumed such a model, without an extragalactic component, and analysed the b^{II}-dependence of the data as a function of hv. The value of τ_p, the absorption optical thickness in the polar directions, is of critical importance; let us treat this as a free parameter. At 280 eV, as we dip away from a pole toward the plane, we must have $z_s > z_H$ to get a flux decrease, and we must *not* have $\tau_p \gg 1$ if the decrease is to be *significant*. On the other hand, if $\tau_p \ll 1$, we will get an increase! We find that $\tau_p \simeq 2$ gives a good fit to the observed decrease by a factor $\simeq 2$ from a pole to plane, provided

$z_s/z_H \simeq 5$, i.e. $z_s \simeq 600$ pc (Figure 3 of Friedman *et al.* in this volume). The value $\tau_p \simeq 2$ is about twice what we would expect from the observed N_H and the absorption calculations of Brown and Gould (1970) for cosmic abundances, which give

$$\sigma(h\nu) \simeq 0.7 \times 10^{-22} (h\nu/1 \text{ keV})^{-3} \quad \text{up to 532 eV} \tag{1}$$

for the cross section per H atom; the main contribution comes from He. Abundances are adjustable, and we cannot reject this model by use of the 280-eV data alone.

Assuming, however, that the *ratio* $\tau_p(280)/\tau_p(680)$ should be as given by Brown and Gould, τ_p is small at 680 eV, and at this energy the same model then predicts a *rise* in the X-ray flux as we go away from the pole, which is not observed (Figure 3 of Friedman *et al.* in this volume). The NRL group conclude that a purely galactic origin for the soft X-rays is excluded. Now for consistency the ratio $\tau_p(280)/\tau_p(680)$ should perhaps also be left free in this discussion, because if we abandon the cosmic abundance ratios, so as to make $\tau_p(280)$ free, we lack a physical model for the absorption. $\tau_p(280)$ is probably due mainly to He, $\tau_p(680)$ to O. But the simplest thing we could do to obtain the large $\tau_p(280) \simeq 2$, viz. adding more He, would *increase* the ratio $\tau_p(280)/\tau_p(680)$ above the Brown and Gould value, pushing $\tau_p(680)$ downwards for fixed $\tau_p(280)$ and strengthening the case made by Davidsen *et al.* (1972).

Rejecting this purely galactic model, these authors consider also a purely extragalactic source, and find that the best fit for such a model implies that τ_p is about $\frac{1}{3}$ the theoretical value. This confirms the result of Bowyer *et al.* (1968), which stimulated all the discussion about 'cloudiness' corrections to the X-ray absorption. But this purely extragalactic model fails, because it cannot explain the sizable residual X-ray intensities near the plane in regions of large N_H (Figure 2). This removes the necessity for 'cloudiness' or other subtractive corrections to τ_p, for when it is recognized that the X-rays on the right-hand side of Figure 2 are of galactic origin, and the exponential-absorption line is drawn to fit the left-hand side only, an absorption $\tau_p(280) \simeq 1$ is obtained, in accord with theory. It still is not clear why these X-rays in the plane were not seen by Bowyer *et al.* (1968). Recent observations have all detected them.

Following through on this two-component model, Davidsen *et al.* find that the best fit (Figure 2) is indeed obtained for $\tau_p(280) \simeq 1$ and $z_s/z_H \lesssim 1$, implying $z_s \lesssim 120$ pc. About $\frac{2}{3}$ of the 280-eV flux received at the poles, i.e. about 200 photons (cm² s sr keV)$^{-1}$, is extragalactic in origin. Interesting information about the sources in the galactic disk can be derived from the data obtained near the plane. The fact that counts are observed in each $10° \times 10°$ box implies that there are many faint sources of soft X-rays rather than a few bright ones. One source per $\lesssim (7 \text{ pc})^3$, or a number density $n \gtrsim 3 \times 10^{-3}$ pc^{-3}, is what is required if they are point sources, but if they are extended the number can be smaller. If the sources are uniformly distributed and there are many in the beam, it is a good approximation to say that the emission is continuously intermixed with the absorbing gas. The specific intensity I_ν received in the plane, where $\tau \gg 1$, is then just the source function $\mathscr{S}_\nu$:

$$I_\nu = \mathscr{S}_\nu = \frac{S_x}{4\pi n_H \sigma}, \tag{2}$$

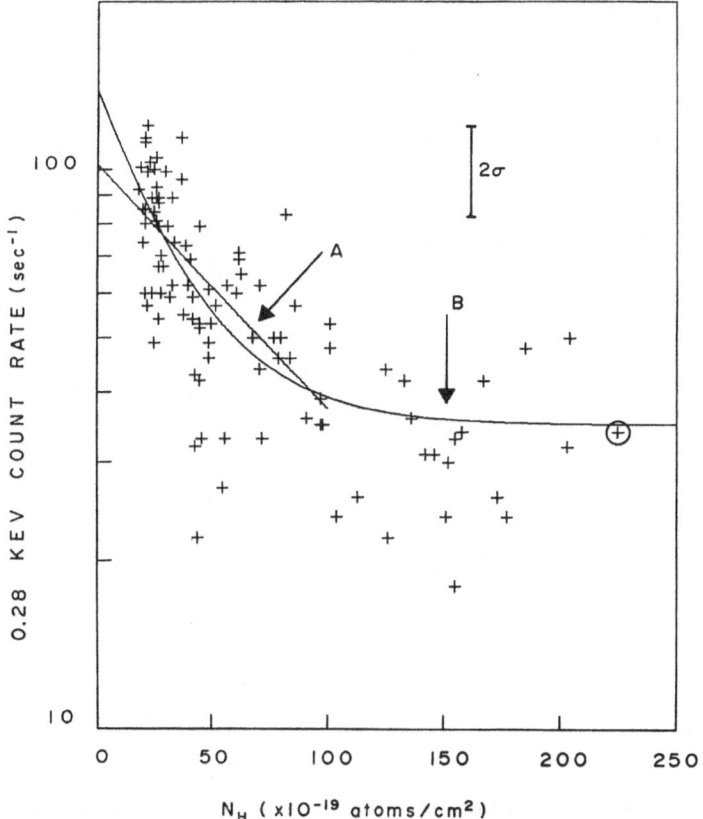

Fig. 2.   Correlation plot of the log of the 280-eV count rate versus the hydrogen column density. The encircled point is the average intensity within 10° of the galactic plane and is plotted at an arbitrary column density. Curve A is an attempt to represent these data by a pure-extragalactic source model; Curve B is for a two-component model (Davidsen *et al.*, 1972).

where S_x is the X-ray emissivity and σ is the absorption cross section per H atom (Brown and Gould, 1970). Since $\sigma(h\nu) \propto \nu^{-3}$ apart from absorption edges, the spectrum of the sources in the plane must be

$$S_x \propto \nu^{-3} I_\nu; \tag{3}$$

the sources are softer than the radiation as received. Assuming thermal bremsstrahlung, Davidsen *et al.* found $T \simeq (2-3) \times 10^6$ K for the disk sources. The extragalactic component is somewhat harder, $T \simeq 4 \times 10^6$ K.

In the NRL paper it is not always clear when the counter windows have been treated as monochromators and when the efficiency curves have been folded into the calculations. Efficiency curves are not given in the preprint, though they are included in the present volume (Friedman *et al.*, p.215). Nevertheless this preprint offers the most coherent picture to date of the soft X-rays.

The one large piece of contrary evidence is the Small Magellanic Cloud (SMC) observation by the Wisconsin group (McCammon *et al.*, 1971). They examined the SMC and a surrounding region of high b^{II} and unusually low galactic obscuration $(N_H \simeq 1 \times 10^{20}\ \mathrm{cm}^{-2} \Rightarrow \tau(280) \simeq 0.3)$. They could not see any drop when they scanned across the SMC (where τ is probably very large) (region 5 in Figure 3A), and they concluded that less than one-fourth of their observed flux of $\simeq 400$ photons $(\mathrm{cm}^2\ \mathrm{s}\ \mathrm{sr}\ \mathrm{keV})^{-1}$ at 280 eV could be coming from beyond the SMC or, roughly speaking, that the observed flux *drops* by at most 100 when the detector is pointed at the SMC.

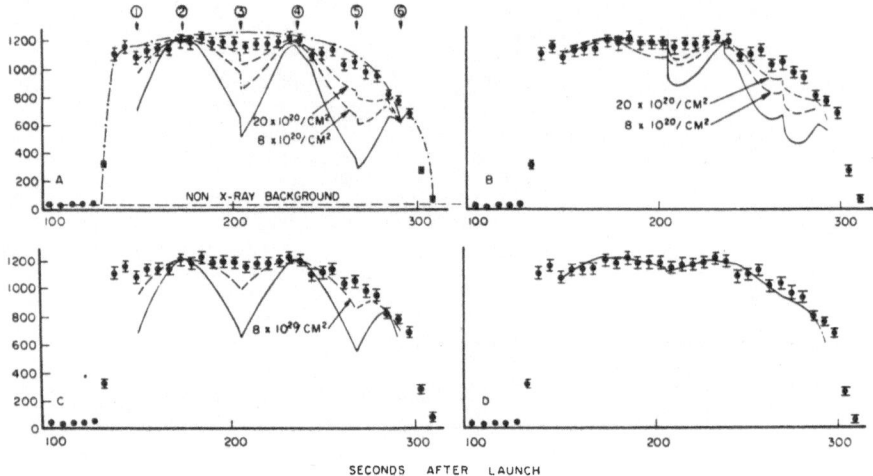

Fig. 3.   Total counting rates in the 120–450 eV range as a function of time during the flight (McCammon *et al.*, 1971). The curves show expected behavior on various models. The models shown in Figure 3a assume that all the flux comes from beyond the SMC. *Solid curve*, absorption predicted if all the gas detected at 21 cm (i.e. the gas associated with the Galaxy-SMC system) is uniformly distributed. *Dashed curves*, cases where the absorbing gas is clumped into clouds of 8 and $20 \times 10^{20}\ \mathrm{cm}^{-2}$ thickness. *Dot-dash curve* includes only the effects of the Earth's atmosphere. For clarity, all models are normalized to the same point. Models in Fig. 3B assume absorption only by gas associated with the SMC, and those in Figure 3C assume absorption only by gas associated with the Galaxy. Curve in Figure 3D is the predicted absorption assuming that only an $E^{-1.4}$ photon spectrum fit to the 2–10 keV data is extragalactic, with the remainder of the observed flux coming from some *local* source.

Does this observation imply that the extragalactic flux is very weak, and is it inconsistent with the picture presented by the NRL group? NRL swept the matter under the rug. When we look at the center of the SMC, the I_v we see should be the source function $\mathcal{S}_v$. In the Galaxy $\mathcal{S}_v$ is $\simeq 120$ in the units above, from the work of Davidsen *et al.* If it were $\simeq 300$ in the SMC, then, since 400-300=100, all 400 units above could be extragalactic, and the counts absorbed in the SMC could be filled in by SMC emission! McCammon *et al.* (1971) noted this and pointed out a bigger difficulty: that their count rate also failed to drop much when they scanned across a region of higher *galactic* absorption (region 3 in Figure 3A). In order to fill *this* in, the *galactic* $\mathscr{S}_v$ in this direction would have to be several times its general value for $b^{II} \simeq 0$. Three comments are in order:

(1) $\mathscr{S}_v$ is the ratio of two quantities which are not necessarily related, and it might vary *systematically* from place to place;

(2) If S_x is due to stars, for example, then there must be *statistical* variation in it. A cone of 8° FWHM looking through 100 pc of the galactic disk $(\tau \simeq 1)$ contains ~ 20 sources if their density is $n \sim 3 \times 10^{-3} \text{ pc}^{-3}$. But half the radiation received comes from the nearest 50 pc of the cone, and the mean number of sources in this portion is only ~ 2, so the statistical fluctuations can be quite large. We might have a lower n and even larger fluctuations.

(3) The discussion is sensitive to the actual value of τ in the regions of lowest obcuration. If $\tau \ll 1$, then the received flux can actually *rise* when we move the line of sight onto a galactic cloud, because we pick up additional (galactic) sources in the cloud, and τ, while it might double, is still small. If He were to be either underabundant or doubly ionized in particular clouds, the absorption would be much reduced.*

We should not try to make much out of little by constructing a general model of soft X-rays in the Galaxy from a few observations in particular directions; in likely cases the lumpiness may frustrate us. The NRL approach of averaging data over large portions of the Galaxy is a good one. As the data accumulate, more can be done, e.g., the X-ray flux should be plotted against N_H for regions at *constant* b^{II} and vice versa; this would help us separate out the effects of galactic sources and perceive attenuation of the extragalactic radiation.

The latest report from the Wisconsin group (Coleman *et al.*, 1972) says that at *low* b^{II} the soft X-ray flux is not correlated with N_H. This is certainly expected, because in the plane we see just the constant $\mathscr{S}_v$, apart from statistical variations.

A preprint by Gorenstein and Tucker (1972) describes work similar to that of Davidsen *et al.* (1972), but the conclusions differ. Nevertheless there are many points of contact. They followed the Wisconsin group in assuming *no* extragalactic component. They looked only at the pole/plane intensity ratio instead of the full b^{II}-dependence, and since their planar value seems to involve only a short span of data, one suspects that their statistics are not as good as NRL's. Having removed the extragalactic flux, they naturally found a thicker disk of galactic sources, $z_s \simeq 800$ pc. As in the work of Davidsen *et al.*, the energy bands are somewhat ill-defined, but Gorenstein and Tucker (1972) are apparently a little more sensitive at low hv. They find that the pole/plane intensity ratio decreases with decreesing hv below 280 eV; NRL found it to decrease with *increasing hv above* 280 eV. Since the X-ray absorption is $\propto \exp(v^{-3})$,

---

* A special word of warning about X-ray absorption calculations: An estimate of the distance to Cygnus X-1 (Gursky *et al.*, 1971) based on N_H and observed X-ray absorption led Kristian *et al.* (1971) to exclude prematurely a candidate that is presently regarded as strong (Bolton, 1972). The distance discrepancy was a factor of 2. I am told that this error involved intrinsic X-ray variation in the source (R. C. Henry, private communication). In any case, it suggests caution. X-ray absorption depends mainly upon He and O, and we should beware of attaching sanctity to fixed abundance ratios for all points in the interstellar medium.

it must be clear that substantial progress in such details will be difficult until we have good spectroscopy with energy resolution $\sim 5\%$, rather than several broad window transmissions. Gorenstein and Tucker (1972) conclude from the observed directional fluctuations that $n \sim 10^{-2} \ pc^{-3}$, and they estimate T of the sources as $\lesssim 10^6$ K. The value for n is not very different from that given by NRL, but the T is significantly smaller, and perhaps points to an additional population of very soft sources.

A recent report by the Livermore group (Palmieri et al., 1972) suggests that their earlier work (Palmieri et al., 1971), which indicated a systematic increase in 280-eV flux from pole to plane, suffered from coarse collimation and ultraviolet contamination. Their new observation reveals a curious 'hump' in the soft X-rays near $l^{II} \simeq 330°$, $b^{II} \simeq 15°$. This hump is 10–15° wide (much larger than the Lupus loop, which lies in the same direction), and its soft X-rays are a factor ~ 2 above neighboring regions. Apparently it is not merely one or two point sources. If at 50 pc distance this might be a diffuse source ~ 15 pc in diameter. One might imagine several kinds of astronomical structures with such a size, but Ilovaisky and Ryter (1971, 1972) have already proposed that old supernova remnants (SNR) of such dimensions could be responsible for the soft X-rays in the plane. This theory has the advantage that SNR are already known to emit soft X-rays; an extrapolation to larger and older remnants gives the desired result, though arguments from available-energy considerations (Tucker, 1971) have tended to suggest that this extrapolation is too generous. Of course a related phenomenon like the 2×10^5 K 'fossil H II regions' of McCray and Schwartz (1972) may contribute. A detailed study of the energetics, heating and cooling is clearly in order. Ilovaisky and Lequeux (1972) claim that the scale height of the old SNR is $\simeq 90$ pc, which would agree with the NRL model of the galactic sources.

Other astronomical theories for the planar flux are available. Strittmatter et al. (1972), in a clever inference from assorted facts, suggest that hot ($T \sim 10^7$ K) coronae of white dwarfs are responsible. The physics in this paper is order-of-magnitude only. Gorenstein and Tucker's criticism that this model is excluded by the short cooling times of the coronae is too hasty; their physical arguments for scale height and electron density are not consonant with those employed in the original paper. Anyway it is not clear that a cooling time shorter than the interval between heat-supplying pulses (implying X-ray pulsations) would be objectionable! Nor is it clear what this characteristic time for heat supply would be if the corona somehow draws its energy from the star's rotation rather than pulsation. Gorenstein and Tucker's other criticism, however, seems well founded: A T as low as $\sim 10^6$ K, suggested by the observed spectrum (3) of X-rays in the plane, conflicts with basic assumptions of the model. The model is good in that it makes several predictions, in particular regarding the observability of individual white dwarfs as sources. It implies a high space density of sources, in agreement with the NRL conclusions, and a large (population II) scale height, rather larger than NRL wish to accept.

Ostriker et al. (1970) suggested that interstellar gas accreting onto $\sim 10^9$ neutron stars in the Galaxy should be heated to $\sim 10^6$ K. The expected X-ray flux is adequate.

Since the accretion is $\propto v^{-3}$, the high-velocity 'runaway' neutron stars at large z (~ 2500 pc) give little contribution, and the effective scale height should be $z_s \leqslant 100$ pc. This agrees with the NRL picture. Recently, when Gorenstein and Tucker (1972) had inferred $z_s \approx 800$ pc from their observations, one of the authors of the Ostriker *et al.* paper is said to have reversed his field and redirected attention to the runaway neutron stars! This is unsettling; a theorist should be able to say what he expects z_s to be before knowing what it is.

There is another possibility involving neutron stars: blackbody radiation at $\sim 10^7$ K from their surfaces, heated by wobble dissipation of rotational energy (Henriksen *et al.*, 1972). I am not competent to discuss the physics in this paper. z_s would presumably be $\geqslant 100$ pc in this case, since the runaways should wobble as much as any.

Without minimizing the possible variety of interpretations, let me advance a few hypotheses I believe are suggested by present data. I hope these will be borne out.

(1) The 280-eV flux at the galactic poles is mainly the transmitted portion of an 'isotropic' extragalactic flux, having an intensity (outside the disk) ≈ 500 photons $(\text{cm}^2 \text{ s sr keV})^{-1}$. (It is possible to account for the b^{II}-dependence of the 280-eV flux by a purely galactic source model, but then the expected b^{II}-dependence at higher $h\nu$ is not in accord with observations. Note that Ilovaisky, in a contributed paper for this Symposium, reached a conclusion contrary to mine, as did Gorenstein and Tucker. He relied heavily on the Livermore data, but as I mentioned, the latest Livermore paper impeached these earlier data somewhat. Dr Hayakawa's conclusion, from his own analysis of data from the Leiden-Nagoya group, is essentially in agreement with mine.)

(2) Failure to observe absorption by the SMC is due to fill-in by sources in the SMC and/or statistical fluctuation in galactic sources in that particular direction. (It will be objected that this is fortuitous, but the alternative of merely abandoning hypothesis (1) would not solve the problem posed by Figure 3C. Perhaps Figure 3D suggests that the soft sources are indeed *extremely* local, e.g. in the outer atmosphere, but I am resisting this conclusion for the moment.)

(3) The X-ray opacity of interstellar gas is approximately as given by Brown and Gould (1970). Any subtractive correction for cloudiness isn't very important in observations to date. (Cloudiness will loom larger at lower $h\nu$ and finer angular resolution.) Incidentally, since interstellar 'clouds' are still not well understood as physical entities, we should not hasten to attribute to the X-ray-absorbing clouds 'known' properties derived from unrelated observations.

(4) At 280 eV in the galactic plane, we see ~ 120 photons $(\text{cm}^2 \text{ s sr keV})^{-1}$, which is the source function resulting from interstellar absorption mixed with a population of galactic sources. These must be either quite numerous ($n \gtrsim 3 \times 10^{-3}$ pc^{-3}) or quite large in size, otherwise they would already have been resolved. Their scale height may be $\lesssim 100$ pc or possibly larger; this requires further work. White dwarfs, neutron stars and SNR are all possibilities; the SNR theory perhaps deserves a slight preference at the moment, because (a) emission has been observed from known resolved SNR, and (b) it contradicts no feature of the background observations. If I_ν is the

received background spectrum in the plane, the intrinsic source spectrum is $\propto \nu^{-3} I_\nu$ at all ν's sufficiently high that many sources are still contributing to I_ν.

3. The Soft Extragalactic Component

What is the origin of the extragalactic excess below 1 keV? It is usual to assume that this is thermal bremsstrahlung, because the exponential spectrum can give a bump at any desired energy if T is chosen appropriately. For 280 eV, $T \sim 10^6$ K is about right. We might postulate hot gas actually within a quasispherical halo ($R \sim 10$ kpc, $n_e \sim 3 \times 10^{-3}$ cm^{-3}) or in the Local group of galaxies ($R \sim 1$ Mpc, $n_e \sim 3 \times 10^{-4}$ cm^{-3}) (Silk, 1970; Rees *et al.*, 1968; Hunt and Sciama, contributed paper for this Symposium). But the standard and much-discussed hypothesis is that of an intergalactic medium having the closure density

$$\varrho_{cl} = \frac{3H_0^2}{8\pi G} \tag{4}$$

and some temperature $T \sim 10^6$ K. If $H_0 = 100$ km (s Mpc)$^{-1}$, the extragalactic flux $\simeq 500$ units at 280 eV suggested by Davidsen *et al.* (1972) would be supplied by a Euclidean sphere of 'cosmological' radius (Felten, 1966)

$$R = \frac{1}{2} R_H = \frac{1}{2} \frac{c}{H_0}, \tag{5}$$

filled with hydrogen plasma at $\varrho = \varrho_{cl}$ and $T \simeq 4 \times 10^6$ K. This is also the T which fits the NRL polar data best. But order-of-magnitude astronomers must be cautious in this problem, because in the real universe T must be a function of epoch, and the medium is quite likely to be hotter at large z (because of adiabatic cooling). The emissivity at 280 eV is quite a strong function of T around 10^6 K. Quite a bit of work has been done on thermal histories of the intergalactic medium, most recently by Bergeron (1969, 1970). This will have to be extended as the soft X-ray data develop.

At present I do not see any fatal objection to a hot dense intergalactic medium, with heat sources placed in z so that the thermal history supplies the observed soft X-rays without transgressing upper limits at $h\nu > 1$ keV. Gunn and Gott (1972) think it likely that $\varrho \ll \varrho_{cl}$, but indirect arguments like theirs are seldom ironclad. It has been suggested that observations of neutral hydrogen in the peripheries of external galaxies enable us to reject the hypothesis of an intergalactic soft X-ray flux. This is incorrect (Felten and Bergeron, 1969), essentially because we do not know the outer structure of external galaxies, or even of our own Galaxy, well enough to employ them as photon counters!

4. 'Mainstream' X-Rays from the Disk, $h\nu = 1$ keV-1 MeV

I pass now to a higher energy band, 1 keV–1 MeV, where the spectrum has usually been represented as a power law. It has been claimed (Seward *et al.*, 1967; Cooke *et al.*, 1969) that in 1–10 keV there is a detectable excess brightness associated with the galactic plane. This has been observed again recently (Bleach *et al.*, 1972) in an interarm region of the disk ($l^{II} \approx 60°$); its line intensity is ≈ 3 photons $(\text{cm}^2 \text{ s rad})^{-1}$ in 2–10 keV. There is no shortage of diffuse-emission theories to explain this line source (e.g. de Freitas Pacheco, 1970; Ipavich and Lenchek, 1970). Time does not permit me to explore these theories; Ilovaisky has reviewed them in a contributed paper for this Symposium and found that they probably are not adequate to produce the observed flux. Anyway they seem a bit superfluous. Recall that Ryter (1970) and Setti and Woltjer (1970) showed that the *weakness* of this unresolved line source permitted us to infer that most of the then-*resolved* sources were intrinsically quite bright and at distances ~ 10 kpc (not much less) – otherwise the unresolved disk would have been much stronger than it is! Then it is clear that this line source *can* in fact be just unresolved galactic sources, e.g. SNR (Ryter, 1970). Bleach *et al.* (1972) argued that, if so, they must be a new family of smaller z_s than the presently resolved sources, and might be more numerous and of lower luminosity, $\sim 10^{33}$ erg s^{-1}. The scale height implied by their data is $z_s \lesssim 600$ pc. This could perhaps be a little larger, since the effective path length to the edge of the galactic disk may be a little larger than the 10 kpc they assumed. They claim that z_s is significantly *thinner* than the z_s for the *resolved* (UHURU) sources, for which they find $z_s \approx 900$ pc. I am not altogether convinced that this discrepancy is significant.

The fact that Clark (paper in this Symposium, p.29) *failed* to observe the enhancement in the plane when looking in a different direction (the longitude band $l^{II} \approx 140$–150°) may suggest that statistical fluctuations in the line density of these sources are quite large and therefore that they are not really very numerous, or it might simply mean that they are numerous in some parts of the Galaxy (interarm?) but not in others. The Leicester group (K. A. Pounds, private communication) found recently that at least 90% of the disk emission they reported earlier can now be accounted for by resolved sources. The enhancement was also seen in 7–12 keV by OSO-3, integrated over a wide range of l^{II}, but it has not yet been checked whether this can now be accounted for by UHURU sources. As usual, more data are needed. The sources contributing to this disk flux could well be the same ones which give the soft X-rays. If these sources are of thermal-bremsstrahlung character, they may then have hot components, $T \sim 8 \times 10^7$ K, but if power-law they must be rather soft, no flatter than $I_v \propto v^{-2.2}$ in energy units (Hudson *et al.*, 1971). Flare stars are a possibility (Edwards, 1971; Cavallo and Horstman, 1972); X-ray lines of iron should then be seen.

5. Mainstream X-Rays: Isotropic Component

Away from the galactic plane, the diffuse flux in 1–100 keV at least is isotropic (within $\lesssim 4\%$ around 10 keV) over large portions of the sky (Schwartz, 1970; Fabian and Sanford, 1971; Schwartz *et al.*, 1971). A 'hole' $\simeq 20°$ across and 10% deep is rumored to have been observed by UHURU in the direction of Draco, but this seems doubtful. If it is real, it may be galactic, and associated with the high-velocity clouds in this direction. The general isotropy suggests an extragalactic origin.

Simple superposition theories are basically attractive. The 1–10 keV luminosity of our Galaxy is $L_X \simeq 5 \times 10^{39}$ erg s^{-1} (a generous estimate). The local density of ordinary galaxies in space is $\simeq 3 \times 10^{-2}$ Mpc^{-3} if $H_0 = 75$ Mpc^{-3} (Sandage, 1965; cf. van den Bergh, 1961; Kiang, 1961).* Using $\frac{1}{2}R_H$ as the effective 'cosmological' superposition radius, we find that the 1–10 keV integrated intensity due to ordinary galaxies should be

$$I \,(\text{erg cm}^{-2}\,\text{s}^{-1}\,\text{sr}^{-1}) \simeq \frac{nL_X c}{8\pi H_0} \simeq 2.4 \times 10^{-9}. \tag{6}$$

The intensity (6) is a factor of $\simeq 25$ below the *observed I*, and this factor has spawned most of the pretentious 'cosmic' theories of the background. But there are other superposition possibilities, of which I will mention several. The 1–10 keV luminosity of 3C 273 is $\simeq 10^{46}$ erg s^{-1} (Kellogg *et al.*, 1971), and the local space density of QSO's may be $\sim 10^{-6}$ Mpc^{-3} (Schmidt, 1970). These values for L_x and n would bring (6) up to four times the observed value! But not all these QSO's are necessarily in the same active state as 3C 273; the strong radio QSO's (QSS's) are 10^2–10^3 times rarer. The rich clusters of galaxies in Virgo, Coma and Perseus (Kellogg *et al.*, 1971; Forman *et al.*, 1972) are extended sources at a level high enough to reduce the shortfall of (6) to a factor ~ 10 if all rich clusters are sources. Seyfert galaxies are another possibility. NGC 4151 is detected, and the present data on NGC 1068 and 1275 permit us to assume that the mean X-ray luminosity of Seyferts is as high as $\sim 2 \times 10^{42}$ erg s^{-1}. Since their space density is $\sim 1\%$ of that of normal galaxies (Burbidge, 1970), this would leave (6) short of the observed intensity by a factor only ~ 6. Finally, we could suppose that the *time-average* luminosity of normal galaxies is $\gg$ the *present* L_X of our own Galaxy, because of high X-ray output by supernovae (Tucker, 1970); the time average is clearly the quantity to be used in (6). We see that superposition theories, even in this simple Euclidean form, are not dead. Setti and Woltjer confirm this in more careful calculations for this Symposium (p. 208).

The high isotropy would put constraints on some superposition theories. Schwartz *et al.* (1971) estimate that the number of sources contributing must be $\gtrsim 4 \times 10^6$. Since the total number of galaxies within $\frac{1}{2}R_H$ is only $\sim 1 \times 10^9$, this limit might be quite tight for QSO or supernova theories, depending upon details. Random fluctuations have not been accounted for explicitly in the argument of Schwartz *et al.* (1971), but I will not spend time on this, as I believe Dr Rees intends to develop it in the next paper.

Turning to the 'metagalactic' theories, I will not go through the familiar 'classical' inverse-Compton explanation for the background radiation (Felten and Morrison, 1966; Setti and Rees, 1970). The apparent sharp 'break' in the spectrum at 40 keV (Figure 1) is a stumbling block for this theory. The break (Schwartz *et al.*, 1970; Horstman and Horstman-Moretti, 1971) is in strong dispute at this Symposium;

---

* Several of the following estimates depend on H_0 in various ways. I will take $H_0 = 75$ to suppress this complication.

I shall assume that it *is* real and discuss the consequences. It will be clear that the existence and shape of this break are of great importance in choosing among theories.

Putting aside any possibility of cosmic far-infrared radiation, I set the photon energy density in intergalactic space equal to

$$\varrho_{bb} \approx a\,(3\,\mathrm{K})^4 \approx 0.4\,\mathrm{eV\,cm}^{-3}. \tag{7}$$

The characteristic time for Compton loss by an electron of energy $\gamma m_e c^2$ is $\propto \gamma^{-1}$. If we want to induce a 'break' in the power-law spectrum of the received Compton radiation, we can do so by postulating another competing loss process with a characteristic time t_L which is, say, the same for all γ, and which therefore acts on the low-energy electrons faster than Compton loss. The resulting break occurs at

$$\varepsilon_0 \sim \frac{2 \times 10^{36}}{\left[t_L\,(\mathrm{s})\right]^2}\,\mathrm{eV} \tag{8}$$

and is 0.5 power (Felten and Rees, 1969). To put this at 40 keV we need $t_L \sim 3 \times 10^8$ yr. There is no physical basis for such a value. For high-energy electrons free in the intergalactic medium, all loss times would be $\gtrsim 10^{10}$ yr. We might suppose that the electrons suffered adiabatic expansion loss with $t_L \sim 3 \times 10^8$ yr in the radio galaxies from which they came. But this is an unreasonably long lifetime for a radio-galaxy outburst, and in any case, confinement this long in the strong magnetic fields of a radio galaxy would cause these electrons to emit much more synchrotron radiation than observations of the background radio brightness will allow. Expanding confinement *near* radio galaxies, in regions of *weak* field, might work; we require $\bar{H} \lesssim 10^{-7}$ G. Perhaps some form of this model still has unexplored possibilities. It can be extended to radio galaxies at large z, and there the constraints are less severe (Bergamini *et al.*, 1967; Felten and Rees, 1969). But further straining at these artifices for producing the break does not seem profitable if the break is indeed sharp, for a reason which will appear shortly.

Brecher and Morrison (1969) obtained a magnificent fit to the bends and curves of the X-ray spectrum (the dashed curve in Figure 1 is theirs) by folding additional complications into the inverse-Compton model. They took normal galaxies rather than radio galaxies as the electron sources, introduced a spectral break arising *in the sources* (and lying at the same energy in each source!) to circumvent the difficulty of the 40-keV break, postulated a distribution of electron spectral indices in the various sources to induce curvature at the two ends of the Compton spectrum, and assumed a very high output of cosmic rays by these 'normal' galaxies, so that our own Galaxy can no longer be taken as an example (Setti and Rees, 1970). Though not all the details in this paper are clear, it is not surprising that a good fit can be obtained with so many degrees of freedom. Notice how well their curve matches the sharp break at 40 keV.

Recent work by Cowsik and Kobetich (1972) reveals a serious difficulty with the Brecher and Morrison spectrum. To make this clear let me first show a figure from Blumenthal and Gould (1970) (Figure 4). The solid curve is the *number* spectrum in energy of photons Compton-scattered by an electron of energy $\gamma m_e c^2$ moving in an isotropic flux of monoenergetic photons, energy E_0. The abscissa is in units of the kinematic maximum energy, $4\gamma^2 E_0$. I have sketched in (dotted curve) the corresponding *energy* spectrum. This is strictly a physical problem, and no astronomical assumptions are involved. Note that the width at half-power points is large, a factor

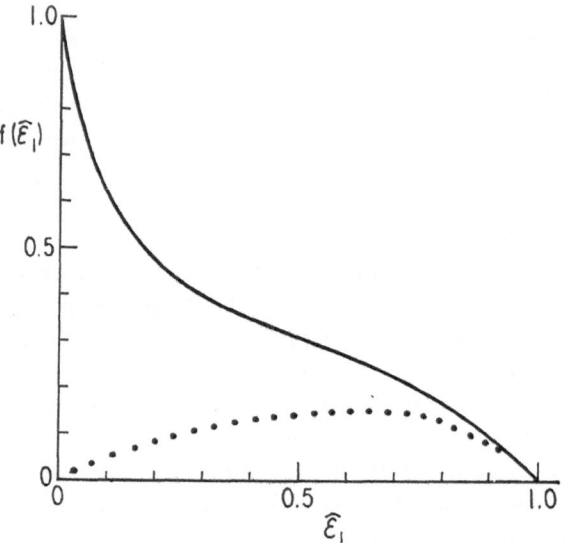

Fig. 4. Differential *number* spectrum of Compton-scattered photons produced by monoenergetic electrons in an isotropic monoenergetic photon flux (Blumenthal and Gould, 1970). Ordinate units are arbitrary. The dotted line (sketched) shows the corresponding energy spectrum. The point 1.0 at right corresponds to the kinematic maximum energy $4\gamma^2 E_0$.

~ 6 in energy, although the *reacting photons and electrons are monoenergetic*. Then when an electron spectrum having a 'break', even a sharp break, is folded with an isotropic photon flux (monoenergetic, or *a fortiori* blackbody) it is clear that a sharp break will *not* be obtained in the radiated Compton spectrum. Figure 5 shows the results of integrations by Cowsik and Kobetich (1972). 'L_{BM69}' shows the Brecher and Morrison (1969) spectrum, with the sharp break given by a delta-function approximation. 'L_2' shows the accurate result for the Compton spectrum radiated by a simple power-law electron spectrum with one abrupt break, interacting with blackbody photons. The width of the knee is seen to be a factor ~ 10. 'L_{ig}' is an even *less* 'angular' Compton spectrum which Cowsik and Kobetich (1972) obtain by introducing additional smearing assumptions, some rather arbitrary, which I will not discuss here. The important point is that an electron power law with a break, even a sharp break, will not produce a Compton spectrum with a break any sharper than L_2, and it is doubtful whether L_2 is an acceptable fit to the 40-keV data. This cuts strongly against all inverse-Compton theories, not just that of Brecher and Morrison

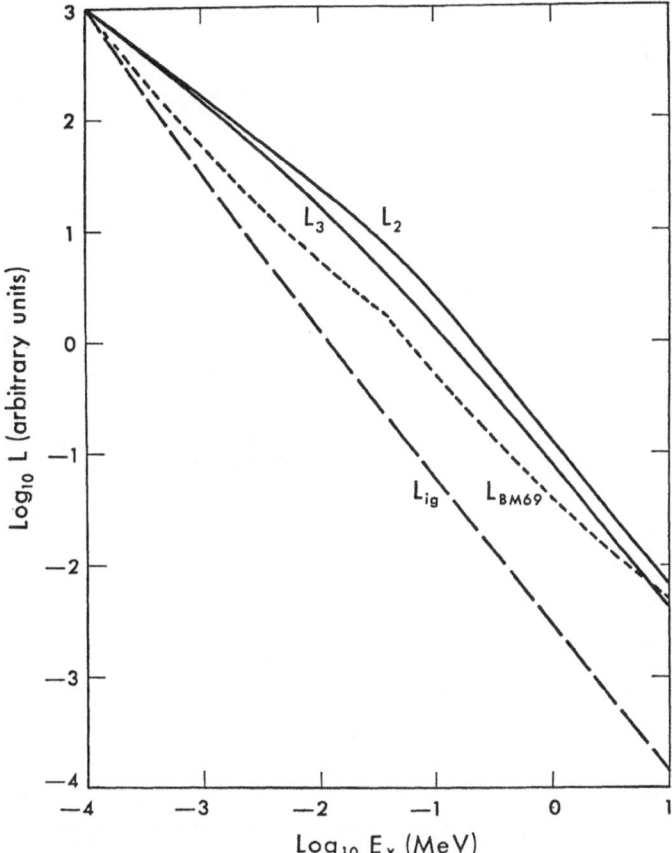

Fig. 5. Shapes of expected 'breaks' in Compton spectra (Cowsik and Kobetich, 1972). See text.

(1969). Of course it would be possible to patch this up by introducing additional components, e.g. a suitably shaped 'hump' in the electron energy distribution. But then the inverse-Compton theory loses its basic appeal of simplicity.

We are left with the problem of explaining the sharp (?) 40-keV break. I will show you one more figure from Cowsik and Kobetich (1972) (Figure 6). Here these authors have plotted the observed diffuse X-rays compared with their smooth (kneeless) Compton spectrum and with a model spectrum for extragalactic γ-rays from white dwarfs, concocted by Cowsik (1971). (This model involves the dubious assumption that most of the galactic *cosmic rays* are also produced by white dwarfs). In their view, the *soft* X-rays come from inverse Compton effect, and the only X-ray band not fitted adequately by these two components is the range from 3 to 100 keV, containing the knee. Therefore, they suggest a third component, namely thermal bremsstrahlung from a very hot intergalactic plasma at $T \simeq 3.3 \times 10^8$ K ($kT \simeq 30$ keV). The advantage of this is that the bremsstrahlung exponential function has an almost

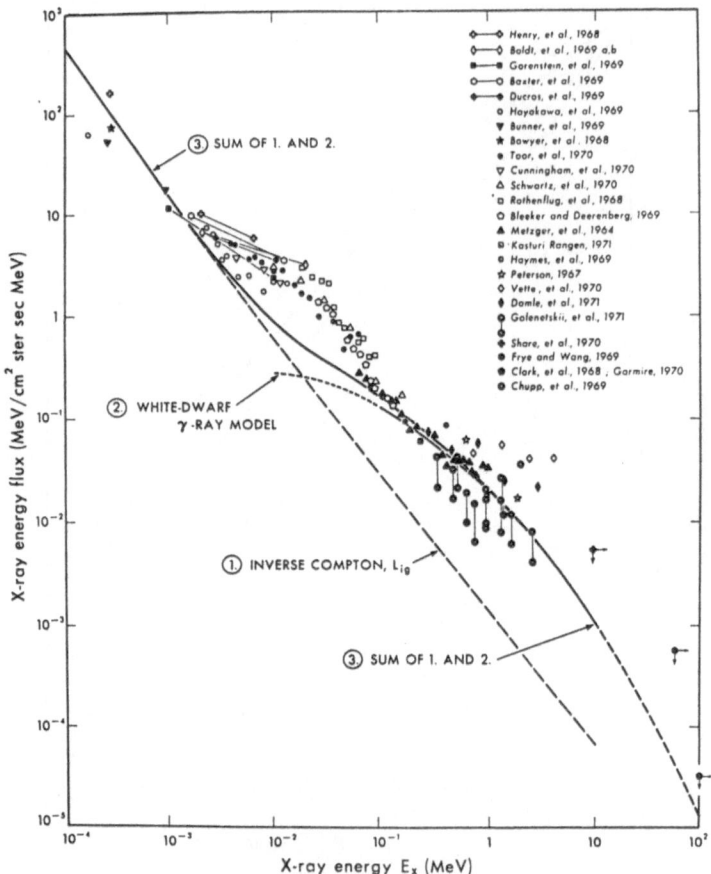

Fig. 6.   Three-component model for the diffuse X-ray background
(Cowsik and Kobetich, 1972). See text.

unique capability of giving the sharp break. If $H_0 = 100$, a density $\varrho = \varrho_{cl}$ would give
~ 10 times more flux than is observed in the break region, but since the product
$\varrho_{cl}^2 R_H \propto H_0^3$ when H_0 is varied, Cowsik and Kobetich (1972) set $H_0 \simeq 55$ as suggested
by recent data (Sandage, 1971) and then find that $\varrho = \varrho_{cl}$ $(n_e \sim 3 \times 10^{-6} \text{ cm}^{-3})$ is
quite consistent with the X-ray flux! Thus something like the 'hot universe' of Gold
and Hoyle (1959) is resuscitated by the new value of H_0, which should not, however,
be taken as firmly established. This new twist to the problem of the hot intergalactic
medium is intriguing. Of course it is no great achievement to fit an observed spectrum
by a three-component model with many adjustable parameters.

6. γ-Rays, $h\nu > 1$ MeV

I have time for only a few remarks about γ-rays. The reality of the turnup near 1 MeV has been disputed (Anand et al., 1970), and is in severe doubt at this Symposium, though an independent group has obtained the same turnup (Vedrenne et al., 1971). Several theoretical models have been proposed (Silk, 1970; Sunyaev, 1970), all extragalactic and some resorting to large z. Indeed the indication now is that these photons *are* isotropic (Damle et al., 1972).

In the 100-MeV range, Cavallo and Gould (1971a) have recalculated the expected flux of '$\pi - \gamma$' photons due to decay of π^0's produced by galactic cosmic rays, assuming that the cosmic-ray density throughout the galactic disk is the same as its local value and using observed neutral-hydrogen column densities. With the recent renormalization of the observational data, they find reasonably good agreement with the observed 'line source' in the plane, except at the galactic center, where an additional source is required. This may be supplied by Compton scattering of the dense infrared radiation at the galactic center by cosmic-ray electrons (Stecher and Stecker, 1970). Since the predicted '$\pi - \gamma$' fluxes are conservative, there may be serious consequences to cosmic-ray theory if the galactic 'line source' turns out not to be there, as a high-resolution observation suggests (Browning et al., 1972).

An attractive feature of the $\pi - \gamma$ process is that its spectrum can be predicted on physical grounds, without astronomical hypotheses. Figure 7 (Cavallo and Gould, 1971a) shows this spectrum (photons s^{-1} MeV^{-1}); it has a flat top and is symmetrical about $\frac{1}{2}m_{\pi^0} c^2 = 67.5$ MeV. An earlier result by Stecker (1970) is also shown. The difference between these curves has become a matter of dispute (Stecker, 1971; Cavallo and Gould, 1971b). It appears that the Cavallo and Gould curve may be more nearly correct, although the matter is not as simple as they tried to make out. A third independent calculation would be useful; Goldsmith and Levy (1971) have undertaken this, at least in part. Observations (Fichtel et al., 1972) indicate that the

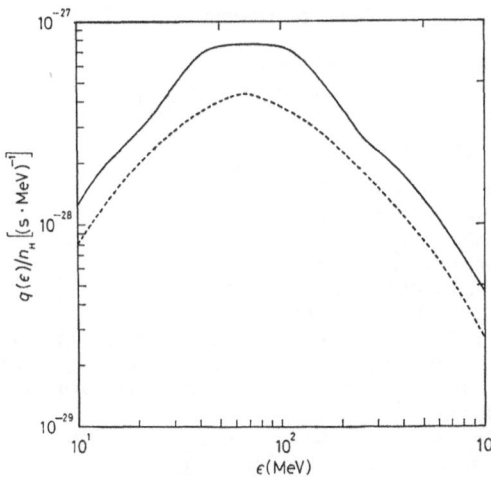

Fig. 7.   Production spectrum of photons in the π-γ process, according to Cavallo and Gould (1971a). The dashed curve is a result given earlier by Stecker (1970).

photons coming from the galactic-center direction are indeed as hard as a $\pi - \gamma$ origin would imply.

Acknowledgements

The reviews by Setti and Rees (1970) and Silk (1970) have been helpful, as have conversations with K. Brecher, P. Morrison, M. J. Rees, P. A. Strittmatter and W. H. Tucker. A referee kindly made the Cowsik and Kobetich figures available to me, together with some comments. Preparation of this manuscript at Steward Observatory has been supported partially by the University Science Development Program of the U.S. National Science Foundation.

References

Anand, K. C., Joseph, G., and Lavakare, P. J.: 1970, *Proc. Ind. Acad. Sci.* **A71**, 225.
Bergamini, R., Londrillo, P., and Setti, G.: 1967, *Nuovo Cimento* **B52**, 495.
Bergeron, J.: 1969, *Astron. Astrophys.* **3**, 42.
Bergeron, J.: 1970, *Astron. Astrophys.* **4**, 335.
Bleach, R. D., Boldt, E. A., Holt, S. S., Schwartz, D. A., and Serlemitsos, P. J.: 1972, preprint.
Blumenthal, G. R. and Gould, R. J.: 1970, *Rev. Mod. Phys.* **42**, 237.
Bolton, C. T.: 1972, *Nature* **235**, 271.
Bowyer, C. S., Field, G. B., and Mack, J. E.: 1968, *Nature* **217**, 32.
Brecher, K. and Morrison, P.: 1969, *Phys. Rev. Letters* **23**, 802.
Brown, R. L. and Gould, R. J.: 1970, *Phys. Rev.* **D1**, 2252.
Browning, R., Ramsden, D., and Wright, P. J.: 1972, *Nature Phys. Sci.* **235**, 128.
Burbidge, G. R.: 1970, *Ann. Rev. Astron. Astrophys.* **8**, 369.
Cavallo, G. and Gould, R. J.: 1971a, *Nuovo Cimento* **B2** (Ser. 11), 77.
Cavallo, G. and Gould, R. J.: 1971b, *Letters Nuovo Cimento* **2**, 1199.
Cavallo, G. and Horstman, H.: 1972, *Nature Phys. Sci.* **235**, 110.
Coleman, P. L., Bunner, A. N., Kraushaar, W. L., McCammon, D., and Williamson, F. O.: 1972, *Bull. Am. Astron. Soc.* **4**, 220.
Cooke, B. A., Griffiths, R. E., and Pounds, K. A.: 1969, *Nature* **224**, 134.
Cowsik, R.: 1971, at the 12th Int. Conf. on Cosmic Rays; *Conference Papers* **1**, 334, Univ. of Tasmania, Hobart.
Cowsik, R. and Kobetich, E. J.: 1972, *Astrophys. J.*, in press.
Damle, S. V., Daniel, R. R., Joseph, G., and Lavakare, P. J.: 1972, *Nature* **235**, 319.
Davidsen, A., Shulman, S., Fritz, G., Meekins, J. F., Henry, R. C., and Friedman, H.: 1972, submitted to *Astrophys. J.*
de Freitas Pacheco, J. A.: 1970, *Astron. Astrophys.* **8**, 420.
Edwards, P. J.: 1971, *Nature Phys. Sci.* **234**, 75.
Fabian, A. C. and Sanford, P. W.: 1971, *Nature Phys. Sci.* **231**, 52.
Felten, J. E.: 1966, *Astrophys. J.* **144**, 241.
Felten, J. E. and Bergeron, J.: 1969, *Astrophys. Letters* **4**, 155.
Felten, J. E. and Morrison, P.: 1966, *Astrophys. J.* **146**, 686.
Felten, J. E. and Rees, M. J.: 1969, *Nature* **221**, 924.
Fichtel, C. E., Hartman, R. C., Kniffen, D. A., and Sommer, M.: 1972, *Astrophys. J.* **171**, 31.
Forman, W., Kellogg, E., Gursky, H., Tananbaum, H., and Giacconi, R.: 1972, preprint.
Gold, T. and Hoyle, F.: 1959, in R. N. Bracewell (ed.), *Paris Symposium on Radio Astronomy*, Stanford Univ. Press, Stanford, p. 583.
Goldsmith, D. W. and Levy, D. J.: 1971, *Bull. Am. Astron. Soc.* **3**, 450.
Gorenstein, P. and Tucker, W. H.: 1972, *Astrophys. J.*, in press.
Gunn, J. E. and Gott, J. R.: 1972, *Astrophys. J.* **176**, 1.
Gursky, H., Gorenstein, P., Kerr, F. J., and Grayzeck, E. J.: 1971, *Astrophys. J. Letters* **167**, L15.
Henriksen, R. N., Feldman, P. A., and Chau, W. Y.: 1972, *Astrophys. J.* **172**, 717.

Horstman, H. and Horstman-Moretti, E.: 1971, *Nature Phys. Sci.* **229**, 148.
Hudson, H. S., Peterson, L. E., and Schwartz, D. A.: 1971, *Nature* **230**, 177.
Ilovaisky, S. A. and Lequeux, J.: 1972, *Astron. Astrophys.* **18**, 169.
Ilovaisky, S. A. and Ryter, Ch.: 1971, *Astron. Astrophys.* **15**, 224.
Ilovaisky, S. A. and Ryter, Ch.: 1972, *Astron. Astrophys.* **18**, 163.
Ipavich, F. M. and Lenchek, A. M.: 1970, *Phys. Rev.* **D2**, 266.
Kellogg, E., Gursky, H., Leong, C., Schreier, E., Tananbaum, H., and Giacconi, R.: 1971, *Astrophys. J. Letters* **165**, L49.
Kiang, T.: 1961, *Monthly Notices Roy. Astron. Soc.* **122**, 263.
Kristian, J., Brucato, R., Visvanathan, N., Lanning, H., and Sandage, A.: 1971, *Astrophys. J. Letters* **168**, L91.
McCammon, D., Bunner, A. N., Coleman, P. L., and Kraushaar, W. L.: 1971, *Astrophys. J. Letters* **168**, L33.
McCray, R. and Schwartz, J.: 1972 in S. P. Maran, J. C. Brandt, and T. P. Stecher (eds.), *The Gum Nebula and Related Problems*, National Aeronautics and Space Administration, in press.
Ostriker, J. P., Rees, M. J., and Silk, J.: 1970, *Astrophys. Letters* **6**, 179.
Palmieri, T. M., Burginyon, G. A., Grader, R. J., Hill, R. W., Seward, F. D., and Stoering, J. P.: 1971, *Astrophys. J.* **169**, 33.
Palmieri, T. M., Burginyon, G. A., Hill, R. W., Seward, F. D., and Scudder, J. K.: 1972, Preprint No. 73682, Lawrence Radiation Laboratory, Livermore.
Peterson, L. E.: 1971, invited paper at AAAS/AAS Symposium, Philadelphia.
Rees, M. J., Sciama, D. W., and Setti, G.: 1968, *Nature* **217**, 326.
Ryter, Ch.: 1970, *Astron. Astrophys.* **9**, 288.
Sandage, A.: 1965, *Astrophys. J.* **141**, 1560.
Sandage, A.: 1971, at Mayall Symposium, Rio Rico, Ariz.
Schmidt, M.: 1970, *Astrophys. J.* **162**, 371.
Schwartz, D. A.: 1970, *Astrophys. J.* **162**, 439.
Schwartz, D. A., Hudson, H. S., and Peterson, L. E.: 1970, *Astrophys. J.* **162**, 431.
Schwartz, D. A., Boldt, E. A., Holt, S. S., Serlemitsos, P. J., and Bleach, R. D.: 1971, *Nature Phys. Sci.* **233**, 110.
Setti, G. and Rees, M. J.: 1970, in L. Gratton (ed.), 'Non-Solar X- and Gamma-Ray Astronomy' *IAU Symp.* **37**, 352.
Setti, G. and Woltjer, L.: 1970, *Astrophys. Space Sci.* **9**, 185.
Seward, F. D., Chodil, G., Mark, H., Swift, C., and Toor, A.: 1967, *Astrophys. J.* **150**, 845.
Silk, J.: 1970, *Space Sci. Rev.* **11**, 671.
Stecher, T. P. and Stecker, F. W.: 1970, *Nature* **226**, 1234.
Stecker, F. W.: 1970, *Astrophys. Space Sci.* **6**, 377.
Stecker, F. W.: 1971, *Letters Nuovo Cimento* **2**, 734.
Strittmaker, P. A., Brecher, K., and Burbidge, G. R.: 1972, *Astrophys. J.* **174**, 91.
Sunyaev, R. A.: 1970, *Sov. Phys.–JETP Letters* **12**, 262 (*Zh. Eksperim. Teor. Fiz. Pis. Red.* **12**, 381).
Tucker, W. H.: 1970, *Astrophys. J.* **161**, 1161.
Tucker, W. H.: 1971, *Science* **172**, 372.
van den Bergh, S.: 1961, *Z. Astrophys.* **53**, 219.
Vedrenne, G., Albernhe, F., Martin, I., and Talon, R.: 1971, *Astron. Astrophys.* **15**, 50.

PART V

PANEL DISCUSSIONS

23. DIFFUSE BACKGROUND OF ENERGETIC X-RAYS*

YASH PAL

Tata Institute of Fundamental Research, Bombay 5, India

Abstract. The experimental situation in regard to the extra-galactic diffuse background of photons of energy above 30 keV is critically reviewed in the light of discussions at this Symposium.

There seems to be some doubt about the spectral break at 40 keV. There is an indication of a small bump around 2 MeV and of a shoulder around 20 MeV. The spectrum below 1 MeV (down to 30 keV) can perhaps be represented as $25\,(E/1\,\mathrm{keV})^{-2.1}$. This spectrum also gives, roughly, the OSO III upper limit at 100 MeV, though it lies well below the two bumps mentioned above. A good deal of the discussion centres on the problem of backgrounds in different types of experiments to measure the diffuse X- and γ-ray fluxes.

The subject of diffuse flux of energetic photons has had a unique distinction; it has been showered with a great deal of attention by theorists, who have valiantly, and often successfully, laboured to explain several of its features which have later turned out to be experimentally questionable. Therefore, there was a strong reaction this time and instead of trying to explain the diffuse background, the panel discussion in this area was organised to clarify what is and what is not.

All the papers submitted for this session are greatly concerned with the question of instrumental and environmental background. I shall not discuss the papers individually or in alphabetical order, but will try to fit them into a single theme. The body of this paper was prepared before the symposium, but substantial additions and corrections have been introduced later.

We will be concerned in this review mainly with the results above ~ 20 keV. The spectrum at lower energies will be referred to only in passing. Also we will mainly talk about the possible cosmic flux, though the cosmic nature may not be so well established for some of the measurements.

1. Features of the Canonical Spectrum of Two Years Ago

The way cosmic photon spectrum looked a couple of years ago is shown in Figure 1. It appeared one had fairly definitive values of flux up to ~ 5 MeV. The important features of the spectrum were widely believed to be:

(a) The possibility of a break in the spectrum around 40 keV.

(b) The possibility of flattening of the spectrum beyond 1 MeV.

(c) The suggestion of a definite flux beyond 100 MeV.

The work during the last year or so, and some of the papers submitted to this Symposium have cast serious doubts on the first two of these features, and added substantially new information in the energy region 1 MeV to 50 MeV.

---

* Dr Pal arranged and led the panel discussion on this topic. The other panel members were: D. Brini, L. Peterson, K. Pinkau, J. Trombka, and P. Lavakare.

Bradt and Giacconi (eds.), X- and Gamma-Ray Astronomy, 279–302. All Rights Reserved.
Copyright © 1973 by the IAU.

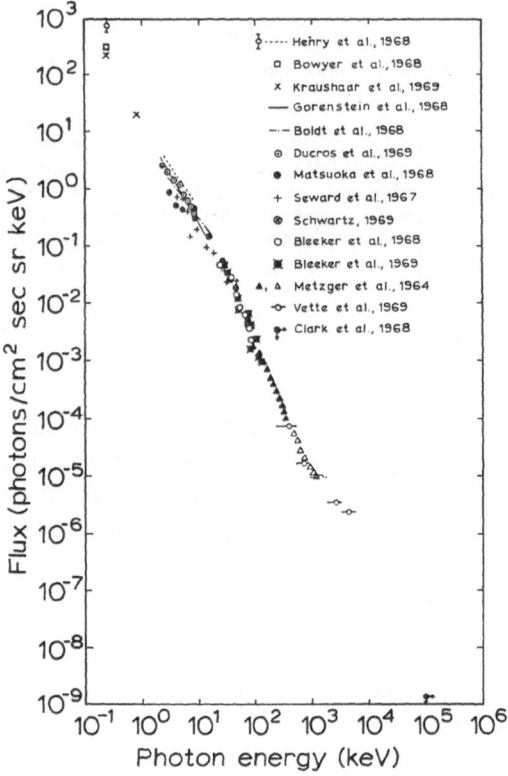

Fig. 1.   Summary of X- and γ-ray diffuse spectrum as of 1970 (Brecher and Burbidge, 1970).

2. Work at Energies below 1 MeV

Measurements in the energy range 20 keV to a few MeV are made with alkali halide crystals, aboard balloons, rockets or spacecraft. Most communications to this conference, as well as the papers published during the last year or so, have been largely concerned with the problems of backgrounds encountered in these measurements.

Following the work of Horstman and Horstman-Morretti (1971) and Makino (1970), several balloon people started taking into account the effect of multiple Compton scattering in the attenuation of the primary X-ray flux. It was pointed out (Manchanda *et al.*, 1971; Kasturirangan and Rao, 1972), that this effect is energy dependent, resulting in an apparent enhancement at lower energies, i.e. ~40 keV, where Compton scattering is more important. Thus Manchanda *et al.* (1971) corrected all the balloon experiments for which the data were available, and suggested that, as earlier indicated by Bleeker and Deerenberg (1970), the balloon X-ray data did not require a spectral break around 40 keV. Kasturirangan and Rao (1972), using a similar reduction of data went to the extent of saying that a single power law exponent may fit all data from 1 keV to 1 MeV. Perhaps, in saying this they do some violence to the rocket data below 10 keV.

In a more recent paper, submitted to this symposium, Manchanda *et al.* (1972) have been even more impressed with the problem of the environmental background at balloon altitudes. The background in these experiments is usually taken out by a straight line extrapolation of the counting rate at depths greater than 10 g cm^{-2}, plotted on a log-log scale. These authors present strong empirical arguments to show that the growth curve may not be linear on a log-log plot but may in fact become flat, even rise, at low depths. This seems to be demanded by the requirement that the attenuation of the deduced primary flux exhibit the same behaviour as the difference between the measured growth curve and the extrapolated background growth curve. This is demonstrated in Figures 2 and 3. The consistent behaviour of the growth curve is indicated by the solid curves at small depths. If this is correct, as is likely in

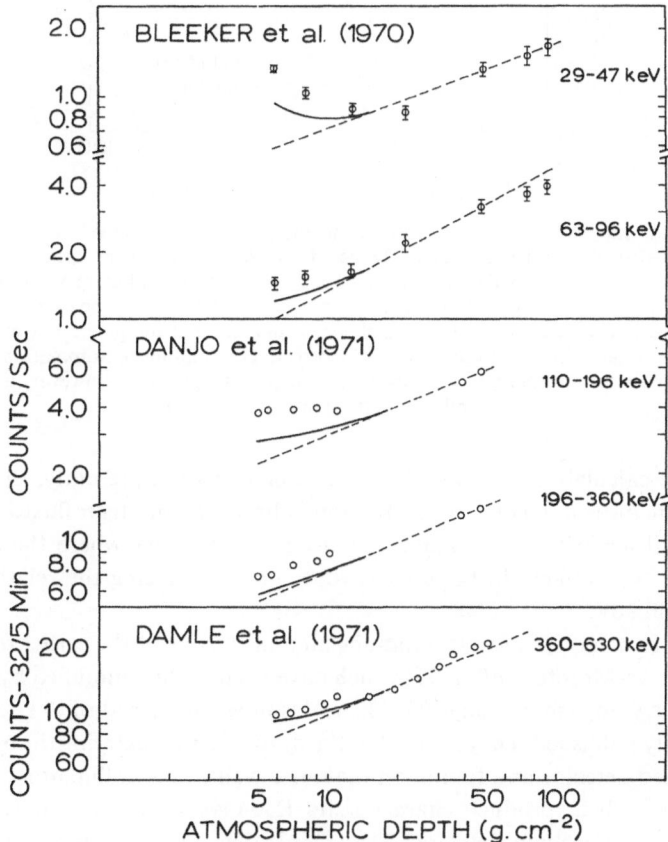

Fig. 2. Growth curves in three balloon experiments to measure primary cosmic photons. The dashed line is the best fit to data points at depths larger than 20 g cm^{-2}. The solid line illustrates the shape of the atmospheric background counting rate required to reproduce the absorption of the primary cosmic X-rays (Manchanda *et al.*, 1972).

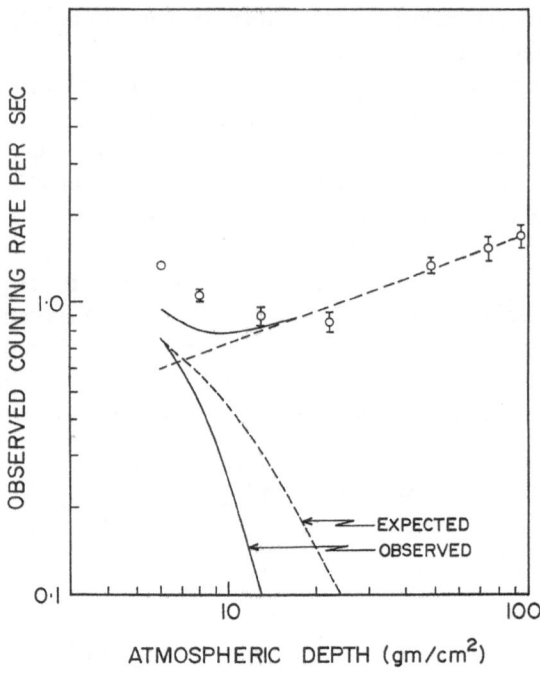

Fig. 3.   Counting rate vs atmospheric depth in the energy channel 29–47 keV in the experiment of Bleeker *et al.* (1970). The dashed line gives the best fit to the data points at large depths. Assuming this to be true representation of the atmospheric background the solid line marked 'observed' gives the attenuation of cosmic X-rays. The dashed line marked 'expected' gives the calculated attenuation of this component. The discrepancy between the two suggests that the power law extrapolation of the atmospheric background to lower depths is incorrect. The solid line extrapolation would yield a correct attenuation for the diffuse component. Its shape and position, however, is not uniquely determined (Manchanda *et al.*, 1972).

view of some calculations by Danjo (1972) to be discussed presently, the fluxes seen in all balloon measurements would be upper limits to the true fluxes. We believe this effect will again be more significant for lower energies where the atmospheric build up effects are likely to be greater, thus further reducing the relative values of flux around 40 keV.

Horstman *et al.* (1972) have recommunicated the results of the earlier rocket experiment (Horstman-Moretti *et al.*, 1971) which have been slightly modified by new corrections. Their new slope in the range 25–200 keV is now 2.0 instead of ∼1.95 which they had previously published. They argue that the diffuse X-rays entering the open aperture of a collimated detector and falling on the inner wall of the collimator can, if photoelectrically absorbed, produce characteristic KX-rays which can contribute to the counting rate of the main detector. They think this can also happen when the collimating shield is active; K absorption probability is largest for incident photons just above the K absorption edge and when the X-ray escapes, the energy left in the shield is too small to veto the count. Thus OSO III measurements, among others, are also subject to this correction. The significant correction due to this effect will be only in their 22–38 keV channel.

There has been comment on the OSO III X-ray measurements from more than one quarter. I suppose people have worried about this observation because this is perhaps the only one which extends a comfortable distance away on either side of the supposed 40 keV break and does indeed show such a break. It is suggested (Dyer and Morfill, 1971) that the corrections due to induced radio activity produced during passage of the spacecraft through the South Atlantic Anomaly may have been inadequate; through experimental and theoretical investigations of spallation effects Dyer and Morfill (1971) and Dyer *et al.* (1972) conclude that a spurious feature at 40 keV would be seen by alkali halide crystals exposed to the particles levels obtaining at the South Atlantic Anomaly or in the interplanetary space.

Dyer and Morfill (1971) have studied the energy spectra observed, in a Cs I crystal, due to the decay of radioactive isotopes produced by bombardment with 155 MeV protons, as a function of time after irradiation. They find that these spectra are in quantitative agreement with the predictions of the semi-empirical formula of Rudstam (1966). Then using the Rudstam formula they calculate the production of various radio-nuclides, and the consequent activity of the crystal, due to exposure to the inner belt protons in the South Atlantic Anomaly, and due to the ambient flux of cosmic-rays at different latitudes. In Figures 4 and 5 are presented the results of their calculation. It is assumed that the passage through the radiation belt corresponds to a 10 min exposure. Also shown in Figure 4 is the expected counting rate from the diffuse X-ray background taking, presumably, an E^{-2} extrapolated spectrum beyond 1 MeV. The authors point out that the main features of their results, namely the marked peaks at 40 keV and 200 keV and the flat spectrum above 1 MeV are all to be found in the observed 'cosmic' spectrum of photons. Further it is seen that the cosmic-ray produced background in the interplanetary space exceeds the diffuse X-ray flux (extrapolated as E^{-2}) at energies greater than 1 MeV. They claim that essentially similar results will apply for Na I crystals. Similar discussion has been made by Fishman (1972).

From this it would seem that the energy range of a few MeV is best studied from low altitude equatorial satellites or balloon platforms near the geomagnetic equator. More about backgrounds in this energy range will be discussed later.

In his analysis of OSO III data, Schwartz (1969) considered mainly the 15 hr half-life Na_{24} activity and determined the corrections from this activity by fitting the decay rates after leaving the South Atlantic Anomaly to a 15 hr half-life. Dyer *et al.* (1972) point out the importance of the background due to a build-up in the activity of long half-life isotopes; according to them such an increased level of counting rate would not be noticed in the Schwartz analysis and will simulate a fictitious flux of diffuse radiation.

In his remarks at the symposium Dr Peterson mentioned that they recognise through their own measurements in OSO 7 that KX-rays from the collimator would contribute a spurious counting rate to the 22–38 keV channel of the OSO III detector

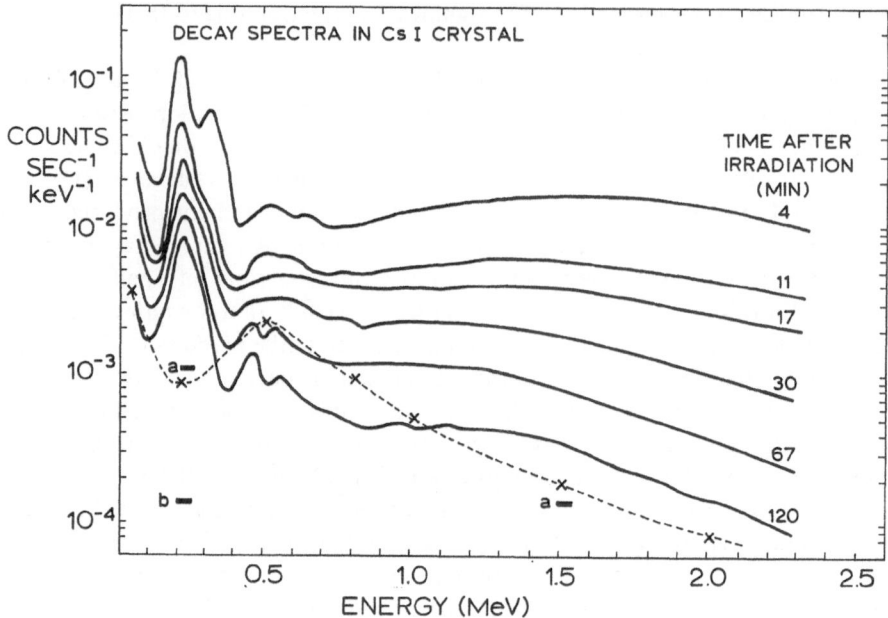

Fig. 4. Predicted spectra (shown solid) for worst case of inner belt traversal, obtained by scaling experimentally observed spectra to 3×10^5 interactions, presented for times up to 2 hr after irradiation. The dotted curve gives the estimated background due to diffuse cosmic X-ray flux (both entry through aperture and leakage through the collimater). 'a' gives the expected level of background due to cosmic-ray damage for a 2 GV cut-off, and 'b' the same for a 14.9 GV cut-off (Dyer and Morfill, 1971).

and "will have to be corrected out." They also see a feature at 60 keV, which is not yet understood, and another feature at about 150 keV which comes and goes as the satellite proceeds through the trapped radiation. The latter feature has a time constant of about 10 days and may be associated with a meta-stable state of I_{128} which becomes activated and decays with roughly a 10 day half life.*

To summarise then, there seems to be growing evidence that the old values of the flux around 40 keV were over-estimated, and that there may not be any break in the spectrum around this energy. Since the proportional counter measurements at lower energies give flatter spectra with a particle spectrum slope of about 1.5–1.7, a change in slope, perhaps gradual, would still be needed at around 20 keV. Some of the

* Commenting on various remarks made by different speakers, Dr Peterson said: "I rather agree with the remarks which have been made by various people which summarise the various kinds of corrections and uncertainties that have crept into these analyses... I have not had the benefit of carefully communicating with Dan Schwartz and we have not reanalyzed the OSO III data in any sense in the light of what we know of the various background effects at this time.... The conclusion I want to make on this is, I think, that there are some corrections necessary for the OSO III. I personally do not see how one can completely end up with a spectrum which has at least no bends in it. It seems to me that at rocket energy, 1–10 keV, (the spectrum) does in fact have a flatter slope. Perhaps the break has been moved down to 20 keV from 40 keV".

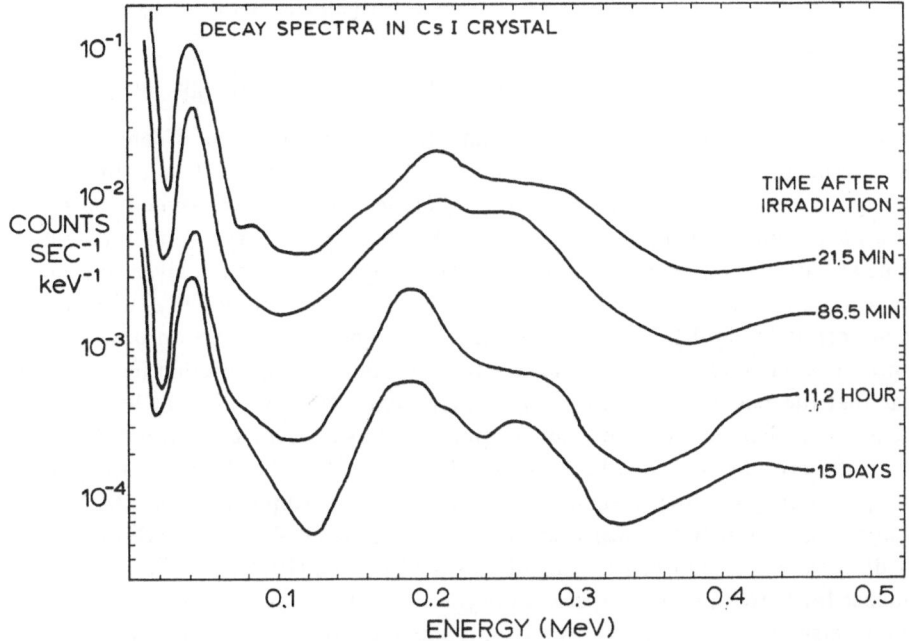

Fig. 5. Predicted spectra due to inner belt spallation for the low energy X-ray region
(Dyer and Morfill, 1971).

authors would like to go right down to 1 keV without any change of slope, but I feel that the integrity of rocket experiments in the 1–10 keV regime should be respected.

The theoretical implications of this may be briefly summarised. Following the work of Brecher and Morrison (1969) it was generally believed that the spectral break around 40 keV reflected the break in the intrinsic spectra of electrons leaking out from discrete sources and interacting with the microwave background radiation to produce X-rays. It had already been pointed out (Cowsik and Pal, 1971) that the break in the X-ray spectrum, if real, could not really be explained in terms of this theory, but was put in as an assumption. Later Cowsik and Kobetich (1971) have made a more detailed calculation of the universal inverse Compton for producing X-rays. They show that when the energy and angular distribution of the microwave photons are taken into account, no such break occurs in the X-ray spectrum. Nor do they find any significant flattening of the calculated spectrum at high energies as was found by Brecher and Morrison. They then proceed to take the difference between the observed (à la 1970) spectrum of X-rays and their calculated spectrum and suggest that this difference in spectrum has a thermal distribution and could have been produced by a hot $(3.3 \times 10^8 \, \text{K})$ intergalactic gas with a density of 3×10^{-6} H atom cm^{-3}. It is clear that a serious reduction in the 40 keV intensity would significantly alter this conclusion; the enhancement in the spectrum may shift to lower energies, and is can probably be accounted for in terms of contributions from discrete sources, at suggested by the UHURU results on various extra-galactic X-ray sources, including galactic clusters.

3. More on Backgrounds and Flux in 1 MeV Range

Several contributions to this Symposium discuss the problem of environmental background in experiments close to the top of the atmosphere. Let us first discuss the experiments of Vedrenne *et al.* (1972). These authors have studied the growth curve of low energy gamma-rays, using a shielded stilbene crystal in three balloon flights at latitudes 62°N, 46°N and 10°N. These growth curves are given in Figure 6. It is seen that at the higher latitudes the curves are fitted well by a p^α law (where p is the atmospheric depth), but at equitorial latitude the curve shows an upturn at low values of p. They treat this as evidence for the existence of a cosmic flux. It is also clear that the background reduces by a large factor when working at low latitudes. The authors also measure the neutron component as a reference, and use it to extrapolate the growth curve at the low latitude. Using the growth curves for different energy bins they obtain a cosmic gamma-ray spectrum from 0.7 and 4.5 MeV which is quite consistent with the measurements of Vette *et al.* (1970). These authors specifically disagree with the measurements of Golenetskii *et al.* (1971), who did not find any evidence for flattening of the spectrum beyond 1 MeV.

Golenetskii's experiment was a Cosmos satellite experiment with a shielded Na I crystal. They tried to analyse the data before entry into the South Atlantic Anomaly, and made use of the latitude effect in the background. They managed to put an upper limit to the cosmic contribution by demanding that the total gamma-ray

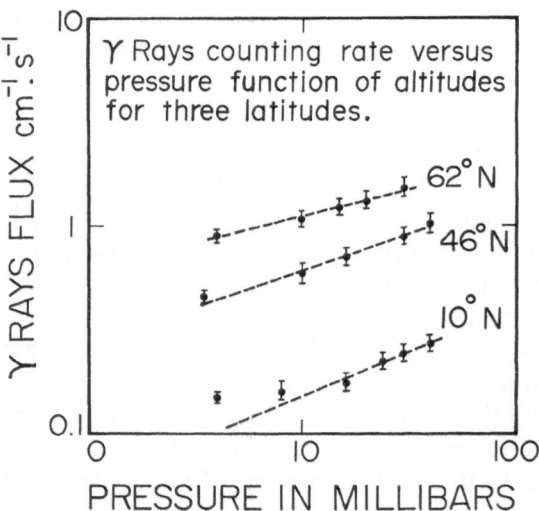

Fig. 6. γ-ray (0.7 to 4.5 MeV) growth curves at three latitudes measured with a $1'' \times 1''$ stilbene crystal surrounded by a plastic anticoincidence jacket (Vedrenne *et al.*, 1972).

atmospheric background vary with latitude in the same manner as the 0.511 MeV
positron annihilation line (see Figure 7). In this way they obtain only upper limits
which, however, lie on a $E^{-2.3}$ extrapolation of the X-ray spectrum and do not
exhibit any flattening. Their energy range was 0.3–3.7 MeV. The results are shown
in Figure 8.

We have a comment about the use of neutrons and 0.511 MeV gamma-rays as
references for the atmospheric background. When these signals are used in studying
growth curves, they may not be physically representative of the growth of the photon
component. If one is measuring neutrons above a threshold, say 1 MeV as in the
experiment of Vedrenne *et al.* (1972), one will miss those neutrons which are produced at around 1 MeV (as the bulk of the neutrons in the atmosphere are) and propagate and degrade in energy due to scattering. On the other hand gamma-rays propagate
through Compton scattering and will merely shift into another energy bin. Similarly
0.511 MeV gamma-rays are removed from the line as soon as they suffer a single
Compton scattering, while the background atmospheric photons are not. The importance of propagation characteristics of atmospheric photons to the shape of their
growth curve will be discussed shortly in connection with the contribution of Danjo
(1972) to this Symposium.

Next we come to the contributions from Damle *et al.* (1971) and Daniel *et al.*

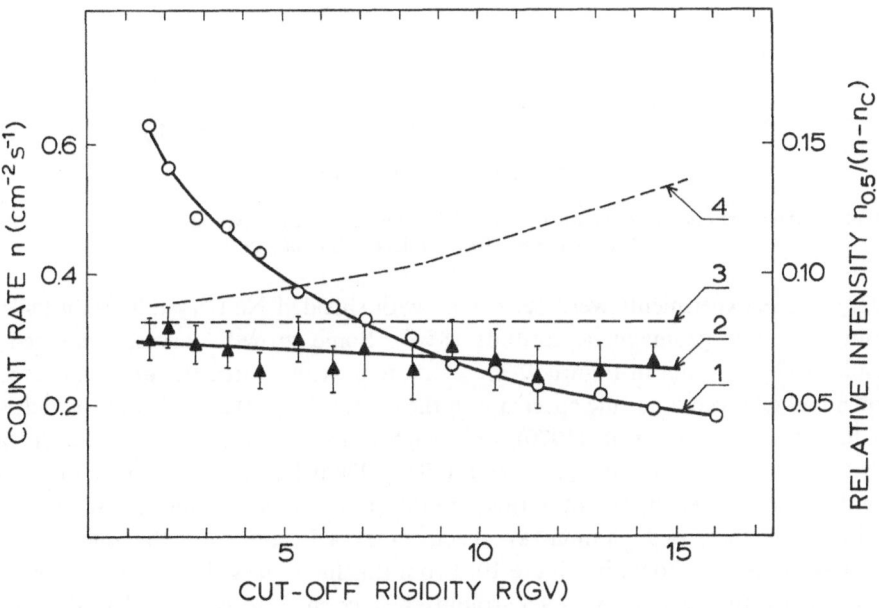

Fig. 7.   γ-ray count rate in the energy range 0.4–2.5 MeV vs cut-off rigidity of primary cosmic rays
(1); observed relative intensity of 0.511 MeV line (2): (3) and (4) represent the relative intensity dependence on rigidity for two different evaluations of diffuse cosmic γ-ray contribution to the total
count rate. Obviously the evaluation (3) is preferred, which gives the upper limits on the diffuse
flux given in the next figure (Golenetskii *et al.*, 1971).

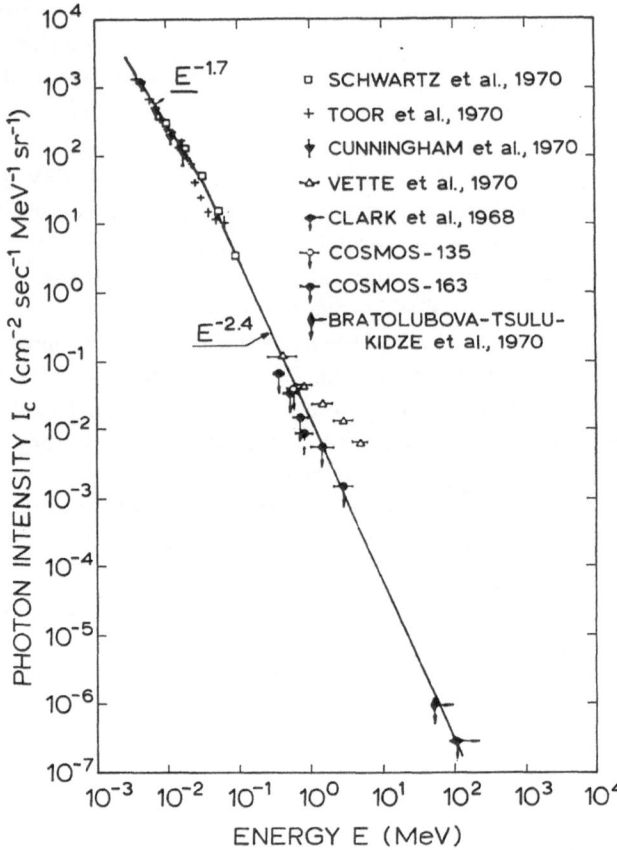

Fig. 8. Some experimental data on the cosmic X-ray and γ-ray fluxes along with the points of
Golenetskii *et al.* (from Golenetskii *et al.*, 1971).

(1972). These experiments were performed with shielded Na I crystals using balloon
flights at a low geomagnetic latitude (8°N). Some growth curves, measured by
Damle *et al.* are given in Figure 9. These authors extrapolated the growth curves as
shown and deduced a cosmic spectrum in the energy range 0.25–4.2 MeV, which was
flat like that of Vette *et al.* (1970), only slightly lower. Their results were, in fact,
very similar to those of Vedrenne *et al.* (1972). Daniel *et al.* (1972) have essentially
repeated the experiment, except in this case they used a much thinner anti-coincidence
shield. Their measured gamma-ray spectrum at 4.7 g cm^{-2} in the energy range
0.2 to ~10 MeV is shown in Figure 10. From this figure it is clear that even without
correcting for the atmospheric background, their count rate is lower than the cosmic
flux supposed to have been measured by Vette *et al.* (1970)! Further a comparison
of their count rate with that of Peterson *et al.* (1970) (also shown in this figure),
obtained with essentially a similar detector but at a latitude of 40°N, shows a large
(a factor of 3) latitude effect and hence the presence of a large atmospheric background

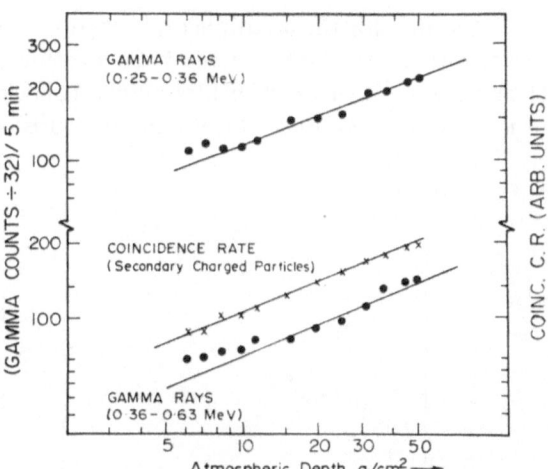

Fig. 9. Growth curves and extrapolations for atmospheric background used by Damle *et al.*
(1971) for two different energies.

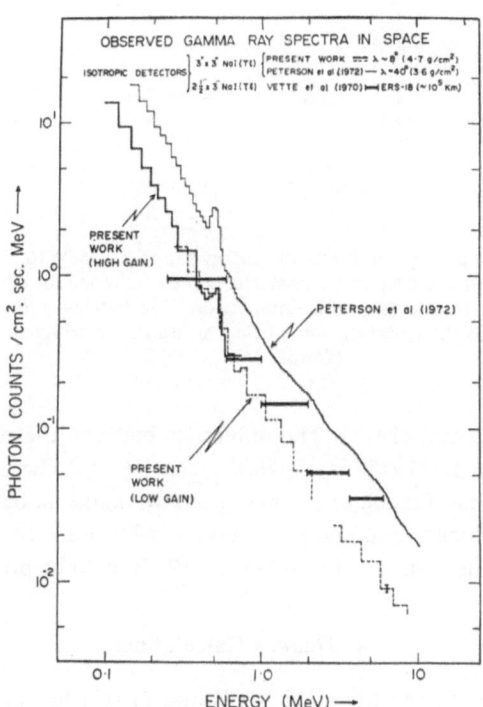

Fig. 10. Comparison of the photon energy loss spectra measured with isotropic detectors, at ceiling
in two balloon experiments and in ERS-18 satellite (Daniel *et al.*, 1972).

in both experiments. Correcting for the atmospheric background using a power law extrapolation of their growth curves, Daniel *et al.* obtain a cosmic spectrum which is shown in Figure 11 along with other experimental points. It is clear that now they do not subscribe to the flattening around 1 MeV and are consistent with the upper

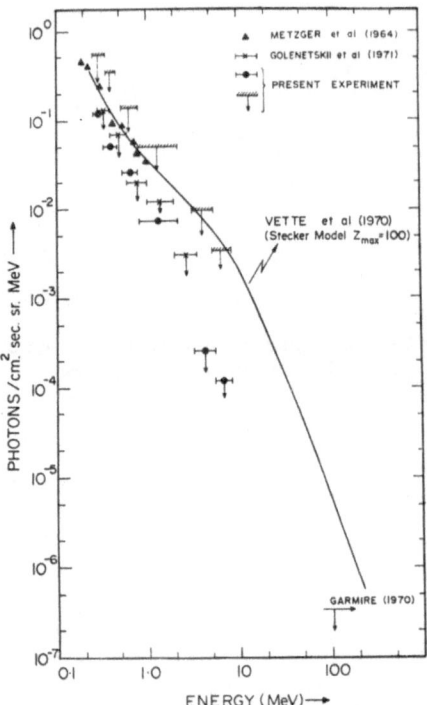

Fig. 11.   The diffuse cosmic photon flux in the energy range 0.2 MeV to 8.5 MeV. The count rates of Metzger *et al.* measured in a deep space probe have been unfolded for detector response. Golenetskii *et al.* (1971) points are also given for comparison. The hatched upper limits are obtained by Daniel *et al.* (1972), if all the counting rate at float altitude is attributed to the cosmic component (Daniel *et al.*, 1971).

limits of Golenetskii *et al.* (1971). The difference between the result of Damle *et al.* (1971) and Daniel *et al.* (1972) is an important lesson. These two detectors differ only in the extent that Damle *et al.* had tried to make a better anti-coincidence shield by making it thicker and ended up having a lot more background. Obviously the lower result, namely that of Daniel *et al.* (1972), is to be preferred.

4. Danjo's Calculations

In a paper submitted to the Symposium, Danjo (1972) has attempted to calculate theoretically the expected shapes for the atmospheric growth curves for gamma rays of different energies. His calculation proceeds as follows:

First he calculates the probability $P(E, x; E_0, x_0)$ of a photon of energy $E(<E_0)$ crossing a depth x, when a single photon of energy E_0 is injected isotropically at a depth x_0. Photo-electric effect, Compton effect and pair creation are taken into account. The calculation is done with a Monte Carlo programme where photons are followed in three dimensions till they are degraded to energies below 50 keV or disappear by photo-electric absorption or escape from the top of the atmosphere. Positrons are assumed to annihilate locally and the resulting gamma-rays are followed further. Compton scattering is determined by the Klein-Nishina formula. Computations are carried out for 5000 photons injected at each of the 12 depths between 2 to 100 g cm^{-2} and for each of 13 injected energies between 0.1 and 7.0 MeV.

$P(E, x; E_0, x_0)$ has now to be folded with the source function of photons $S(E_0, x_0)$. Danjo assumes that photons are all generated through bremsstrahlung of electrons. So he needs an electron source function $J(E_e, x)$. He tries to deduce this function for Hyderabad (16.9 GV cut-off), by calculating the electrons arising from the knock on process and from $\pi \to \mu \to e$ and $\pi^\circ \to 2\gamma \to 4e$ processes. In this he uses the results of Perola and Scarsi (1966) and Beurmann (1971) and makes appropriate corrections for the difference in the primary cut-off rigidity. The total flux of electrons at any depth is obtained by integrating over all angles in the upper hemisphere, assuming that the flux is a function only of the slant depth in the atmosphere. (Note that there might be some approximation here particularly for the flux of electrons from muon decay which may depend on the zenith angle). The source function of X-rays is supposed to be isotropic and is given by

$$S(E_0, x_0) \, dE_0 = dE_0 \int J(E_e, x) \, \sigma(E_e, E_0) \, dE_e,$$

where $\sigma(E_e, E_0)$ is the bremsstrahlung cross-section.

The growth curves for atmospheric X-rays from all directions $I_X(E, x)$ are then given by

$$I_X(E, x) = \int_E^\infty \int_0^\infty S(E_0, x_0) \, P(E, x; E_0, x_0) \, dE_0 \, dx_0.$$

The growth curves for 0.2–0.3 MeV and 0.55–0.75 MeV are shown in Figures 12 and 13. Here the contributions from various depths in the atmosphere are individually shown. It is seen that "the resultant growth curves can be approximated by a power law function for depths greater than 10 g cm^{-2}, grossly reflecting the behaviour of $S(E_0, x_0)$". But for smaller depths the curves flatten due to the build up effect of photons of higher energies generated at higher depths.

Because of these calculations Danjo strongly stresses that flattening of the growth curve at small depths cannot by itself be taken as evidence for the presence of a cosmic component, as has been hitherto assumed. This result has to be viewed in conjunction with the finding of Manchanda et al. (1972), at slightly lower energies, that they do need a flattening of the atmospheric background growth curves in order

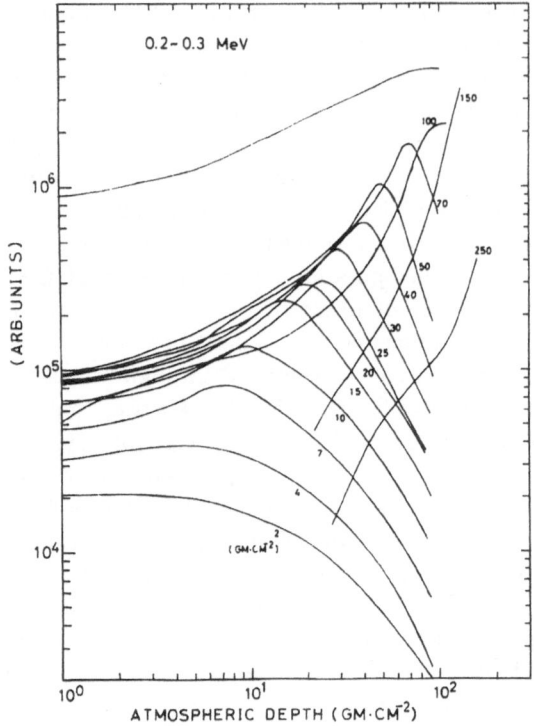

Fig. 12. Calculated growth curves for omni-directional atmospheric photons in the energy range 0.2–0.3 MeV. Curves showing the contributions from various depths are also shown (Danjo, 1972).

that their deduced cosmic component exhibit a proper attenuation behaviour. Though there might be arguments about the details of Danjo's calculation, it seems to us that his main thesis is irrefutable. Danjo concludes that there is so far no definite evidence from a balloon experiment for the existence of cosmic photons of more than 0.2 MeV!

Danjo has also shown that the growth curve of 0.511 positron annihilation gamma-rays would show a steep attenuation with depth, in agreement with observation, because photons on this line are not subject to the build up effect. Furthermore, the downward moving photons are also much less subject to build up than the upward moving photons; the latter are the bulk of photons at small atmospheric depths.

Peterson remarked in this connection, "We have obtained balloon results for many years and we always regarded trying to obtain the spectrum of the diffuse component from balloon results as being a rather dangerous process at best. We feel that the use of a source function such as has been done by Danjo is a much better process than to generate a growth curve with the counting rates. We have generated source functions not based upon a calculation starting from a cosmic ray input but one which is empirically derived from the gamma-ray data itself; this will appear in the literature shortly".

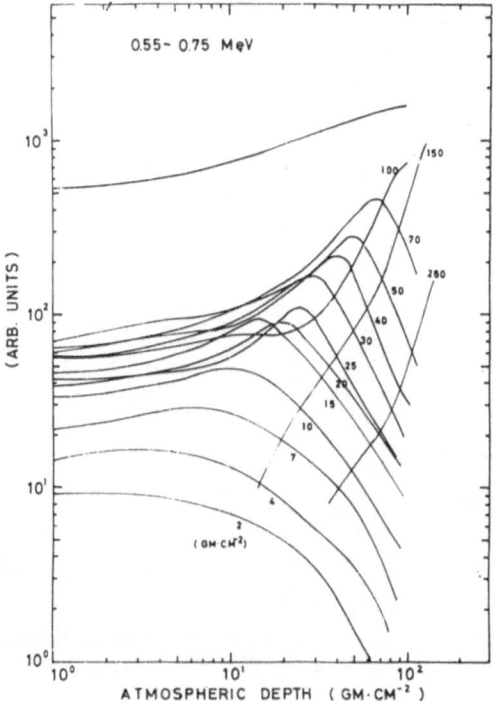

Fig. 13.   The same as Figure 12 for the energy range 0.55–0.75 MeV (Danjo, 1972).

5. Positron Annihilation Line at 0.511 MeV

Peterson reported a definite flux of 2×10^{-2} photons $cm^{-2} s^{-1}$ from their Apollo experiments, though he pointed out that the effect of 8 kg of matter in the vicinity of the detector on the extended boom have not been properly evaluated. Lavakare, however puts an upper limit of 3×10^{-2} photons $cm^{-2} s^{-1}$ to such a flux in their balloon measurements from Hyderabad. This has to be viewed in the light that Haymes has reported a flux of 2×10^{-3} photons $cm^{-2} s^{-1}$ at 470 keV from the galactic centre.

6. New Results at Energies Beyond 5 MeV

Several new results have been communicated to this Symposium at energies beyond 5 MeV. Three of these are spark chamber experiments at balloon altitudes, while the fourth is based on an experiment on Apollo 15 and 16. The Apollo experiments also give results in the sub-MeV region.

Share *et al.* (1972) from NRL have made two balloon flights, one over Texas (4.5 GV cut-off) and the other over Argentina (11.9 GV cut-off), in which they observe electron pairs produced in emulsion layers in spark chambers placed below. The threshold energy is quoted as 10 MeV. The growth curves for these events are shown

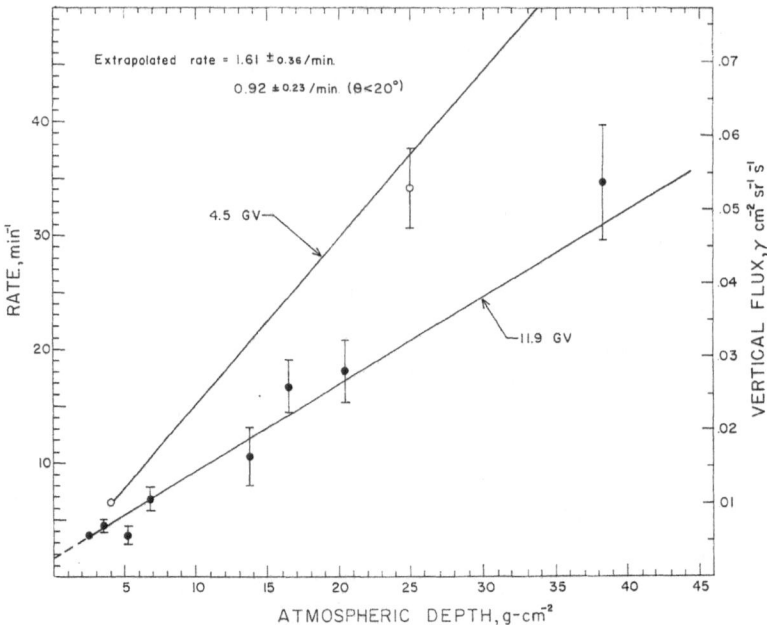

Fig. 14.   Growth curves for electron pairs observed in two balloon flights, one over Texas (4.5 GV
cut-off) and the other over Argentina (11.9 GV cut-off) by Share *et al.*, 1972).

in Figure 14. The Argentina flight suggests non-zero extrapolated counting rate at
the top of the atmosphere. However, when the authors study the azimuthal distribution of the pairs observed at ceiling, they find a strong east-west effect with a (west/
east) ratio of 1.42 ± 0.22 for zenith angles greater than 15°. They calculate an expected
ratio of 1.40 for charged particles at this location. It is likely that the east-west
effect in gamma rays may be caused by particle interactions in the overlaying material
of the instrument. The authors suggest, however, that the excess from the West may
also be due to the fact that the galactic plane lies in that direction. They quote an
upper limit to the integral intensity >10 MeV as 4×10^{-3} γ cm^{-2} s^{-1} sr^{-1}. However,
they prefer to convert this into a limit of 5.5×10^{-5} γ cm^{-2} s^{-1} sr^{-1} MeV^{-1} on the
differential flux at 27 MeV, because at this energy the response of their telescope is
not very sensitive to the assumed spectral index. This is the limit shown in Figure 21.

   Mayer-Hasselwander *et al.* (1972) of the Max Planck Institute have reported the
results of two balloon flights in which a gamma-ray astronomy spark chamber was
carried to residual pressures of 1.7 g cm^{-2} and 2.2 g cm^{-2} respectively. Figure 15
shows the growth curves for clearly visible electron pairs in these flights. The excess
above the power law extrapolation is clearly evident. It is to be noted that the discussion of the build up effect due to Danjo does not apply to this experiment because
of their high energy, 30–50 MeV, and also because in this experiment one is dealing
with downward moving photons. The authors also study the atmospheric growth
curve of the opening angle distribution of the pairs recorded by them and find

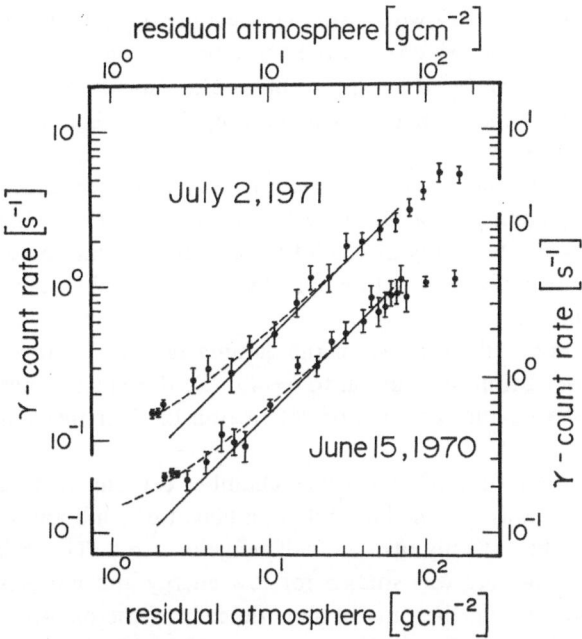

Fig. 15.   Growth curve for clearly visible electron pairs in two balloon flights of a γ-ray astronomy spark chamber (Mayer-Hasselwander *et al.*, 1972).

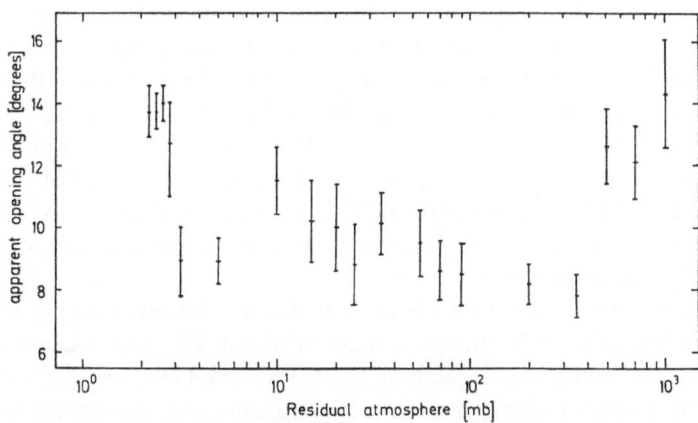

Fig. 16.   Apparent opening angle distribution of electron pairs vs atmospheric depth (Mayer-Hasselwander *et al.*, 1972).

(Figure 16) that the spectrum becomes harder (as expected) when one approaches the Pfotzer maximum, but begins to get soft again as one nears the top of the atmosphere. Such an effect has also been noticed by Share *et al.* (1972). This is supposed to show that the observed signal is not of atmospheric origin alone, but is mixed with a soft spectrum of celestial gamma-rays. If some one could find a way of generating a build up of downward moving gamma-rays of this energy range and at this altitude, then this argument would not be a strong support for a celestial contribution, because the build up spectrum would also be soft. However, as mentioned earlier, it is unlikely that such a build-up is possible for these directional detectors at these energies.

Taking into account their response function Mayer-Hasselwander *et al.* find that the flux of celestial gamma rays at 30–50 MeV should be more intense than $0.012 \times E^{-2.3}$ interpolation between the X-ray and OSO III gamma-ray data by a factor of 10. If they take their maximal response function this factor is reduced to 4. Their results are shown in Figure 21.

They do not find any dependence of the gamma ray flux on the galactic latitude, although their observations extend out to $b = 40°$. So their flux is presumably cosmic. The authors emphasise that accelerator calibration of their instrument has not yet been carried out.

Frye discussed the results of their spark chamber experiment to observe gamma-rays. The instrument was switched on only at a height of 3 mb and hence the growth curves could be looked at only over a small range of depth. The ceiling altitude was 1.2 mb. The counting rate was studied for low energy and high energy pairs both on the way up and during the time of slow descent of the balloon. The results are shown in Figure 17. While for the low energy (~ 30 MeV) channel both the growth curves show a non-zero intercept, the high energy (> 100 MeV) curves behave differently on the way up and on the way down. This latter behaviour which is not yet understood, has lead Frye to view his positive flux at low energies with some suspicion. This experience may also be relevant to the other experiments. If however, the positive intercept at low energies is taken seriously, Frye obtains a flux at 30 MeV which is consistent with the result quoted by the Max Planck group.

Finally we come to the results from the flights of Apollo 15 and Apollo 16. Along with the astronauts each of these vehicles also carried a Na I crystal with a plastic anti-coincidence shield. The experiment was conducted by Trombka *et al.* (1973). The preliminary results of this have been published in the Proceedings of Apollo 15 Conference held early this year.

The instrument was located on a boom which could be extended out to a distance of 8 m from the spacecraft, though a mass of about 8 kg was located next to the instrument on the boom. This amount of matter was about the same as for the ERS-18. The energy loss spectra obtained from the instrument were almost exactly the same in Apollo 15 and Apollo 16 experiments and also agreed with the measurements of ERS-18 and Ranger III (Metzger *et al.*, 1964) upto 1 MeV, though above 2 MeV the Apollo points fall considerably below the ERS-18 measurements. The Apollo 15 result (with detector extended) is shown in Figure 18 along with the points from ERS-18 and Ranger III. There is a clear peak at 0.5 MeV, due to positron annihilation, in the Apollo experiment. The general shape of the energy loss spectrum indicates a broad hump around 2 MeV. The procedure for converting this spectrum to a photon spectrum was described at some length by Trombka, (Arnold *et al.*, 1972),

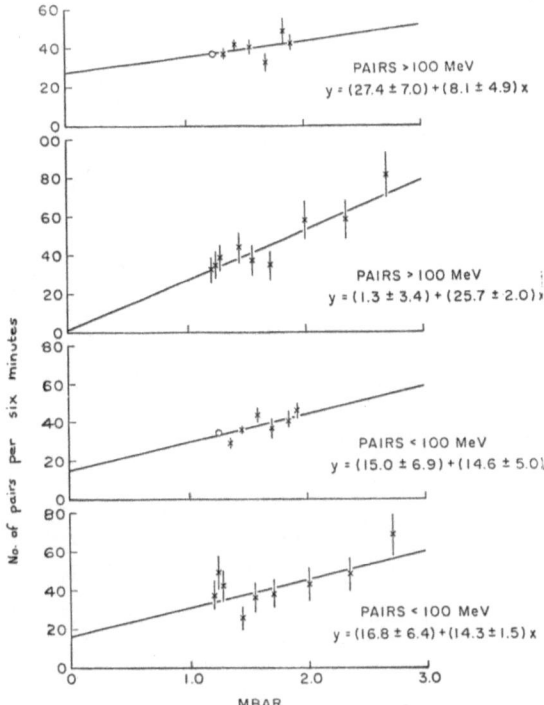

Fig. 17. Growth curves for low energy and high energy pairs reported by Frye (this Symposium). For both energies the curves have been obtained during ascent as well as descent. Though both the low energy curves show a non-zero intercept, the two curves for the high energy (100 MeV) pairs are inconsistent. The equations in the figure give the coordinate representations of the straight line fits drawn.

where he discussed the manner in which the contributions of discrete lines due to spacecraft activity and of spallation products in the crystal to the pulse height spectrum were estimated. Specifically the authors used the exponential form suggested by Fishman (1972) for the pulse height distribution of the spallation products.*

The resulting photon spectrum obtained by these authors is shown in Figure 19. Trombka stressed that the low energy part of this spectrum should not be taken too seriously as the corrections are many and not all complete, but the results above 8 MeV should be good. I tend to more than agree with this; the corrections are just too many and too complex and one cannot be sure that everything conceivable has been taken into account. In particular the spallation contribution is expected to be maximum in the energy region around 2 MeV and a factor of 2 uncertainty in this could wipe out the hump in that place. Also as the authors say, the effect of the 8 kg of matter next to the detector has not been properly evaluated so far.*

* Though the results presented at the Symposium had used Fishman (1972) corrections, the curves later submitted (and included here) by Peterson and Trombka have estimated the spallation effects on the basis of the work of Dyer and Morfill (1971). As a result of this, and possibly some other refinements, the hump around 2 MeV seen in Figures 19, 20 and 21 is slightly less pronounced than was indicated in the preliminary results available at the Symposium.

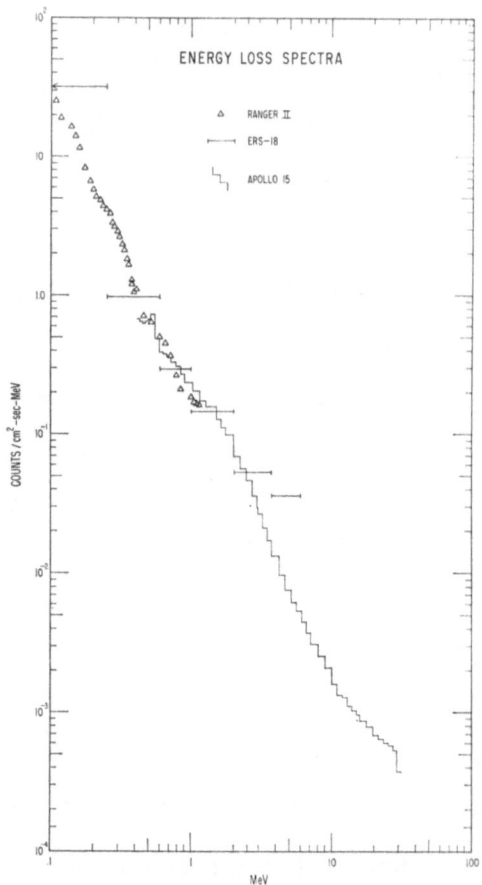

Fig. 18. Energy loss spectrum observed in the Apollo 15 experiment (Arnold *et al.*, 1972). This spectrum (with boom extended and anti-coincidence operational) below 1 MeV is almost exactly the same as that measured in ERS-18 and Ranger III using similar detectors. Above 2 MeV the ERS-18 points lie well above this spectrum.

The value of flux quoted by Trombka is 5×10^{-5} photons cm^{-2} s^{-1} sr^{-1} MeV^{-1} at 30 MeV. This is consistent with the flux of Pinkau (Mayer-Hasselwander *et al.*, 1972) and upper limits of Share and Frye. There is some reason for complacency here, but such coincidences may however be dangerous.

7. Region of 100 MeV

There are no new results in the 100 MeV range and we are left with only one point which comes from the classic OSO III measurement (see Clark *et al.*, 1971). This is sometimes given as a definite flux, though a downward pointing arrow is often added by more conservative people, including the authors of this point.

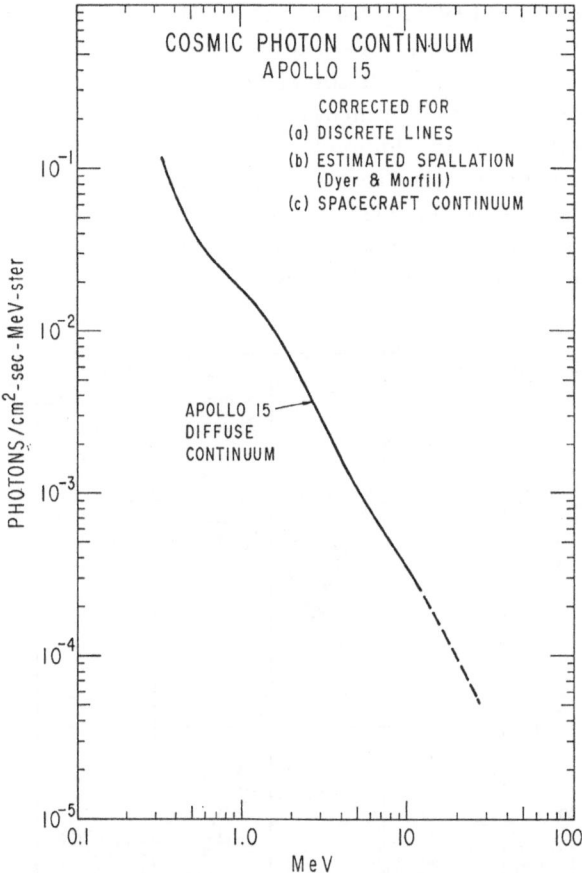

Fig. 19. Photon spectrum deduced from the Apollo experiments (Trombka, this Symposium).

8. The Energetic Photon Spectrum à la this Symposium

Taking congnisance of the discussion about various backgrounds, I would like to bet in favour of the following spectrum from 20 keV to 1 MeV:

$$N(E) = 25 \, (E/1 \text{ keV})^{-2.1} \, (\text{cm}^2 \text{ s sr keV})^{-1}.$$

I believe most of the authors who have communicated to this symposium would compromise on this. This spectrum along with some experimental points is exhibited in Figure 20. It is seen that we have opted against the 40 keV break, but we leave open the manner of transition from energies well below 20 keV to those above 20 keV.*

* However, Dr Schwartz comments: "Data contained in the reference Schwartz (1969) show that the suggested further corrections are not large compared to the discrepancy of OSO-III data and the dashed spectrum (in Figure 20). Again, the data (in Figure 20) do not *show* that there is no break at 40 keV, but are merely of poor enough statistical quality that they do not *require* one (even if it really exists)".

The data beyond 200 keV is given in Figure 21. In this we have excluded the Ranger III and ERS-18 data, and show instead the tentative photon spectrum from the Apollo flights presented by Peterson and Trombka at this meeting; this is done because the raw pulse height spectrum measured in Ranger III, ERS-18 and the Apollo flights was about the same. In addition there are the values and limits of Daniel *et al.* (1972), Golenetskii *et al.* (1971), Share *et al.* (1972), Mayer-Hasselwander *et al.* (1972) and OSO III. The Apollo shoulder around 2 MeV, as mentioned earlier is rather uncertain though it is partly supported by Vedrene *et al.* (1972) and not strongly contradicted by Daniel *et al.* (1972). The discrepancy comes in with respect to the two

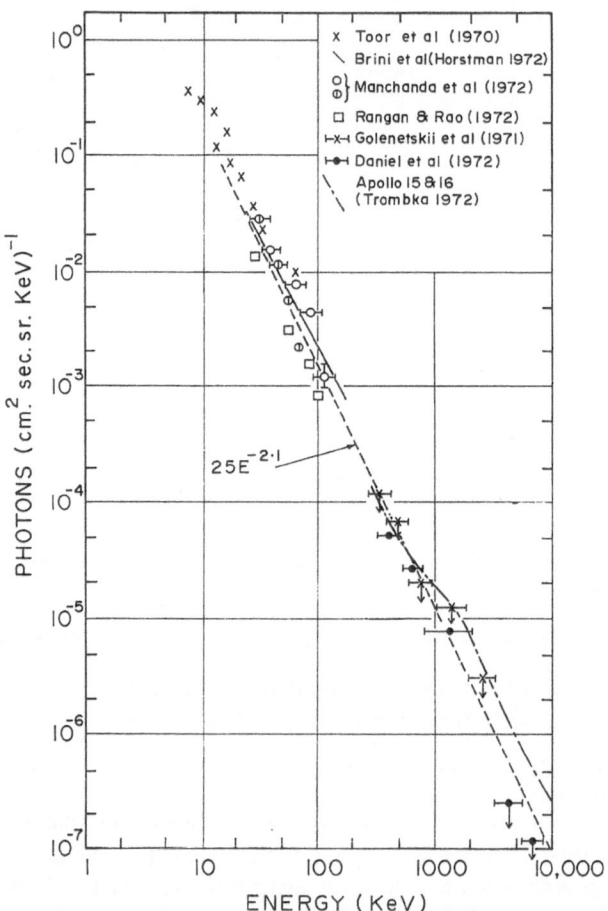

Fig. 20.   Spectrum of photons from 20 keV to 1 MeV. It was agreed that Schwartz (1969) points need further correction and hence are not included. Of the others only a few recent representative measurements have been shown in this slide. In particular, from the balloon experiments of Machanda *et al.* (1971), only the points from two higher flights are taken. The doted curve is $N(E) = 25(E/1$ keV$)^{-2.1}$ (cm^2 s sr keV)$^{-1}$. Note that there is no break around 40 keV. At higher energies the Apollo results are shown while Ranger III and ERS-18 data are excluded (see text).

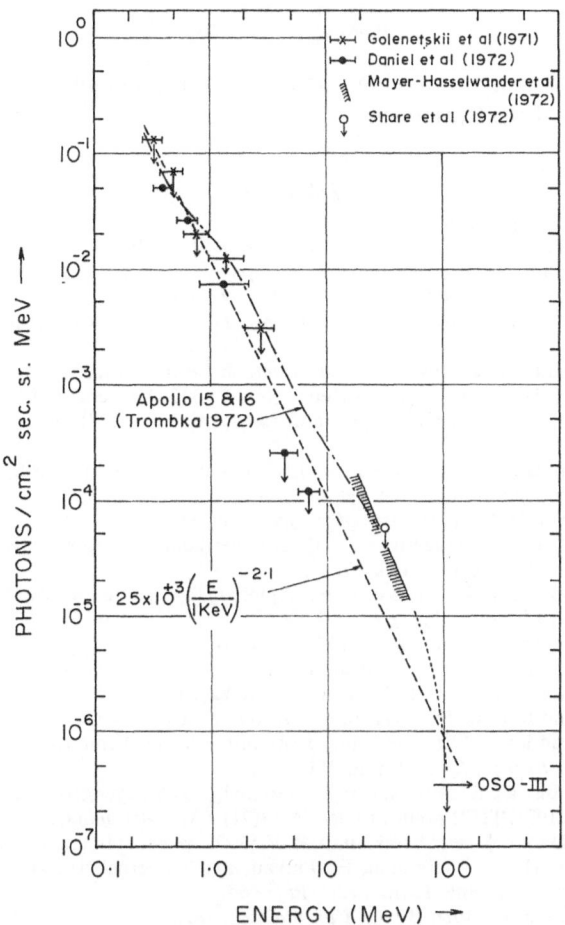

Fig. 21. Selected experimental data beyond 200 keV discussed at this Symposium.
See text for discussion.

highest energy (3–6 MeV) limits of Daniel *et al.* and the Apollo measurements. It appears to us that the favourite cosmological epoch of Stecher, $z \sim 100$, is a bit out of luck. On the other hand the Max Planck and NRL (and possibly Frye, though he was very cautious about claiming a definite flux inspite of an experimental result which appeared at least as definitive as any other) result, combined with the OSO III measurement at 100 MeV, would seem to demand a shoulder at about 25 MeV. This would give the pride of place to an epoch around $z = 2$–3 if the production is via π° decay; however, the phenomenologists would be advised to wait a while before theorising about this.

I would like to emphasise that even though the measurements of Mayer-Hasselwander and OSO III are made with directional detectors there is relatively poor direct evidence so far, for the cosmic (meaning extragalactic) nature of all the flux beyond 1 MeV.

302 YASH PAL

Acknowledgements

I am indebted to various authors for sending me their preliminary results two to three weeks in advance of the Symposium.

References

Beuermann, K. P.: 1971, preprint.
Bleeker, J. A. M. and Deerenberg, A. J. M.: 1970, *Astrophys. J.* **159**, 215.
Brecher, K. and Burbidge, G. R.: 1970, *Comm. Astrophys. Space Sci.* **2**, 75.
Brecher, K. and Morrison, P.: 1969, *Phys. Rev. Letters* **23**, 802.
Clark, G., Garmire, G., and Kraushaar, W.: 1971, *Proc. 12th Int. Conf. Cosmic Rays*, (ed. by Hobart), Vol. I, p. 91.
Cowsik, R. and Kobetich, E. J.: 1971, *Proc. Int. Conf. on Cosmic Rays*, Vol. **1**, p. 38.
Cowsik, R. and Yash Pal: 1971, Technical Report No. 71-113, University of Maryland.
Damle, S. V., Daniel, R. R., Joseph, George, and Lavakare, P. J.: 1971, *Astrophys. Space Sci.* **14**, 473.
Daniel, R. R., Joseph, G., and Lavakare, P. J.: 1972, paper submitted to this Symposium.
Danjo, A.: 1972, paper submitted to this Symposium.
Dyer, C. S. and Morfill, G. E.: 1971, *Astrophys. Space Sci.* **14**, 243.
Dyer, C. S., Engel, A. R., and Quenby, J. J.: 1972, paper submitted to this Symposium.
Fishman, G. J.: 1972, *Astrophys. J.* **163**, 171.
Golenetskii, S. V., Mazets, E. P., Ilinskii, V. N., Aptekar, R. L., Dredov, M. M., Guruyan, Yu. A., and Panov, V. N.: 1971, *Astrophys. Letters* **9**, 69.
Horstman, H.: 1972, paper submitted to this Symposium.
Horstman-Moretti, E., Fuligini, F., and Brini, D.: 1971, *Nuovo Cimento* **6**, 68.
Horstman, H. and Horstman-Moretti, E.: 1971, *Nature* **229**, 148.
Kasturirangan, K. and Rao, U. R.: 1971, preprint; to appear in *Astrophys. Space Sci.*
Kasturirangan, K. and Rao, U. R.: 1972, paper submitted to this Symposium.
Makino, F.: 1970, *Astrophys. Space Sci.* **8**, 251.
Manchanda, R. K., Biswas, S., Agarwal, P. C., Gokhale, G. S., Iyengar, V. S., Kunte, P. K., and Sreekantan, B. V.: 1971, TIFR preprint CR-XA-13(71). (Also *Astrophys. Space Sci.* (1972) **15**, 272.)
Manchanda, R. K., Danjo, A., and Sreekantan, B. V.: 1972, paper submitted to this Symposium.
Mayer-Hasselwander, H. A., Pfefferman, E., Pinkau, K., Rothernel, H., and Sommer, M.: 1972. Max-Planck Institute preprint, *Extraterrest. Phys.* **64**.
Perola, G. C. and Scarsi, L.: 1966, *Nuovo Cimento* **46A**, 781.
Peterson, L. E., Gruber, D., Matteson, J. L., and Vette, J. I.: 1970, preprint UCSD-Sp-70-05.
Rudstam, G.: 1966, *Z. Naturforch.* **21A**, 1027.
Schwartz, D. A.: 1969, Thesis, U.C.S.D.
Schwartz, D. A., Hudson, H. S., and Peterson, L. E.: 1970, *Astrophys. J.* **162**, 431.
Share, G. H., Kinzer, R. L., and Seeman, N.: 1972, paper submitted to this Symposium.
Trombka, J. I., Metzger, A. E., Arnold, J. R., Matteson, J. L., Reedy, R. C., and Peterson, L. E.: 1973, *Astrophys. J.,* in press.
Vedrenne, G., Albernhe, F., Talon, R., and Martin, I. M.: 1972, paper submitted to this Symposium.
Vette, J. I., Gruber, D., Matteson, J. L., and Peterson, L. E.: 1970, *Astrophys. J.* **160**, L161.

24. HIGH-ENERGY DISCRETE SOURCES*

G. G. FAZIO

*Smithsonian Astrophysical Observatory and Harvard College Observatory,
Cambridge, Mass. 02138, U.S.A.*

Abstract. The origin of the gamma-radiation from the galactic plane and the region near the galactic center is still uncertain. However, during this meeting, several groups reported evidence for discrete sources of cosmic gamma-rays. Most of the sources are located near the galactic plane, and some are associated with X-ray sources. The galactic gamma-radiation may be due to these previously unresolved sources. Other sources detected may be associated with variable radio galaxies.

The Crab Nebula still remains the most investigated source at gamma-ray energies. Pulsed emission from NP 0532 was detected in the 10 to 30 MeV region, but no continuous emission was observed. At the highest energies, pulsed emission was reported at $\sim 10^{12}$ eV. Continuous emission from the Crab Nebula was observed at $\sim 10^{11}$ eV; the radiation may be time variable.

The recent gamma-ray experiments on Apollo 15 and 16 and the ESRO satellite TD-1 are described, as well as future experiments on the satellites SAS-B, COS-B, and HEAO-B.

1. Introduction

Although gamma-ray astronomy has advanced rapidly, it has not enjoyed the spectacular success that X-ray astronomy has. Gamma-ray astronomy is a difficult field of research for several reasons. The incident flux is very low – e.g., at 100 MeV, it is $\approx 10^{-5}$ photon cm^{-2} s^{-1} – and what is even more important, this flux must be detected in a background radiation of charged cosmic-ray particles that is $\sim 10^4$ times greater. There are also basic technical difficulties associated with gamma-ray detectors. Because of the relatively low absorption cross section, gamma-ray detection requires rather massive experiments. Also, owing to the low incident flux, the area-time factor for a detector must be large, again leading to large detectors and long exposure times. In addition to these problems, the detectors must be placed above the atmosphere in high-altitude balloons or satellites. However, at energies above 10^{11} eV, ground-based detection of cosmic gamma-rays, through their interaction in the atmosphere, becomes feasible.

In the late 1950s, theoretical predictions of the flux of cosmic gamma-rays were very optimistic, and these predictions encouraged considerable experimental activity. However, gamma-ray astronomy has become the only field of space science where the detected fluxes were considerably lower than the theoretical predictions. Unfortunately, there have been no surprises.

Development of more sensitive detectors has advanced rapidly, and over the last 10 yr, the minimum detectable flux has been lowered by more than a factor of 10^3. Evidence for discrete sources of cosmic gamma-rays has appeared, and major advances will soon be made by means of several satellite experiments.

* Dr Fazio arranged and led the panel discussion on this topic. The other panel members were: B. Agrinier, G. Frye, H. Helmken, R. Hillier, G. Hutchinson, D. Kniffen, J. Kurfess, K. Pinkau, G. Share and T. C. Weekes.

This discussion on the present status of the identification of discrete sources has been divided into the following sections: (a) Crab Nebula and the pulsar NP 0532; (b) Galactic center region; (c) Galactic plane in the Cygnus-Cassiopeia region; (d) Galactic plane in the anticenter region; and (e) Future satellite experiments.

2. Crab Nebula and the Pulsar NP 0532

Before this meeting, the continuous flux of X-rays from the Crab Nebula had been measured up to energies of the order of 400 keV (Figure 1), while the pulsar NP 0532 has been detected up to energies of the order of a few MeV. Hillier *et al.* (1970) first reported evidence for pulsed emission in the 0.6 to 9 MeV region that was comparable to or exceeded in intensity the extrapolated total spectrum. In the X-ray region of the spectrum, the percentage of the total emission that was pulsed varied from ∼2% at

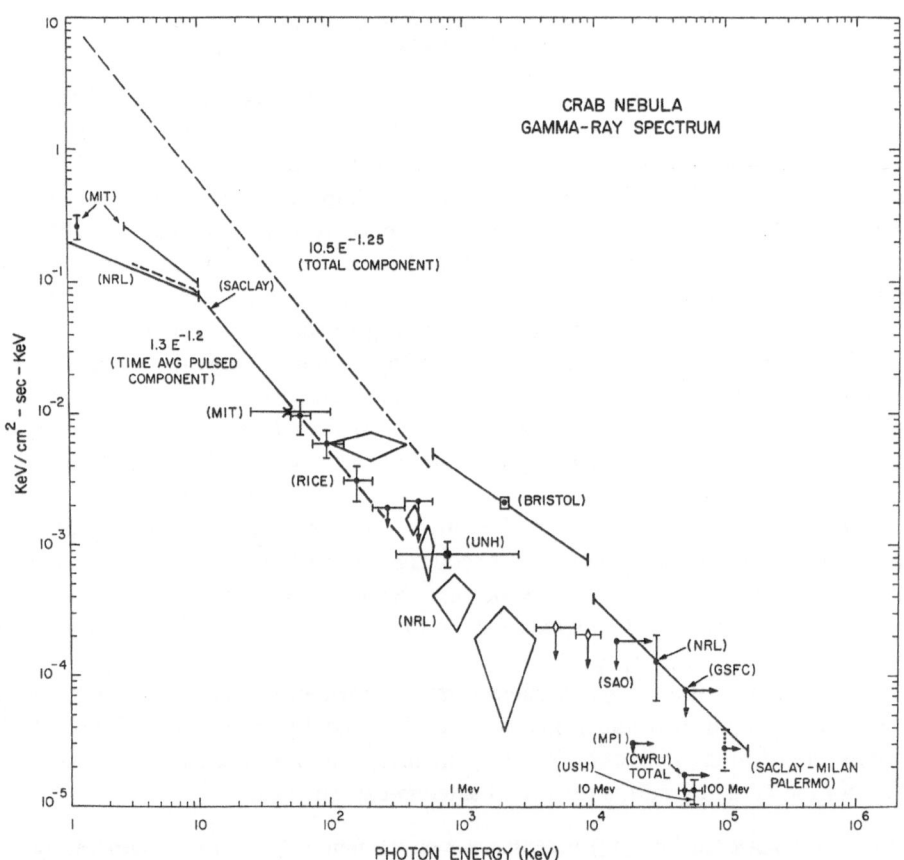

Fig. 1.   Gamma-ray spectrum of the Crab Nebula. The graph summarizes information available before this symposium. The spectrum for the 'Total Component' is taken from Peterson (Figure 7, this volume).

1 keV to ~15% in the 100 keV region. Kurfess (1971) reported evidence for pulsed emission up to 1 MeV, but in the 100 to 400 keV region the pulsed emission was $42 \pm 12\%$ of the total emission, and the ratio of the secondary peak to the primary peak was 2.3 ± 0.2. The pulsed-emission spectrum assumed was $10 E^{-2.2}$ photon cm^{-2} s^{-1} keV^{-1}. Orwig et al. (1971) simultaneously reported results consistent with Kurfess' in the 0.25 to 2.3 MeV region and indicated their results are an order of magnitude lower than those of Hillier et al. (1970).

During the discussion, J. D. Kurfess reviewed his results, emphasizing both the changes in the pulse shape that occurred in going from the X-ray to the gamma-ray region and the implications for the higher energy spectrum. He also noted that, owing to background corrections, his quoted flux in the 100 to 400 keV region may be over-estimated by 20%. A discrepancy still exists with the data of Hillier et al. (1970). Hillier's results may be too high because of two factors: (i) A hard spectrum was assumed that tends to overestimate the flux, and (ii) the quoted energy range of the detector was too high. In defense of his results, Hillier noted that there was an onboard energy calibration and the atmospheric background rate was consistent with other experiments. A time-dependent flux is one solution to this problem; however, Hillier suggested more data are needed before this problem can be solved. It is also interesting to note that Kurfess has not determined the relative phase of his X-ray data with respect to the optical pulse of NP 0532.

In the region above 10 MeV, some confusion has existed as to whether a pulsed emission had been detected from NP 0532. Most of the recent results were reported at the 12th International Conference on Cosmic Rays held in August 1971 in Hobart, Tasmania, Australia. They have been summarized in the rapporteur paper by Frye (1971) and are given here in Table I. Only one positive result was reported, by the University of Southampton group (Browning et al., 1971), but this was at marginal significance.

The most significant new result reported during the discussion was by the Case-Melbourne team, represented by G. M. Frye, Jr. They obtained an energy flux (time-averaged pulsed emission) of $(6 \pm 3) \times 10^{-5}$ keV cm^{-2} s^{-1} keV^{-1} at 20 MeV. Evidence was based on a 5σ peak within 1 ms of the main pulse among events analyzed with the

TABLE I

Hobart Conference results on the Crab Nebula

Group	Energy (MeV)	Pulsed flux (photon cm^{-2} s^{-1})	Average flux (photon cm^{-2} s^{-1})	Reference[a]
SAO	>15	$<1.4 \times 10^{-4}$	$<4.7 \times 10^{-4}$	OG 15
NRL	>20	$<1 \times 10^{-4}$	$<2 \times 10^{-4}$	OG 13, 14
MPI	>20	$<3 \times 10^{-5}$		OG 16
Southampton	>50	$=1.3 \times 10^{-5}$		OG 17
Saclay-Milan-Palermo	>100	$=3 \times 10^{-5}$		OG 16

[a] References refer to papers presented at the 12th International Conference on Cosmic Rays, Hobart, Tasmania, Australia, August 1971.

pulsar period (Figure 2). For the continuous integral flux, only an upper limit could be placed: $\leqslant 6 \times 10^{-4}$ photon cm^{-2} s^{-1} above 10 MeV and $\leqslant 2.4 \times 10^{-4}$ above 20 MeV. This would indicate that the pulsed emission in the 10 to 30 MeV region is $\geqslant 25\%$ of the continuous emission of the Nebula. The ratio of the interpulse to the main pulse was estimated to be less than 1.7, which is a significant change from the value of 2.3 reported in the region of several hundred keV. The width of the main pulse was 2 ms. Above 50 MeV, Frye reported an upper limit to the pulsed emission of 4×10^{-5} photon cm^{-2} s^{-1}.

L. Scarsi, representing the Saclay-Milan-Palermo group, also reported detection of pulsed gamma-ray emission above 20 MeV from NP 0532. The integral flux above

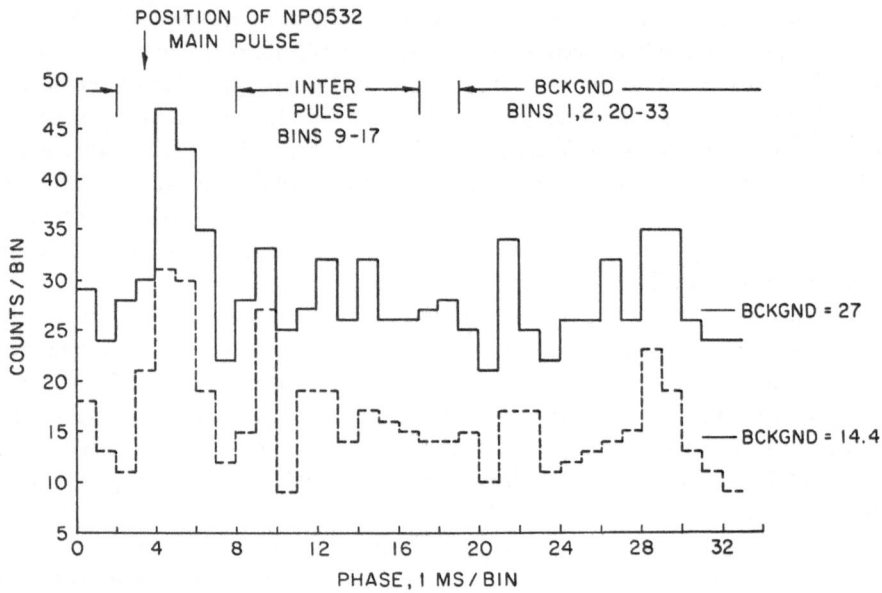

Fig. 2.   Gamma-ray photon count from NP 0532 as a function of the pulsar phase in the energy interval 10 to 30 MeV (solid line) and > 10 MeV (dotted line), as presented by G. M. Frye, Jr.

20 MeV was (6 to 9) $\times 10^{-5}$ photon cm^{-2} s^{-1}, in agreement with the Case-Melbourne results. The data were based on the results of six balloon flights of a spark-chamber detector during 1969 to 1971. A 2.6σ effect was observed in the primary pulse, which was in phase with the optical, radio, and X-ray pulses. The events were summed over a gamma-ray arrival direction of 12° half-angle centered on the Crab Nebula, and the shape of the pulse as a function of phase was consistent with the results of Kurfess (Figure 3).

The above results on NP 0532 do not agree with the upper limit to the pulsed flux previously published by the Max Planck Institute in Munich (Kettenring et al., 1971). K. Pinkau reviewed how these results were obtained and stated that he hoped that the additional data now being reduced will resolve this discrepancy.

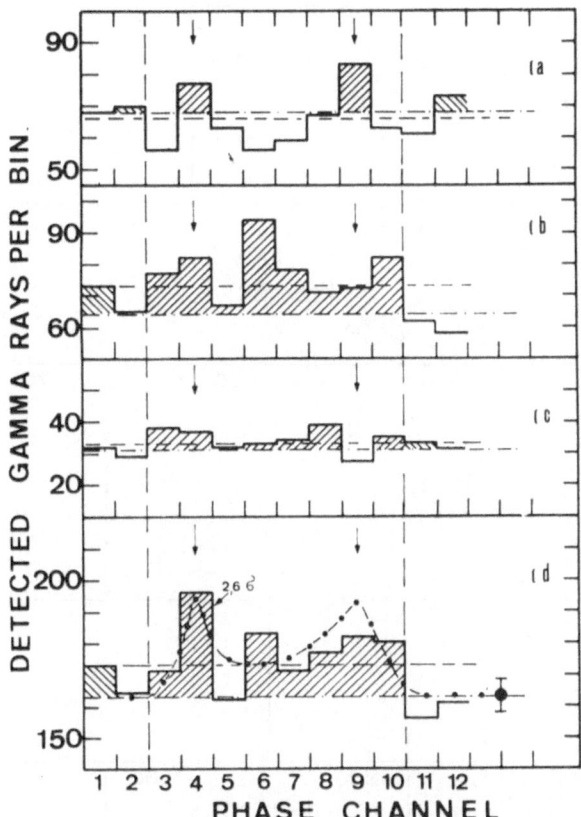

Fig. 3. Gamma-ray photon count from NP 0532 as a function of the pulsar phase in the energy region > 20 MeV from a series of six flights, summed in pairs (a), (b), and (c) and the total data (d). The data in (d) are compared with the 100 to 400 keV X-ray data. Presented by L. Scarsi.

The Laboratory for Cosmic Ray Physics, U.S. Naval Research Laboratory (NRL), had previously reported possible evidence for pulsed gamma-ray emission from NP 0532 above 10 MeV. G. Share, commenting on these results obtained with a nuclear emulsion spark-chamber detector, noted that of the 52 gamma-ray events previously detected in the spark chamber only and occurring near both optical peaks, only one photon came $\leqslant 2°$ from the Crab Nebula. This null result places an upper limit of 6×10^{-5} photon cm^{-2} s^{-1} for energies above 15 MeV. One interpretation of this is that the previously reported pulsation was due to low-energy events, $\leqslant 15$ MeV. The NRL group also set an upper limit to the total flux from the Crab Nebula of 10^{-4} photon cm^{-2} s^{-1} at energies above 15 MeV.

Above 10^{11} eV, cosmic gamma-rays interact within the atmosphere, producing Čerenkov radiation that can be detected with ground-based instruments. Cosmic-ray nuclei produce similar radiation; however, gamma-rays can be distinguished by their anisotropy in the direction of the suspected source. Using this technique, T. C. Weekes

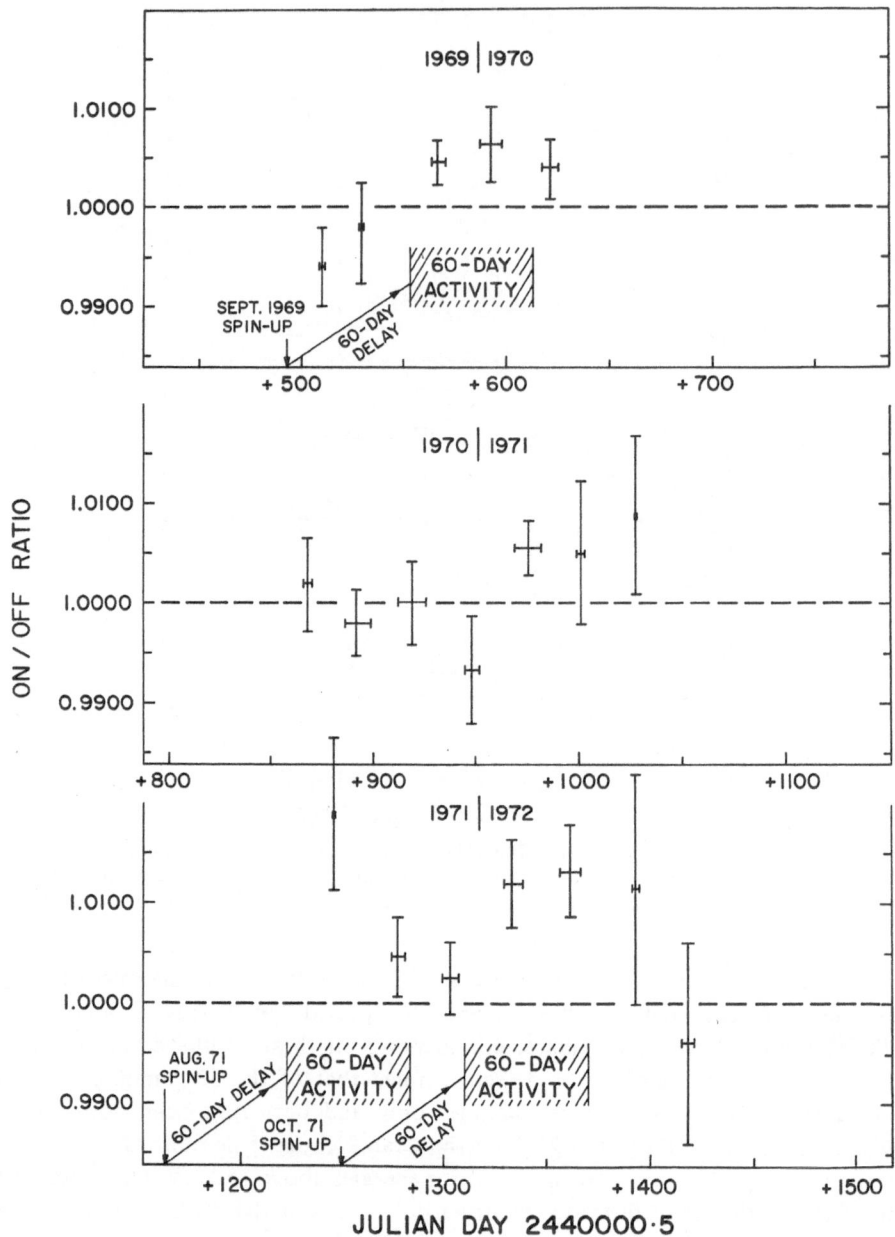

Fig. 4.   Ratio of the air-shower counts recorded from the direction of the Crab Nebula to those from an arbitrary reference direction. A ratio significantly > 1 indicates a flux of 2.5×10^{11} eV gamma rays from the Nebula. Presented by T. C. Weekes.

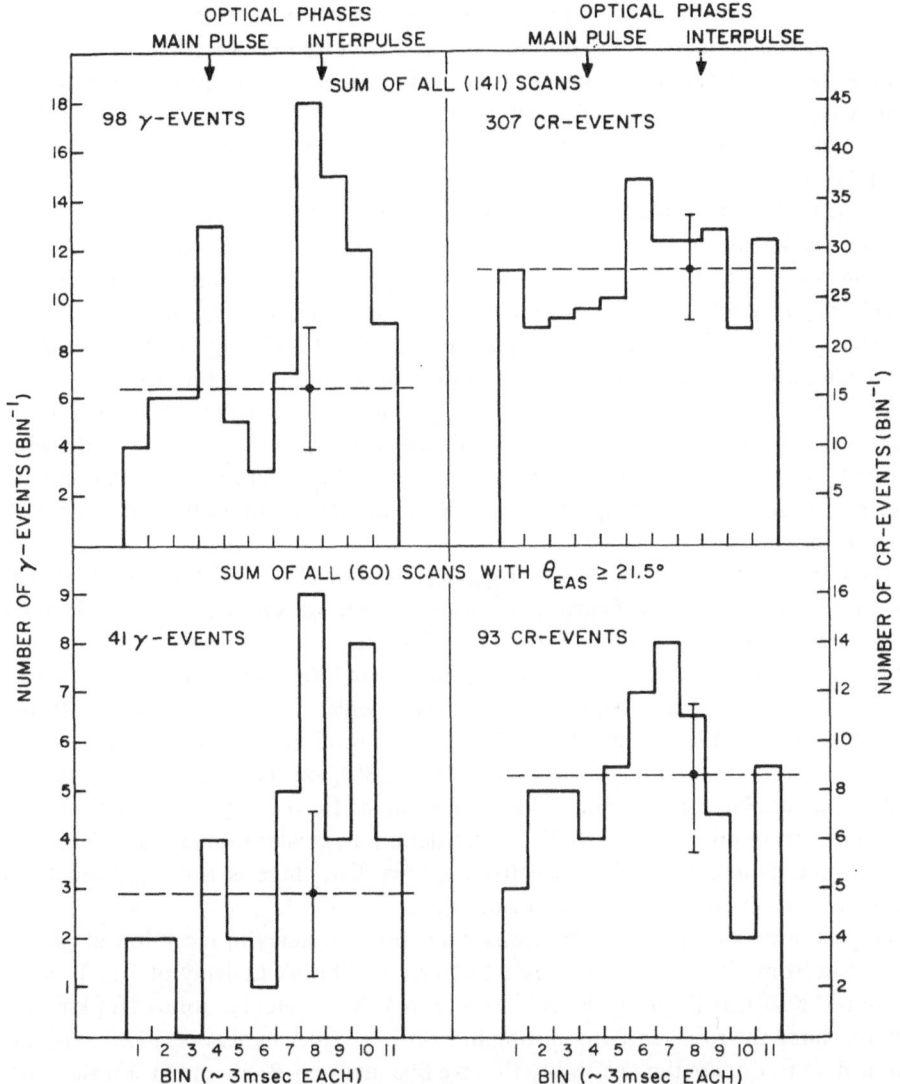

Fig. 5. Phase analysis of total gamma-ray data ($> 6 \times 10^{11}$ eV) from NP 0532. Data are divided according to the median zenith angle of NP 0532 for energy-spectrum determination. Data from Grindlay (1972).

reported the detection, by the group at the Smithsonian Astrophysical Observatory (SAO), of gamma-rays with energy greater than 2.5×10^{11} eV from the Crab Nebula. Observations of the Nebula over a 3-yr period indicate an average flux of $(4.4 \pm 1.4) \times 10^{-11}$ photon cm^{-2} s^{-1} at the 3.1σ level. This flux corresponds to an emission of 6×10^{33} ergs s^{-1}. However, the gamma-ray flux may vary with time, with the most significant flux $((1.21 \pm 0.24) \times 10^{-10}$ photon cm^{-2} s$^{-1})$ occurring 60 to 120 days after a major spin-up of the pulsar NP 0532. This increase was observed on three different occasions, and if only this flux is used, the effect is at the 5σ level (Figure 4). The total gamma-ray energy observed on each occasion was $\sim 10^{41}$ ergs, an energy approximately equal to that of the pulsar spin-up.

Grindlay (1971) had previously reported evidence for the detection of pulsed gamma-rays of $\sim 10^{12}$ eV from NP 0532 and has recently confirmed these results (Grindlay, 1972). The sum of his data now yields a 5.5σ peak at the interpulse and a 3.5σ peak at the primary pulse, consistent with the pulsed flux ($\geqslant 6.8 \times 10^{11}$ eV) of 1.25×10^{-11} photon cm^{-2} s^{-1} (Figure 5). The ratio of the interpulse to the primary pulse is $\sim 3.5:1$, which follows the trend observed in the low-energy gamma-ray region but is opposite to the effect reported at this meeting in the 10 to 100 MeV region. The gamma-ray spectrum appears consistent with an extrapolation from the X-ray region.

Weekes also described the results from the teams at University College, Dublin, and U.K.A.E.R.E., Harwell, who reported evidence for the detection of a pulsed flux above 2×10^{12} eV. A significant (3.8σ) peak was observed, but it dit not occur in phase with the optical primary peak. This observation gives an upper limit to the flux of 2×10^{-12} photon cm^{-2} s^{-1}.

The interesting question as to whether the continuous flux, which was detected from the Crab Nebula by the SAO group, is pulsed or not, was dicussed by H. F. Helmken of SAO. Periodic analysis of 122 hr of data, collected over 3 yr, showed no significant effects. For energies above 1.5×10^{11} eV, the upper limit to the flux was 1.2×10^{-11} photon cm^{-2} s^{-1} for an assumed primary pulse width of 1.3 ms. These results agree with previously published results by SAO. Based on the continuous flux reported, the absence of a pulsed flux places an upper limit of 30% on the fraction contributed by the pulsed flux. Grindlay's results, when extrapolated from 6×10^{11} eV to 1.5×10^{11} eV with a spectrum proportional to $E^{-1.5}$, predict a flux greater by a factor of 7 than this upper limit. Unless the pulsed spectrum is very flat, there is a disagreement with Grindlay's results that needs to be resolved.

Two tentative and probably spurious indications of delay in reception of gamma-ray pulses from NP 0532 were reported by Frye at 100 MeV (a delay of 1 ms) and the Dublin-U.K.A.E.R.E. group at 10^{12} eV (17 ms). R. C. Henry, Johns Hopkins University, stated the results are interesting in light of a paper by Rawls (1972), who used pulsar data to discuss the possibility that we live in a non-Riemannian 'Finsler' space (Rund, 1959). Rawls used data up to 1 MeV to show that the fundamental length associated with our space, if Finsler, is $\leqslant 1.9 \times 10^{-18}$ cm. The delays at higher gamma-ray energies would imply smaller lengths, and the fact that the effect is a delay would mean the length is imaginary. Henry commented that this possibility must not be taken seriously but is worth keeping in mind.

3. Galactic Center Region

The scintillation detector experiment on the OSO 3 satellite produced convincing evidence for gamma-ray emission (> 100 MeV) from the galactic plane, with maximum intensity from the region of the galactic center (Clark *et al.*, 1971; Kraushaar *et al.*, 1972). Fichtel *et al.* (1972) verified these results at the 4σ level, using a balloon-borne spark chamber. In general, these experiments have been interpreted as being consistent with a line source in the galactic plane with an intensity near the galactic center of $\sim 1.2 \times 10^{-4}$ photon cm^{-2} s^{-1} rad^{-1}. Fichtel *et al.* also reported no significant excess of gamma-rays observed in the 50 to 100 MeV interval, supporting the evidence for a π^0-decay origin for the radiation. Frye *et al.* (1969) found no evidence for a line source in the region of the galactic center, with an upper limit of 3×10^{-5} photon cm^{-2} s^{-1} rad^{-1} for a $\pm 6°$ line source, but they detected several discrete sources. Whether the gamma radiation from the galactic center results from a diffuse interstellar source or from a number of discrete sources remains an interesting problem.

From a superposition of the results of three flights in 1969, Frye *et al.* (1971) had already reported evidence for several sources: (i) Sgr $\gamma - 1$, (ii) Gγ 2 + 3 and Gγ 341 + 1, which are coincident with known X-ray sources, and (iii) Lib $\gamma - 1$, a time-variable

Point Sources of Cosmic Υ-Rays

PER Υ-1

234-P:	$N_O = 26$	$N_B = 15.0$	
620-P:	$N_O = 65$	$N_B = 41.6$	$\sigma = 4.6$

$F = 2.3 \times 10^{-5}$ Υ's cm^{-2} sec^{-1}

$a = 65°$, $\delta = +36°$

CEN Υ-1

429-A:	$N_O = 42$	$N_B = 24.2$	
467-A:	$N_O = 26$	$N_B = 16.8$	$\sigma = 4.2$

$F = 2.5 \times 10^{-5}$ Υ's cm^{-2} sec^{-1}

$a = 212°$, $\delta = -38°$

G Υ 327+2

429-A:	$N_O = 49$	$N_B = 34.1$	
467-A:	$N_O = 22$	$N_B = 8.5$	$\sigma = 4.4$

$F = 3.0 \times 10^{-5}$ Υ's cm^{-2} sec^{-1}

$a = 234°$, $\delta = -52°$

Fig. 6.   New point sources of cosmic gamma rays (> 50 MeV) as reported by G. M. Frye, Jr.

source seen in November 1969 but not in Febraury 1969. This last may be associated with the radio source AP Lib (PKS 1514–24), a BL Lac-type object. During the Symposium, Frye also presented evidence for two new sources at the 4σ level in this region: Cen $\gamma-1$ with an intensity of 2.5×10^{-5} photon cm^{-2} s^{-1} and Gγ 327+2 with an intensity of 3×10^{-5} photon cm^{-2} s^{-1}, both for energies above 50 MeV. This latter source may be identified with the highly variable X-ray source 2U1516−56. A summary of the positions of these sources is given in Figure 6. A number of $\geqslant 3\sigma$ peaks are seen in the fluxes measured from a region of 2 sr centered on the galactic center, but the distribution of all these sources, when analyzed as a latitude distribution, is not consistent with a $\pm 6°$ line source in the plane (Figure 7). Frye's results could still be consistent with the OSO 3 experiment since the angular resolution of that detector was $\pm 15°$.

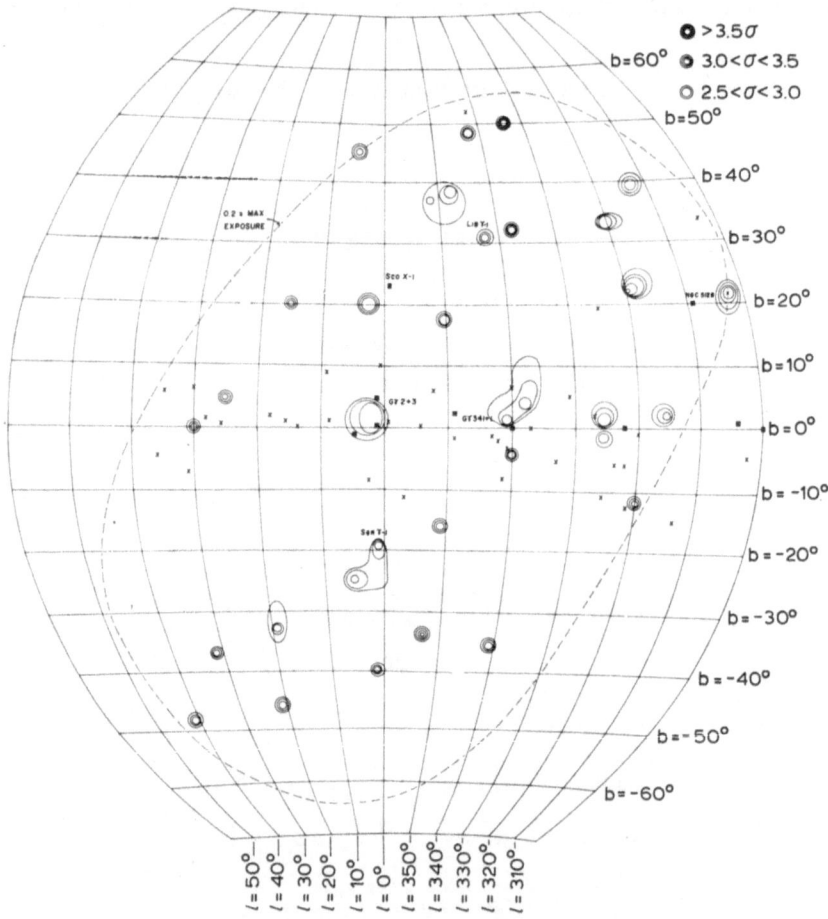

Fig. 7.   A map, in galactic coordinates, showing a number of gamma-ray sources in the galactic-center region. Presented by G. M. Frye, Jr.

D. A. Kniffen, of Goddard Space Flight Center (GSFC), summarized the results of Fichtel *et al.* (1972) and discussed the interpretation of their flux measurements on the galactic center. Although the data have been interpreted in terms of a line source, the possibility that they consist of an accumulation of several unresolved discrete sources cannot be ruled out. However, these sources must be within $\pm 6°$ of the galactic plane. This interpretation would be consistent with recent results of the University of Southampton group, but it is in clear contradiction with Frye's findings. Kniffen also pointed out that a discrete-source interpretation must be reconciled with the results of their spectral measurments (π^0-decay-type spectrum) and those of NRL.

G. K. Rochester submitted a paper by the Imperial College, London, group verifying the flux intensities of OSO 3 and Fichtel *et al.* in the 200 MeV to 10 GeV region. Results were presented from two balloon flights in Zambia that used a scintillation detector. An excess number of counts was also observed in the region of Sgr XR-2, which accounted for almost half the excess counts from the line source.

G. Hutchinson, from the University of Southampton, presented some new data on the origin of the gamma-ray (> 100 MeV) flux in the region of the galactic center. Their balloon-borne spark chamber, launched from Palestine, Texas, in September 1971, observed the galactic center at relatively large zenith angles. They detected three sources above the 3σ level; each was associated with a galactic X-ray source listed in

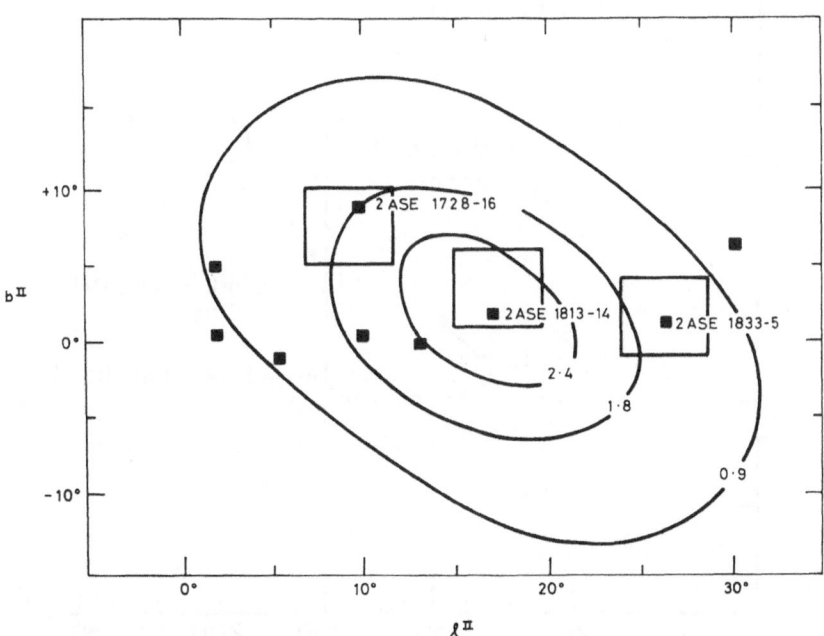

Fig. 8. Gamma-ray point sources observed by the University of Southampton group in the region of the galactic center. The black squares show the position of known X-ray sources, and the open squares indicate regions of anomalously high gamma-ray emission. Exposure-value contours are in units of 10^6 cm^2 s. Reported by G. Hutchinson.

the UHURU catalog (Figure 8). The source strengths were of the order of 2×10^{-5} photon cm^{-2} s^{-1}. Other than these sources, no other significant flux was observed within $\pm 3°$ of the galactic equator, giving an upper limit to a line source of 8×10^{-5} photon cm^{-2} s^{-1} rad^{-1}. If the three sources were viewed by a detector that could not resolve them and the result interpreted as a line source, the flux would be 1.5×10^{-4} photon cm^{-2} s^{-1} rad^{-1}, in agreement with Clark *et al.* (1971) and Fichtel *et al.* (1972). The area of the sky observed in this experiment only partially overlapped the region investigated by Frye.

Preliminary results on the galactic-center region obtained during a recent series of balloon flights from Argentina were presented by Share and Helmken.

Share quoted evidence for a possible source near Gγ 341+1 at energies > 10 MeV with an intensity of $\sim 2 \times 10^{-4}$ photon cm^{-2} s^{-1}. Search for a line source of radiation (> 10 MeV) gave only an upper limit to the flux of 3×10^{-4} photon cm^{-2} s^{-1} rad^{-1}, implying that the spectrum observed above 100 MeV must flatten at lower energies, which is characteristic of a π^0-decay mechanism (Figure 9). No gamma-rays > 15 MeV were detected from Lib γ-1 at an intensity expected from Frye's results.

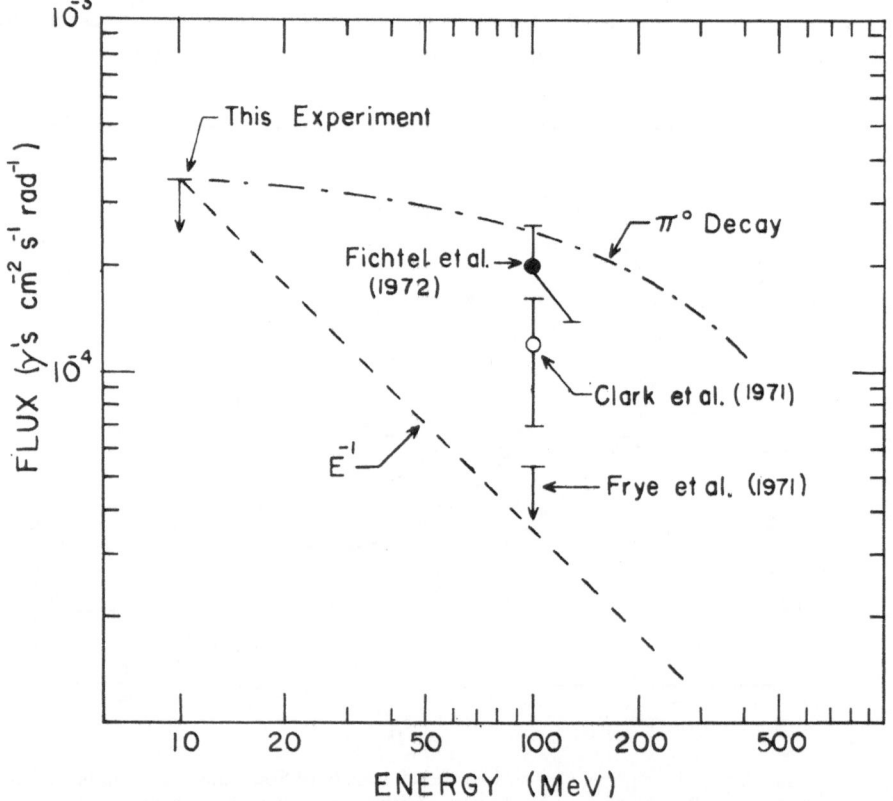

Fig. 9.   Upper limit to the flux of a gamma-ray line source in the galactic plane near the center ($328.5° < l^{II} < 325°$, $-12.5° < b^{II} < 12.5°$) as determined by the NRL group and reported by G. Share.

Using a gas-Čerenkov detector with an energy threshold of 10 MeV, Helmken observed the galactic center on each of two flights from Argentina. A positive excess flux was observed on both flights, with a total value at the 2.6σ level. Preliminary analysis of the data gives an integral flux above 12 MeV of $(1.9\pm0.7)\times10^{-4}$ photon cm^{-2} s^{-1}. Comparison of this value with the results of Fichtel *et al.* shows that it is also consistent with a line source in the plane in which the source mechanism consists of equal parts of Compton scattering and π^0 decay. In Figure 10, Helmken's data point is plotted along with the low-energy gamma-ray spectrum from the galactic-center region observed previously by Johnson *et al.* (1972).

In this lower energy range, Johnson *et al.* also found evidence for a spectral feature at 0.5 MeV, the electron-positron annihilation line.

There is no doubt that the galactic plane and the region near the galactic center are sources of gamma radiation. What the source of this radiation is – i.e., whether it is diffuse and interstellar in origin or a group of discrete sources – is still uncertain. At this meeting, no new and significant evidence was presented for a true diffuse source of gamma-rays in the galactic plane. The experimental results to date, however, are

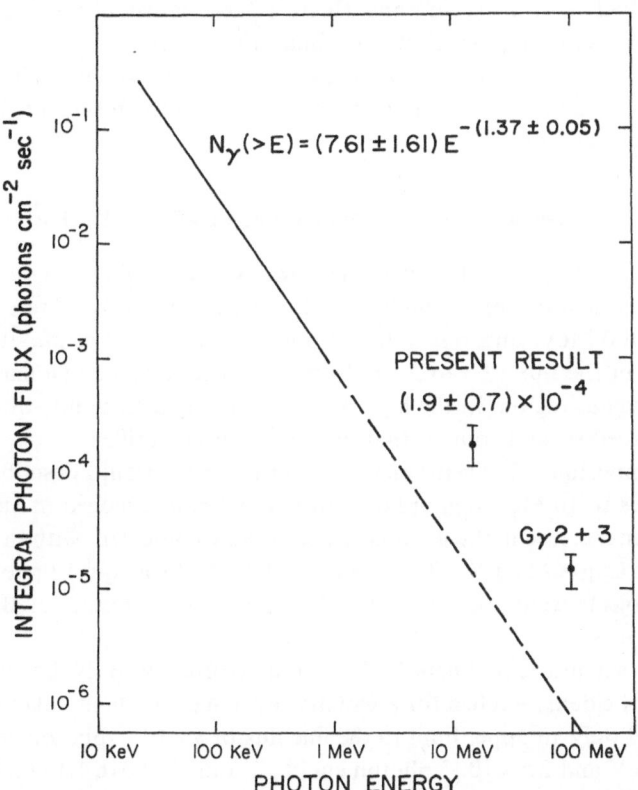

Fig. 10. Integral gamma-ray flux from the galactic center as measured by H. Helmken, compared to an extrapolation of the flux measured by W. N. Johnson *et al.* (1972) at lower energies.

not in conflict with the diffuse flux predicted from π^0 meson decay. The mesons are produced by the interaction of a uniform source of cosmic rays on atomic hydrogen. Cavallo and Gould (1971) predict a flux of $\sim 3 \times 10^{-5}$ photon cm^{-2} s^{-1} rad^{-1} in the plane and in the region of the galactic center due to this mechanism. Stecker (1970) predicts a flux a factor of 2 lower. The OSO 3 results in the galactic plane, away from the center, are consistent with the predicted flux, and the upper limits to a diffuse source in the plane as determined by Frye *et al.* (1969) are a factor of 2 (center) to 3 (anticenter) above this predicted flux.

In regard to the multitude of discrete sources reported, no source has been seen at a high statistical significance by two different groups. Owing to the rather small number of balloon flights, this may be an indication of time variability. Frye has proposed the following criteria for acceptance of a source: (i) a 5σ effect from a single observation or (ii) a 4σ effect from two observations. Further, he recommends that anything less than this should be listed as a 'possible' source and promulgated only as information for other groups. Hutchinson requested that groups exchange their sky maps to compare 'possible' source positions. Helmken has suggested that to compare data, a common method of analysis should be followed, e.g., the calculation of upper limits as proposed by Fichtel *et al.* (1969) and Hearn (1969). It would also be helpful if the value of σ was given on all suspected gamma-ray sources.

What gamma-ray astronomy needs is more sensitive experiments with good angular resolution. Perhaps the answer will come this year with the results of the TD-1A and SAS-B satellite experiments.

4. Galactic Plane in the Cygnus-Cassiopeia Region

The University of Southampton group (Browning *et al.*, 1972) recently reported evidence at the $\sim 3\sigma$ level for a number of discrete gamma-ray sources in the energy range above 100 MeV, and researchers from Toulouse (Niel *et al.*, 1972) found an enhancement in the flux (2.5 to 3.5σ) from the galactic plane in a narrow longitude zone in the Cygnus region. In the history of gamma-ray astronomy, this region of the sky has produced several sources that have never been verified.

At this Symposium, A. Dean, of the Milan-Palermo group, describing an experiment in the 0.8 to 10 MeV region flown from a balloon launched in Sicily, presented evidence for an increase in the gamma-ray count rate coincident with the transit of the X-ray sources Cyg X2 and X4. The flux above 0.8 MeV was ~ 0.12 photon cm^{-2} s^{-1}. Both these objects were possible 100 MeV sources in the results of Browning *et al.* (1972).

An OGO 5 satellite experiment by P. A. J. de Korte and B. N. Swanenburg, at the University of Leiden, searched for a gamma-ray line emission at galactic longitude of $60°$ but placed only an upper limit to the line flux of 5×10^{-5} photon cm^{-2} s^{-1} rad^{-1} above 400 MeV and 2.5×10^{-5} photon cm^{-2} s^{-1} rad^{-1} above 1.15 GeV.

At the Hobart Conference, Vladimirsky *et al.* (1971), of the Crimean Astrophysical Observatory, reported possible detection of a high-energy gamma-ray ($>10^{12}$ eV) source in the Cygnus region of the galactic plane. Weekes reported that SAO had searched for this source but was unable to detect it at 10^{11} eV. The Crimean group presented a paper at this meeting stating that they had searched the region a second time, and they, too, were unable to detect the source. Perhaps we are again dealing with time-variable sources.

The Crimean team (Stepanyan *et al.*, 1972) also reported a new source in Cassiopeia in the energy region above 2×10^{12} eV. Observed during three drift scans on September 29 and October 14 and 15, 1971, this source had a flux of 10^{-10} photon cm^{-2} s^{-1} at the 3.9σ level. Its position was $\alpha = 01^h 11^m \pm 6^m$ and $\delta = 62° \pm 1°$. It was not observed on three previous drift scans, on September 22, 25, and 27, 1971.

5. Galactic Plane in the Anticenter Region

Other than the Crab Nebula, one of the most spectacular sources reported in this region of the sky has been the variable Seyfert galaxy 3C120, labeled Tau γ–1. This source was reported by Volobuyev *et al.* (1971), of the Moscow Physical Engineering Institute, based on the results of scintillation-counter experiments on the satellites Cosmos 251 and 264. The flux is rather large, 5×10^{-4} photon cm^{-2} s^{-1}, a factor of 20 greater than any other source reported. These measurements were made during a period when the radio intensity of 3C120 was at a maximum. Kirillov-Ugryumov pointed out in a paper submitted to the Symposium the similarity between this object and the source reported by Frye, PKS 1514–24 (AP Lib), both variable radio galaxies with star-like nuclei. Other galaxies with similar characteristics that should be investigated for gamma-ray emission are VRO 42.22.01 (BL Lac) and ON 231 (W Com), which have also been suggested by Frye, as well as B2 1215+30, 3C273, and OJ 287. At this meeting, Frye reported no evidenc for OJ 287 being a gamma-ray source (Figure 11) and had previously published an upper limit to the flux from 3C273 (Frye and Wang, 1969).

Frye reported one new source in this region of the sky, observed on each of two flights with a combined effect of 4.6σ. This source, Per γ–1, was located at $\alpha = 65°$, $\delta = 35.5°(l = 164°, b = -10°)$ and had an intensity of 2.3×10^{-5} photon cm^{-2} s^{-1}. No source was reported at the position of 3C120; however, a 4σ source was observed within the $\pm 10°$ error box on 3C120, located at $\alpha = 76°$, $\delta = 1°$ with a flux of 5×10^{-5} photon cm^{-2} s^{-1}. No evidence for a line source in the galactic plane has been observed (Figure 11).

Three balloon-flight experiments by the Saclay-Milan-Palermo group to search for gamma radiation in the anticenter region were described by B. Agrinier, of the Center for Nuclear Studies, Saclay. Seven discrete sources were reported at the 2 to 3σ level. The Crab Nebula was observed at the 2σ level, giving a flux (> 20 MeV) of $(6.2 \text{ to } 9.0) \times 10^{-5}$ photon cm^{-2} s^{-1}. The point sources detected are in the vicinity of the galactic disk, and hence the OSO 3 line source in this region could be accounted for by the discrete sources reported in this experiment (Figure 12).

The NRL group (Share) reported only an upper limit to the line emission (> 15 MeV) from the plane near the anticenter ($\leqslant 3.5 \times 10^{-4}$ photon cm^{-2} s^{-1} rad^{-1}).

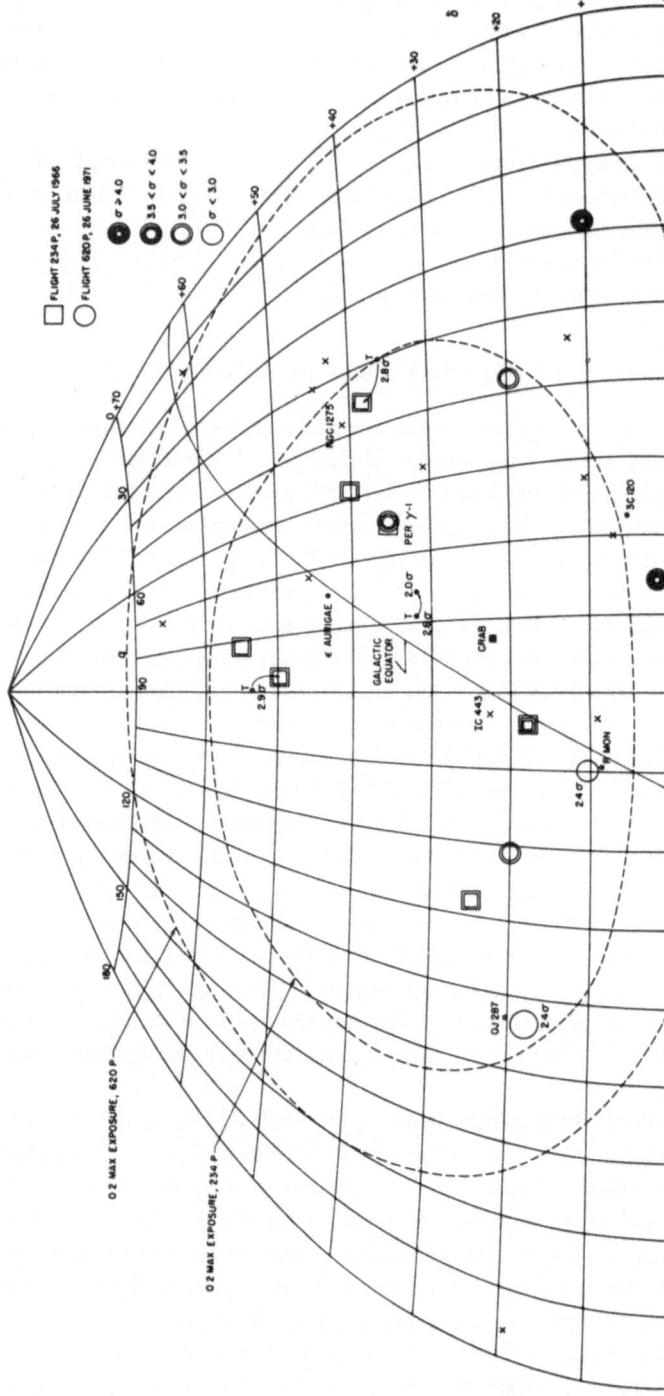

Fig. 11. A map, in celestial coordinates, showing the position of gamma-ray sources in the galactic anticenter region. Presented by G. M. Frye, Jr.

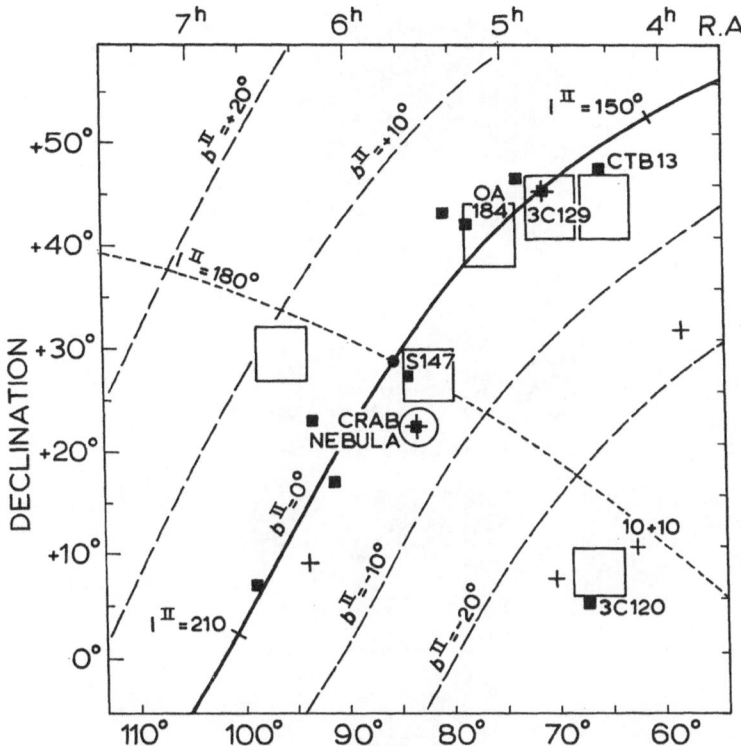

Fig. 12. A map of the sky in the region of the galactic anticenter, presented by B. Agrinier for the Saclay-Milan-Palermo group, showing gamma-ray sources as open squares or circles. The crosses indicate X-ray sources, and the black squares, supernova remnants.

6. Future Satellite Experiments

The Apollo 15 and 16 flights carried a gamma-ray spectrometer consisting of a 3″ × 3″ Na I (Tl) crystal with an anticoincidence scintillator and a 512-channel pulse-height analyzer. Measurements of the diffuse gamma-ray flux with this detector were reported earlier. During the last 30 hr of transearth coast of Apollo 16, the detector was extended to 2 m and the spacecraft arranged to rotate. The spacecraft acts as an occulting disk of about 0.25 sr solid angle. A strong source would be indicated by a reduction in count rate. Hence, a map of the sky in the 0.1 to 4.0 MeV region with about 40 resolution elements can be obtained. L. Peterson of the University of California in San Diego reported that these data are now being reduced.

In March 1972, the ESRO satellite TD-1A was launched. It contains an experiment, called MIMOSA, that uses a vidicon spark chamber to search for cosmic gamma-rays above 50 MeV (Figure 13). The sensitive area of the detector is 130 cm², the area-solid-angle factor is 28 cm² sr, and the angular resolution is ~3°. The data rate is 1 video frame per 5 s. The experiment is functioning and the gamma-ray count rate being

Fig. 13. The gamma-ray spark-chamber experiment (S133) on the ESRO TD-1A satellite.

recorded is ~1 event per 5 min. The detector always points at the zenith during the satellite orbit. The experiment is a cooperative effort of the Center for Nuclear Studies, Saclay, the Max Planck Institute, Munich, and the University of Milan.

Pinkau also reported on the status of the ESRO satellite COS-B, which is a single experiment devoted entirely to gamma-ray astronomy in the region above 30 MeV. The detector is a wire spark chamber with an energy calorimeter (Figure 14). The sensitive area is ~576 cm² and the area-solid-angle factor, ~70 cm² sr. The energy resolution will be about 50% at 100 MeV. The satellite will be spin-stabilized and, in contrast to other experiments, placed in a highly eccentric orbit. The main advantages of this orbit are reduction of Earth albedo, reduction of radiation-belt effects, minimum occultation by the Earth, and adequate ground-station coverage. The satellite, due to be launched in 1974, is a cooperative program among the groups at Leiden, ESTEC, Saclay, Milan, Palermo, and Munich.

The SAS-B satellite will also be a single gamma-ray experiment (see Figure 15) conducted by NASA/GSFC (C. E. Fichtel, R. C. Hartman, and Kniffen). The experiment will consist of a wire spark chamber sensitive to gamma-ray energies above 25 MeV. The sensitive area of the experiment is ~500 cm²; the area-solid-angle factor,

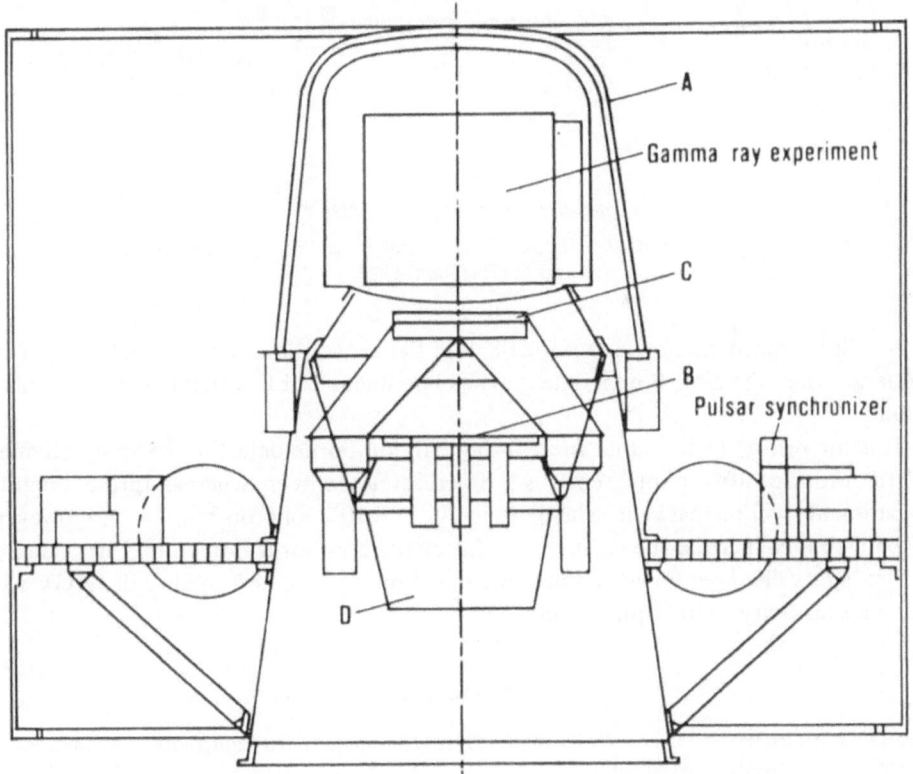

Fig. 14.   The COS-B satellite configuration, showing the gamma-ray spark chamber and calorimeter.

125 cm² sr; and the angular resolution, ∼2° at 100 MeV. The satellite will be placed in a 550-km circular equatorial orbit in October 1972.

The HEAO-B gamma-ray experiment is a joint undertaking of the groups at NASA/GSFC, Stanford University, Grumman Aerospace Corporation, and Max Planck Institute. The design for this experiment is not yet complete, but it includes a large, digitized spark chamber and a cesium iodide crystal spectrometer to allow differential energy measurments over the interval from 20 to 10^4 MeV. The detector

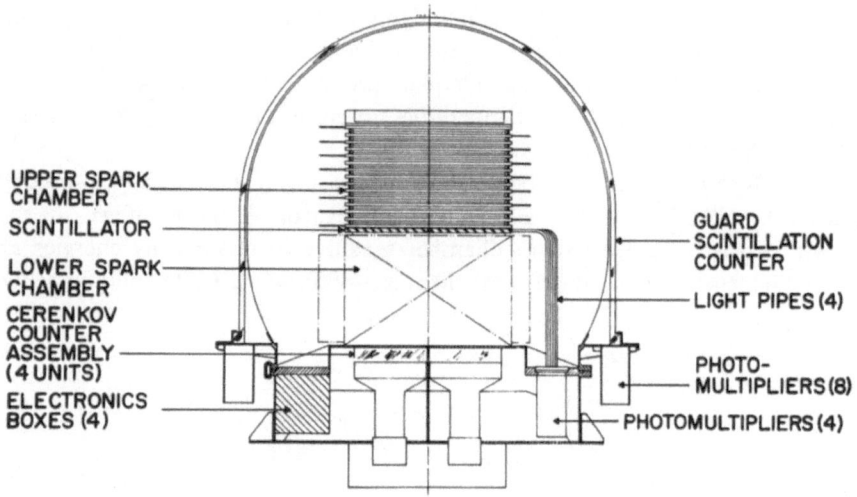

SAS-B GAMMA RAY EXPERIMENT

Fig. 15.   Schematic of the digitized spark chamber used as a gamma-ray telescope on the satellite SAS-B.

area will be approximately a factor of 8, and the area-solid-angle product a factor of 16 larger than the SAS-B experiment. The experiment will be operated in a scanning mode.

It is interesting to note that present-day balloon-borne detectors have sensitivities of the order of 10^{-5} photon cm^{-2} s^{-1} for discrete sources, whereas future satellite experiments will increase the sensitivity to 10^{-6} to 10^{-7} photon cm^{-2} s^{-1}. However, it is possible to increase the sensitivity of future balloon-borne detectors. For example, a 1-m² spark chamber flown at a pressure of 0.5 mb and a cutoff rigidity of 12 GV will have a sensitivity ∼10^{-6} photon cm^{-2} s^{-1}.

References

Browning, R., Ramsden, D., and Wright, P. J.: 1971, 12th International Conference on Cosmic Rays, Hobart, Tasmania, Paper OG-17.
Browning, R., Ramsden, D., and Wright, P. J.: 1972, *Nature* **235**, 128.
Cavallo, G. and Gould, R. J.: 1971, *Nuovo Cimento* **2**, 77.

Clark, G. W., Garmire, G. P., and Kraushaar, W. L.: 1971, 12th International Conference on Cosmic Rays, Hobart, Tasmania, Paper OG-29.

Fichtel, C. E., Kniffen, D. A., and Ögelman, H. B.: 1969, *Astrophys. J.* **158**, 193.

Fichtel, C. E., Hartman, R. C., Kniffen, D. A., and Sommer, M.: 1972, *Astrophys. J.* **171**, 31.

Frye, G. M., Jr.: 1971, Rapporteur Paper on Gamma-Ray Astronomy, 12th International Conference on Cosmic Rays, Hobart, Tasmania.

Frye, G. M., Jr., Staib, J. A., Zych, A. D., Hopper, V. D., Rawlinson, W. R., and Thomas, J. A.: 1969, *Nature* **223**, 1320.

Frye, G. M., Albats, P. A., Zych, A. D., Staib, J. A., Hopper, V. D., Rawlinson, W. R., and Thomas, J. A.: 1971, 12th International Conference on Cosmic Rays, Hobart, Tasmania, Paper OG-24.

Frye, G. M. and Wang, C. P.: 1969, *Astrophys. J.* **158**, 925.

Grindlay, J. E.: 1971, *Nature* **234**, 153.

Grindlay, J. E.: 1972, *Astrophys. J. Letters* **174**, L9.

Hearn, D. R.: 1969, *Nucl. Instr. Methods* **70**, 200.

Hillier, R. R., Jackson, W. R., Murray, A., Redfern, R. M., and Sale, R. J.: 1970, *Astrophys. J. Letters* **162**, L177.

Johnson, W. N., III, Harnden, F. R., Jr., and Haymes, R. C.: 1972, *Astrophys. J. Letters* **172**, L1.

Kettenring, G., Mayer-Hasselwander, H. A., Pfeffermann, E., Pinkau, K., Rothermel, H., and Sommer, M.: 1971, 12th International Conference on Cosmic Rays, Hobart, Tasmania, Paper OG-16.

Kraushaar, W. L., Clark, G. W, Garmire, G. P., Borken, R., Higbie, P., Leong, C., Thorsos, T.: 1972, *Astrophys. J.*, in press.

Kurfess, J. D.: 1971, *Astrophys. J. Letters* **168**, L39.

Niel, M., Vedrenne, G., Claverie, A., and Bouigue, R.: 1972, Centre d'Étude Spatiale des Rayonnements, Toulouse, Report 71-270, March.

Orwig, L. E., Chupp, E. L., and Forrest, D. J.: 1971, *Nature* **231**, 171.

Rawls, J. M.: 1972, *Phys. Rev. D* **5**, 487.

Rund, H.: 1959, *The Differential Geometry of Finsler Spaces,* Springer-Verlag, Berlin.

Stecker, F. W.: 1970, *Astrophys. Space Sci.* **6**, 377.

Stepanyan, A. A., Vladimirsky, B. M., and Fomin, V. P.: 1972, preprint.

Vladimirsky, B. M., Povlov, I. V., Stepanyan, A. A., and Fomin, V. P.: 1971, 12th International Conference on Cosmic Rays, Hobart, Tasmania, Paper OG-25.

Volobuyev, S. A., Galper, A. M., Iyudin, A. F., Kirillov-Ugryumov, V. G., Luchkov, B. I., and Ozerov, Yu. V.: 1971, 12th International Conference on Cosmic Rays, Hobart, Tasmania, Paper OG-19.